Lecture Notes in Computer Sci

Edited by G. Goos, J. Hartmanis and J. van

T0250747

Springer
*Berlin
Heidelberg
New York
Barcelona
Hong Kong
London
Milan
Paris
Singapore
Tokyo*

Dongho Won (Ed.)

Information Security and Cryptology – ICISC 2000

Third International Conference
Seoul, Korea, December 8-9, 2000
Proceedings

 Springer

Series Editors

Gerhard Goos, Karlsruhe University, Germany
Juris Hartmanis, Cornell University, NY, USA
Jan van Leeuwen, Utrecht University, The Netherlands

Volume Editors

Dongho Won
Sungkyunkwan University, School of Electronical and Computer Engineering
300 Chunchun-dong, Jangan-gu, Suwon, Kyonggi, Korea
E-mail: dhwon@simsan.skku.ac.kr

Cataloging-in-Publication Data applied for

Die Deutsche Bibliothek - CIP-Einheitsaufnahme

Information security and cryptology : third international conference ;
proceedings / ICISC 2000, Seoul, Korea, December 8 - 9, 2000. Dongho
Won (ed.). - Berlin ; Heidelberg ; New York ; Barcelona ; Hong Kong ;
London ; Milan ; Paris ; Singapore ; Tokyo : Springer, 2001
  (Lecture notes in computer science ; Vol. 2015)
  ISBN 3-540-41782-6

CR Subject Classification (1998): E.3, G.2.1, D.4.6, K.6.5, F.2.1, C.2, J.1

ISSN 0302-9743
ISBN 3-540-41782-6 Springer-Verlag Berlin Heidelberg New York

Springer-Verlag Berlin Heidelberg New York
a member of BertelsmannSpringer Science+Business Media GmbH

http://www.springer.de
© Springer-Verlag Berlin Heidelberg 2001
Printed in Germany

Typesetting: Camera-ready by author, data conversion by PTP-Berlin, Stefan Sossna
Printed on acid-free paper   SPIN: 10782183   06/3142   5 4 3 2 1 0

Preface

I would like to welcome all the participants to the 3rd International Conference on Information Security and Cryptology (ICISC 2000). It is sponsored by the Korea Institute of Information Security and Cryptology (KIISC) and is being held at Dongguk University in Seoul, Korea from December 8 to 9, 2000. This conference aims at providing a forum for the presentation of new results in research, development, and application in information security and cryptology. This is also intended to be a place where research information can be exchanged.

The Call for Papers brought 56 papers from 15 countries and 20 papers will be presented in five sessions. As was the case last year the review process was totally blind and the anonymity of each submission was maintained. The 22 TPC members finally selected 20 top-quality papers for presentation at ICISC 2000.

I am very grateful to the TPC members who devoted much effort and time to reading and selecting the papers. We also thank the experts who assisted the TPC in evaluating various papers and apologize for not including their names here. Moreover, I would like to thank all the authors who submitted papers to ICISC 2000 and the authors of accepted papers for their preparation of camera-ready manuscripts. Last but not least, I thank my student, Joonsuk Yu, who helped me during the whole process of preparation for the conference.

I look forward to your participation and hope you will find ICISC 2000 a truly rewarding experience.

December 2000   Dongho Won

Organization

The 3rd International Conference on Information Security and Cryptology (ICISC 2000) is organized and sponsored by the Korea Institute of Information Security and Cryptology (KIISC).

Executive Committee

Kil Hyun Nam   General Chair (Presient of KIISC, Korea)
Dong Ho Won   TPC chair (Sungkyunkwan University, Korea)
Jae Ho Shin   Organizing Chair (Dongguk University, Korea)

Technical Program Committee

Dong Ho Won	Sungkyunkwan University, Korea
Zongduo Dai	Academia Sinica, P.R.C.
Ed Dawson	Queensland University of Technology, Australia
Chul Kim	Kwangwoon University, Korea
Kwang Jo Kim	Information and Communications University, Korea
Kaoru Kurosawa	Tokyo Inst. of Tech., Japan
Kwok-Yan Lam	National University of Singapore, Singapore
Kyoung Goo Lee	KISA, Korea
Pil Joong Lee	Pohang University of Sci. and Tech., Korea
Chae Hoon Lim	Future System Inc., Korea
Masahiro Mambo	Tohoku University, Japan
Jong In Lim	Korea University, Korea
Chris Mitchell	University of London, U.K.
Sang Jae Moon	Kyungpook National University, Korea
Kaisa Nyberg	Nokia Research Center, Finland
Eiji Okamoto	Toho University, Japan
Tatsuaki Okamoto	NTT, Japan
Choon Sik Park	ETRI, Korea
Sung Jun Park	BCQRE CO., LTD, Korea
Bart Preneel	Katholieke Universiteit Leuven, Belgium
Heung Youl Youm	Soonchunhyang University, Korea
Moti Yung	CertCo, U.S.A.
Yuliang Zheng	Monash University, Australia

Organizing Committee

Jae Ho Shin	Dongguk University, Korea
Chee Sun Won	Dongguk University, Korea
Sang Kyu Park	Hanyang University, Korea
Ha Bong Chung	Hongik University, Korea
Dong Hoon Lee	Korea University, Korea
Jae Jin Lee	Dogguk University, Korea
Howang Bin Ryou	Kwangwoon University, Korea
Seok Woo Kim	Hansei University, Korea
Yong Rak Choi	Taejon University, Korea
Hyun Sook Cho	ETRI, Korea
Hong Sub Lee	KISA, Korea
Seung Joo Han	Chosun University, Korea
Min Surp Rhee	Dankook University, Korea
Seog Pal Cho	Seonggul University, Korea
Kyung Seok Lee	KIET, Korea
Joo Seok Song	Yonsei University, Korea
Jong Seon No	Seoul National University, Korea
Tai Myoung Chung	Sungkyunkwan University, Korea

Sponsoring Institutions

MIC (Ministry of Information and Communication), Korea
IITA (Institute of Information Technology Assessment), Korea
KISA (Korea Information Security Agency, Korea
Dongguk University, Korea
The Electronic Times, Korea

Table of Contents

A Note on the Higher Order Differentail Attack of Black Ciphers with Two-Block Structures

Ju-Sung Kang, Seongtaek Chee, and Choonsik Park

Section 8100, Department of Basic Technology, NSRI
161 Kajong-Dong, Yusong-Gu, Taejon, 305-350, KOREA
{jskang,chee,csp}@etri.re.kr

Abstract. We study on the security against higher order differential attack on block ciphers with two-block structure which have provable security against differential and linear cryptanalysis. The two-block structures are classified three types according to the location of round function such as C(Center)-type, R(Right)-type, and L(Left)-type. We prove that in the case of 4 rounds encryption function, these three types provide an equal strength against higher order differential attack and that in the case of 5 or more rounds, R-type is weaker than C-type and L-type. Moreover, we show that these facts also hold similarly for probabilistic higher order differential attack.

Key words : DC, LC, provable security, (probabilistic) higher order differential attack, two-block structure

1 Introduction

The most well-known methods of analyzing block ciphers today is differential cryptanalysis(DC) and linear cryptanalysis(LC). DC, proposed by Biham and Shamir[2, 3] in 1990, is a chosen plaintext attack in which the attacker chooses plaintexts of certain well-considered differences. LC, published by Matsui[11] in 1993, is a known plaintext attack. In the attack based on LC, the attacker finds some effective linear expressions which are called linear approximations.

There are some measures in order to evaluate the security of block ciphers against DC and LC. When DC was proposed for the first time, the security of given block ciphers was evaluated by its maximum differential characteristic probability, which was found heuristically. However, Lai et al.[10] pointed out that block cipher designers should use the maximum average of differential probability instead of the maximum differential characteristic probability for evaluating security against DC. This situation of LC is similar to that of DC. When LC was first presented, the security against LC was evaluated by the maximum linear approximation probability. Nyberg[15] claimed that designers should use the maximum average of linear approximation probabilities, which is called the linear hull probability, instead of the maximum linear approximation probability.

Nyberg and Knudsen[16] showed that Feistel ciphers are provably secure against DC and LC, which means that the upper bounds of the maximum average of differential and linear probabilities of them are sufficiently small. Moreover they proposed the block cipher which is called $\mathcal{KN}$ cipher and provably secure against DC and LC. The notion of provable security against DC and LC was also researched by Matsui[12], Aoki and Ohta[1], and Kaneko et al.[7].

Meanwhile, block ciphers with provable security against DC and LC do not guarantee their security against other attacks. For example, Jakobsen and Knudsen[6] showed that the $\mathcal{KN}$ cipher can be broken by the higher order differential attack with much less complexity than DC and LC. The complexity of DC(or LC) for the $\mathcal{KN}$ cipher with 6 rounds is about 2^{60}, but that of higher order differential attack is only about 2^9. Therefore, it is clear that higher order differential attack provides a security evaluation aspect different from that of provable security against DC and LC.

In this work we investigate the relationship between the security of block ciphers against higher order differential attack and the overall structures of block ciphers. The overall structures of block ciphers with provable security against DC and LC are mainly two-block structures in which the plaintext is divided into two equal sub-blocks and one sub-block is transformed by a round function and then exclusive-ored with the other sub-block in each round. The two block structures are classified as three types according to the location of round function such as C(Center)-type, R(Right)-type, and L(Left)-type. We prove that in the case of 4 rounds encryption function, these three types provide the equal strength against higher order differential attack and that in the case of 5 or more rounds, R-type is weaker than C-type and L-type. Further, we also discuss the security of block ciphers with two-block structures against probabilistic higher order differential attack[4].

2 Two-block structures with provable security against DC and LC

Let F be a round function of a block cipher with n input and output bits. That is, $F : Z_2^n \to Z_2^n$. For any given Δx, Δy, a, $b \in Z_2^n$, define differential and linear probability of F by

$$DP^F(\Delta x \to \Delta y) = \frac{\#\{x \in Z_2^n \ : \ F(x) \oplus F(x \oplus \Delta x) = \Delta y\}}{2^n}$$

and

$$LP^F(a \to b) = \left(\frac{\#\{x \in Z_2^n \ : \ <a, x> = <b, F(x)>\}}{2^{n-1}} - 1 \right)^2 ,$$

respectively, where $< \alpha, \beta >$ denotes the parity(0 or 1) of bitwise product of α and β. DP^F and LP^F for a strong cryptographic function F should be small for any $\Delta x \neq 0$ and $b \neq 0$. So we define parameters represent immunity of F against DC and LC as follows:

$$DP^F_{\max} = \max_{\Delta x \neq 0, \ \Delta y} DP^F(\Delta x \to \Delta y)$$

and

$$LP_{\max}^F = \max_{a,\ b \neq 0} LP^F(a \to b)\ .$$

Assume that E is a key-dependent encryption function with $2n$ input and output bits and K be the set of all possible key values. For any $k \in K$, $E^{(k)}$ denotes an one-variable function from Z_2^{2n} to Z_2^{2n} with the fixed key k. Then we can define $DP^{E^{(k)}}(\Delta x \to \Delta y)$ and $LP^{E^{(k)}}(a \to b)$ alike to the definitions for F, where Δx, Δy, a, $b \in Z_2^{2n}$. The averages of differential and linear probability of E are defined by

$$DP^E(\Delta x \to \Delta y) = \frac{1}{\#K} \sum_{k \in K} DP^{E^{(k)}}(\Delta x \to \Delta y)$$

and

$$LP^E(a \to b) = \frac{1}{\#K} \sum_{k \in K} LP^{E^{(k)}}(a \to b)\ ,$$

respectively. If DP^E and LP^E are sufficiently small for any Δx, Δy, a, $b \in Z_2^{2n}$ such that $\Delta x \neq 0$ and $b \neq 0$, we say that E is provably secure against DC and LC. Equivalently, E has provable security against DC and LC if

$$DP_{\max}^E \overset{def}{=} \max_{\Delta x \neq 0,\ \Delta y} DP^E(\Delta x \to \Delta y)$$

and

$$LP_{\max}^E \overset{def}{=} \max_{a,\ b \neq 0} LP^E(a \to b)$$

are low enough.

The two-block structures with provable security against DC and LC are classified as three types according to the location of round function F. Three possible positions of the round function F are expressed in Fig. 1.

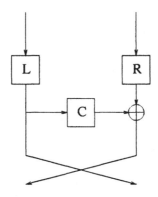

Fig. 1. Possible positions of the round function

We say that a two-block structure is C-type if the round function F is situated in the center of the two blocks. Similarly, we define that a two-block structure is R-type(or L-type) if the round function F is located in the right(or left) hand side of the two blocks. See Fig. 2.

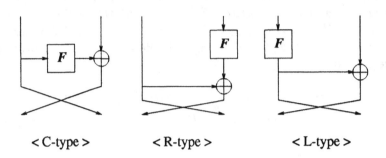

< C-type >   < R-type >   < L-type >

Fig. 2. Three types of two-block structures

The two-block structure of C-type is well-known as Feistel structure and applied to the overall structure of DES. Nyberg and Knudsen[16] proved that if $DP_{\max}^F \leq p$ and $LP_{\max}^F \leq q$, then $DP_{\max}^E$(or $LP_{\max}^E$) of the block cipher with C-type structure can be evaluated as $DP_{\max}^E \leq 2p^2$(or $LP_{\max}^E \leq 2q^2$), where the number of rounds is 4(or 3 if F is bijective) or more. Aoki and Ohta[1] improved the results of Nyberg and Knudsen as follows: $DP_{\max}^E \leq p^2$ and $LP_{\max}^E \leq q^2$ if F is bijective and the block cipher of C-type has more than 3 rounds.

Matsui[12] introduced another structure of block ciphers with provable security against DC and LC which is different from Feistel structure. It was the two-block structure of R-type and applied to the overall structure of block cipher MISTY[13]. The R-type structure has a merit that it realizes parallel computation of the round functions, but its round functions must be a bijection for decryption process. Matsui[12] proved that $DP_{\max}^E \leq p^2$ and $LP_{\max}^E \leq q^2$ if F is a bijection and the block cipher of R-type has more than 3 rounds. That is, block ciphers of C-type and R-type have the same provable security against DC and LC.

On the other hand, Kaneko et al.[7] showed that the block cipher of L-type and R-type have the same provable security against DC and LC by using the duality between probabilities of differential and linear hull. Skipjack[14] is a block cipher of four-block structure, but it seems that the overall structure of Skipjack is an extension of the L-type two-block structure.

Three different types, C-type, R-type, and L-type, have the same provable security against DC and LC, but this does not guarantee that their security against other attacks is also the same. Especially, it seems that their immunities against higher order differential attack are not in accord with each other. So we investigate the security against higher order differential attack of three different types. Throughout this paper we assume that the round functions in the R-type and L-type are bijective.

3 Algebraic degree and higher order differential attack

Let $x = (x_1, \cdots, x_n) \in Z_2^n$ be an n-dimensional binary vector and $f(x)$ be a Boolean function. Then $f(x)$ can be represented as the algebraic normal form:

$$f(x) = a_0 \oplus a_1 x_1 \oplus \cdots \oplus a_n x_n$$
$$\oplus a_{12} x_1 x_2 \oplus \cdots \oplus a_{n-1,n} x_{n-1} x_n$$
$$\cdots$$
$$\oplus a_{12\ldots n} x_1 x_2 \cdots x_n .$$

The algebraic degree of f, $\deg(f)$, is defined as the degree of the highest degree term of its algebraic normal form. If F is a vectorial Boolean function such as $F(x) = (f_1(x), \cdots, f_m(x))$, then the degree of F is defined as

$$\deg(F) = \max_{1 \le i \le m} \deg(f_i) .$$

Lai[9] introduced the notion of higher order derivative and proposed some useful facts for analyzing block ciphers.

Definition 1 *For a vectorial Boolean function $F : Z_2^n \longrightarrow Z_2^m$, the derivative of F at point $a \in Z_2^n$ is defined as*

$$\Delta_a F(x) = F(x \oplus a) \oplus F(x) ,$$

the i-th derivative of F at $(a_1, \cdots, a_i) \in (Z_2^n)^i$ is defined by

$$\Delta_{a_1, \cdots, a_i}^{(i)} F(x) = \Delta_{a_i} \left(\Delta_{a_1, \cdots, a_{i-1}}^{(i-1)} F(x) \right) ,$$

where $\Delta_{a_1, \cdots, a_{i-1}}^{(i-1)} F(x)$ is $(i-1)$-th derivative of F at $(a_1, \cdots, a_{i-1})$ and 0-th derivative of F is defined to be F itself.

Knudsen[8] extended the notion of classical differentials into higher order differentials which can be inferred from the definition of higher order derivatives.

Definition 2 *Let $F : Z_2^n \longrightarrow Z_2^m$ be a vectorial Boolean function. The differential of order i is an $(i+1)$-tuple $(a_1, \cdots, a_i, b) \in (Z_2^n)^i \times Z_2^m$ such that*

$$\Delta_{a_1, \cdots, a_i}^{(i)} F(x) = b , \quad x \in Z_2^n .$$

Various higher order differential attacks[6, 8, 17, 18] are based on the following propositions shown by Lai[9].

Proposition 1
$$deg(\Delta_a F(x)) \le deg(F(x)) - 1 .$$

Proposition 2 *Let $\mathcal{L}[a_1, \cdots, a_i]$ denote the subspace generated by $\{a_1, \cdots, a_i\}$. Then*

$$\Delta_{a_1, \cdots, a_i}^{(i)} F(x) = \bigoplus_{c \in \mathcal{L}[a_1, \cdots, a_i]} F(x \oplus c) .$$

Proposition 3 *If a_i is linearly dependent of $a_1, \cdots, a_{i-1}$, then $\Delta_{a_1, \cdots, a_i}^{(i)} F(x) = 0$.*

Knudsen[8] showed that the attacks on 5 rounds cipher of C-type using higher order differentials are much more efficient than conventional DC. Also Jakobsen and Knudsen[6] pointed out that the higher order differential attack is effective in attacking the $\mathcal{KN}$ cipher because the algebraic degree of the round function F is low. Later, Shimoyama et al.[17] improved this attack by reducing the required chosen plaintexts and running time.

4 Higher order differential attack on block ciphers of two-block structures

In this section we investigate the security of two-block structures against higher order differential attack. Let $x = (x^L, x^R)$ be the plaintext, where x^L and x^R denote the left and right sub-blocks of x, respectively. Similarly, let

$$E(x) = y = (y^L, y^R) = (E^L(x), E^R(x))$$

be the corresponding ciphertext and $x_i = (x_i^L, x_i^R)$ denote the corresponding result of i iterative rounds encryption. We say that the round function is $F : Z_2^n \longrightarrow Z_2^n$ with $\deg(F) = d$ and

$$F_{k_i}(x) \stackrel{\triangle}{=} F(x \oplus k_i) \, ,$$

where k_i is the i-th round subkey of n-bit. Further, we assume that $d \leq n$ and round subkeys are always exclusive-ored just prior to the round function F.

4.1 Higher order differential attack on 4 rounds block ciphers with two-block structures

Firstly, we assume that the number of rounds is 4 in each two-block structure. The 4 rounds block cipher of C-type structure is in Fig. 3. We obtain that

$$x_3^L = x^R \oplus F_{k_1}(x^L) \oplus F_{k_3}(x^L \oplus F_{k_2}(x^R \oplus F_{k_1}(x^L))) \, ,$$
$$x_3^R = x^L \oplus F_{k_2}(x^R \oplus F_{k_1}(x^L)) \tag{1}$$

from Fig. 3. It suffices to know only the value of x_3^R in order to analogize the value of x_3, since $x_3^L = y^L$. Say

$$x_3^R = T_{[k_1, k_2]}^C(x) = x^L \oplus F_{k_2}(x^R \oplus F_{k_1}(x^L)) \, ,$$

because x_3^R depends upon k_1 and k_2 by (1). By the proposition 1, 2, and 3, for any linearly independent vectors $a_1, \cdots, a_d$ such that $a_j = (0, a_j^R) \in Z_2^n \times Z_2^n$, $1 \leq j \leq d$, and for some constant c,

$$\Delta_{a_1, \cdots, a_d}^{(d)} T_{[k_1, k_2]}^C(x) = \bigoplus_{v \in \mathcal{L}[a_1, \cdots, a_d]} T_{[k_1, k_2]}^C(x \oplus v) = c \tag{2}$$

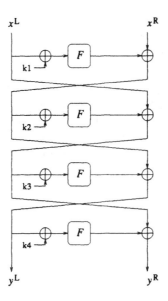

Fig. 3. C-type structure with 4 rounds

holds for all $x \in Z_2^{2n}$ and $k_1, k_2 \in K$. Thus we can obtain that

$$c = \bigoplus_{v \in \mathcal{L}[a_1, \cdots, a_d]} T_{[0,0]}^C(v) . \tag{3}$$

Moreover in the C-type structure $x_3^R = F_{k_4}(y^L) \oplus y^R$ holds, hence

$$T_{[k_1,k_2]}^C(x) = F_{k_4}(E^L(x)) \oplus E^R(x) . \tag{4}$$

Consequently, from (2), (3), and (4), we obtain that

$$\bigoplus_{v \in \mathcal{L}[a_1, \cdots, a_d]} F_{k_4}(E^L(v)) \oplus E^R(v) = \bigoplus_{v \in \mathcal{L}[a_1, \cdots, a_d]} T_{[0,0]}^C(v) \tag{5}$$

By using (5), the process of recovering the last round subkey value k_4 is as follows:

1. On Z_2^{2n}, choose linearly independent vectors $a_1, \cdots, a_d$ such that

$$a_j = (0, a_j^R) \in Z_2^n \times Z_2^n , \quad \forall 1 \le j \le d .$$

2. Assume the value of k_4.
3. For each $v \in \mathcal{L}[a_1, \cdots, a_d]$, do the following:
   (a) Obtain $E^L(v)$ and $E^R(v)$.
   (b) Compute $T_{[0,0]}^C(v)$.
   (c) Compute $F_{k_4}(E^L(v))$.
4. If (5) holds, then k_4 is the right key. Otherwise, go to step 2.

From the above process, we can obtain the fact that the required number of chosen plaintexts is 2^d and the maximum running time is about $4 \cdot 2^d + 2^{d+n}$ times of the computation of round function F for finding the last round key.

On the other hand, 4 rounds R-type and L-type structures are expressed in Fig. 4 and 5, respectively. Remember that the round function F should be invertible for decryption in R-type and L-type structures. Assume that in the last round of R-type and L-type structures, more round subkey is exclusive-ored just after the round function F. This assumption is resonable since without such a process, the round function F in the last round has no cryptographic meaning.

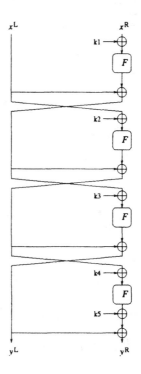

Fig. 4. R-type structure with 4 rounds

On the 4 rounds R-type structure, we obtain that

$$
\begin{aligned}
x_3^L &= x^L \oplus F_{k_1}(x^R) \oplus F_{k_2}(x^L) \oplus F_{k_3}(x^L \oplus F_{k_1}(x^R)) \,, \\
x_3^R &= x^L \oplus F_{k_1}(x^R) \oplus F_{k_2}(x^L)
\end{aligned}
\tag{6}
$$

by figure 4. By (6),

$$
\tilde{x}_3^R = T^R_{[k_1,k_2,k_4]}(x) = x^L \oplus F_{k_1}(x^R) \oplus F_{k_2}(x^L) \oplus k_4 \,,
$$

where $\tilde{x}_3^R = x_3^R \oplus k_4$, and

$$\bigoplus_{v \in \mathcal{L}[a_1, \cdots, a_d]} F_{k_5}^{-1}(E^L(v) \oplus E^R(v)) = \bigoplus_{v \in \mathcal{L}[a_1, \cdots, a_d]} T_{[0,0,0]}^R(v) \qquad (7)$$

since $\tilde{x}_3^R = F_{k_5}^{-1}(y^L \oplus y^R)$. Therefore we know the fact that the required chosen plaintexts and running time of C-type and R-type structures are the same.

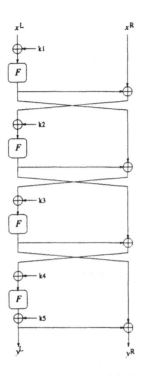

Fig. 5. L-type structure with 4 rounds

Similarly, we can obtain that by Fig. 5,

$$x_3^L = F_{k_2}(x^R \oplus F_{k_1}(x^L)) \oplus F_{k_3}(F_{k_1}(x^L) \oplus F_{k_2}(x^R \oplus F_{k_1}(x^L))) ,$$
$$x_3^R = F_{k_3}(F_{k_1}(x^L) \oplus F_{k_2}(x^R \oplus F_{k_1}(x^L))) . \qquad (8)$$

We must know the value of x_3^L to recover x_3 since $x_3^R = y^L \oplus y^R$. At this point, it is easy to have misunderstanding that the algebraic degree of the function representing x_3 is at least d^2 by taking a quick glance at (8). However in the L-type structure, if we recover (x_3^L, x_3^R), then we can also find the value of x_2^R since $x_2^R = x_3^L \oplus x_3^R$. So the attacker can use the formula

$$x_2^R = F_{k_2}(x^R \oplus F_{k_1}(x^L)) \qquad (9)$$

instead of (8). Now let $\tilde{x}_3^L = x_3^L \oplus k_4$, $\tilde{x}_2^R = \tilde{x}_3^L \oplus x_3^R$, and

$$\tilde{x}_2^R = T_{[k_1,k_2,k_4]}^L(x) = F_{k_2}(x^R \oplus F_{k_1}(x^L)) \oplus k_4 .$$

Then by choosing linearly independent vectors $a_1, \cdots, a_d$ such that $a_j^L = 0$, $1 \le j \le d$, we can obtain the formula which similar to (5) of C-type and (7) of R-type that

$$\bigoplus_{v \in \mathcal{L}[a_1,\cdots,a_d]} F_{k_5}^{-1}(E^L(v)) \oplus E^L(v) \oplus E^R(v) = \bigoplus_{v \in \mathcal{L}[a_1,\cdots,a_d]} T_{[0,0,0]}^L(v) ,$$

since

$$\tilde{x}_2^R = F_{k_5}^{-1}(y^L) \oplus y^L \oplus y^R .$$

From this, we obtain the fact that the required chosen pliantexts and running time for the higher order differential attack of 4 rounds L-type structure are the same to those of C-type and R-type structures with 4 rounds.

Theorem 1 *Let the algebraic degree of round function F be deg(F)=d. Then the required chosen plaintexts and maximum number of computations of round function F for the higher order differential attack of 4 rounds C-type, R-type, and L-type structures in order to recover the last round keys are all 2^d and $4 \cdot 2^d + 2^{d+n}$, respectively.*

4.2 Higher order differential attack on 5 or more rounds block ciphers with two-block structures

Now we consider the security of 5 or more rounds of three different types against higher order differential attack. Note that we should pay attention to the equation of x_4^R in the 5 rounds C-type and R-type structures in order to analyze the security against higher order differential attack. But in the case of 5 rounds L-type structure, we must observe that the equation of x_3^R. Concerning equations are as follows:

$$\text{C-type}: x_4^R = x^R \oplus F_{k_1}(x^L) \oplus F_{k_3}(x^L \oplus F_{k_2}(x^R \oplus F_{k_1}(x^L))) , \quad (10)$$
$$\text{R-type}: x_4^R = x^L \oplus F_{k_1}(x^R) \oplus F_{k_2}(x^L) \oplus F_{k_3}(x^L \oplus F_{k_1}(x^R)) , \quad (11)$$
$$\text{L-type}: x_3^R = F_{k_3}(F_{k_1}(x^L) \oplus F_{k_2}(x^R \oplus F_{k_1}(x^L))) . \quad (12)$$

By (10) and (12), it is easy to know that the algebraic degrees of functions representing x_4^R of C-type and x_3^R of L-type are at least d^2. On the contrary, from (11), we obtain the fact that in the 5 rounds R-type structure, the algebraic degree of function representing x_4^R can be decreased down to d if we choose linearly independent vectors $a_1, \cdots, a_d$ such that $\forall a_j^R = 0$, $1 \le j \le d$.

We investigate the relationship between the number of rounds of three types and the security against higher order differential attack to recover the last round keys. In the cases of C-type and L-type structures, the minimal algebraic degree for the attack is d where the number of rounds is 4, and that is d^2, d^3, $\cdots$

for 5 rounds, 6 rounds, $\cdots$, so on, respectively. However, in the case of R-type structure, the minimal algebraic degree for the attack after 4 rounds is increasing as d times per two rounds. We summarize these results in the following theorem.

Theorem 2 *Assume that r denotes the number of rounds with $r \geq 4$ and the algebraic degree of the round function F is d. Then the required plaintexts and running time for the higher order differential attacks of r rounds C-type, R-type, and L-type structures in order to recover the last round keys are as Table 1. In Table 1, $\lceil x \rceil$ denotes the greatest integer not exceeding x, and the running time represents the number of computations of round function F.*

Table 1. Higher order differential attack of three types

Types	# of rounds	# of chosen plaintexts	Running time
C-type	$r \geq 4$	$2^{d^{r-3}}$	$r \cdot 2^{d^{r-3}} + 2^{d^{r-3}+n}$
R-type	$r \geq 4$	$2^{d^{\lceil (r-3)/2 \rceil}}$	$r \cdot 2^{d^{\lceil (r-3)/2 \rceil}} + 2^{d^{\lceil (r-3)/2 \rceil}+n}$
L-type	$r \geq 4$	$2^{d^{r-3}}$	$r \cdot 2^{d^{r-3}} + 2^{d^{r-3}+n}$

4.3 Probabilistic higher order differential attack on the 4 or more rounds block ciphers with two-block structures

Recently, Iwata and Kurosawa[4] showed that a Feistel type block cipher is broken where the round function is approximated by a low degree vectorial Boolean function. They called this attack a probabilistic higher order differential attack because it was a generalization of the higher order differential attack to probabilistic one. In this subsection, we study the security against probabilistic higher order differential attack for block ciphers of two-block ciphers. At first, we introduce the notion of (d, μ)-expression in [5, 4].

Definition 3 *A vectorial Boolean function F is (d, μ)-expressible if there exists a vectorial Boolean function F' such that $deg(F) \leq d$ and*

$$Pr_x(F(x) = F'(x)) \geq \mu .$$

Now we assume that the round functions of C-type, R-type, and L-type structures are all (d, μ)-expressible. Then by the similar process to above two subsections and the proof of Theorem 4.1 of [4], we can obtain the following theorem.

Theorem 3 *Assume that r denotes the number of rounds with $r \geq 4$ and the round function F is (d, μ)-expressible. Then the required plaintexts and running time for the probabilistic higher order differential attacks of r rounds C-type, R-type, and L-type structures in order to recover the last round subkeys are as*

Table 2. In Table 2, N denotes the number of running times Algorithm 1 of [4] and the success probability is given by

$$\sum_{1 \le i \le N} \binom{N}{i} p^i (1-p)^{N-i} \left(\sum_{0 \le j \le i-1} \binom{N}{j} 2^{-nj}(1-2^{-n})^{N-j} \right)^{2^n - 1},$$

where $p = 1 - 2^{d+1}n(1-\mu)$.

Table 2. Probabilistic higher order differential attack of three types

Types	# of rounds	# of chosen plaintexts	Running time
C-type	$r \ge 4$	$N2^{d^{r-3}}$	$N(r \cdot 2^{d^{r-3}} + 2^{d^{r-3}+n})$
R-type	$r \ge 4$	$N2^{d^{\lceil (r-3)/2 \rceil}}$	$N(r \cdot 2^{d^{\lceil (r-3)/2 \rceil}} + 2^{d^{\lceil (r-3)/2 \rceil}+n})$
L-type	$r \ge 4$	$N2^{d^{r-3}}$	$N(r \cdot 2^{d^{r-3}} + 2^{d^{r-3}+n})$

A merit of the R-type structure is that it realizes parallel computation of the round functions without losing provable security against DC and LC. Thus from a view point of computational efficiency, R-type structure is better than C-type and L-type structures. However, by Theorem 2 and 3, we can obtain the fact that the R-type is weaker than C-type and L-type from a view point of security against (probabilistic) higher order differential attack. Here we comprehend again that it is hard to the security is compatible with computational efficiency.

On the other hand it is generally believed that designing a cryptographically good random function is easier than designing a good random permutation. Using this and based on our result, we can conclude that C-type is the best choice from these two points of view.

5 Conclusion

In this work we investigated the security against higher order differential attack for two-block structures of three different types which are provably secure against DC and LC. We basically used the relationship between the algebraic degree of round function and higher order differential attack. In the case of 4 rounds encryption, we showed that C-type, R-type, and L-type structures all have the same security against higher order differential attack and probabilistic higher order differential attack. Further, we proved that in the case of 5 or more rounds encryption, C-type and L-type are stronger than R-type in the view point of security against higher order differential attack and probabilistic higher order differential attack.

References

1. K. Aoki and K. Ohta, "Strict evaluation of the maximum average of differential probability and the maximum average of linear probability", *IEICE TRANS. FUNDAMENTALS*, No. 1, 1997, pp. 2-8.
2. E. Biham and A. Shamir, "Differential cryptanalysis of DES-like Cryptosystems", *Advances in Cryptology - CRYPTO'90, LNCS 537*, Springer-Verlag, 1990, pp. 2-21.
3. E. Biham and A. Shamir, "Differential cryptanalysis of DES-like Cryptosystems", *Journal of Cryptology*, 1991, No. 4, (1), pp. 3-72.
4. T. Iwata and K. Kurosawa, "Probabilistic higher order differential attack and higher order bent functions", *Advances in Cryptology - ASIACRYPT'99, LNCS 1716*, Springer-Verlag, 1999, pp. 62-74.
5. T. Jakobsen, "Cryptanalysis of block ciphers with probabilistic non-linear relations of low degree", *Advances in Cryptology - CRYPTO'98, LNCS 1462*, Springer-Verlag, 1998, pp. 212-222.
6. T. Jakobsen and L. R. Knudsen, "The interpolation attack on block ciphers", *Fast Software Encryption'97, LNCS 1267*, Springer-Verlag, 1997, pp. 28-40.
7. Y. Kaneko, F. Sano, and K. Sakurai, "On provable security against differential and linear cryptanalysis in generalized Feistel ciphers with multiple random functions", *Proceedings of SAC'97*, 1997, pp. 185-199.
8. L. R. Knudsen, "Truncated and higher order differentials", *Fast Software Encryption'95, LNCS 1008*, Springer-Verlag, 1995, pp. 196-211.
9. X. Lai, "Higher order derivatives and differential cryptanalysis", *Communications and Cryptography*, Kluwer Academic Press, 1994, pp. 227-233.
10. X. Lai, J. Massey, and S. Murphy, "Markov Ciphers and Differential Cryptanalysis", *Advances in Cryptology - Eurocrypt'91, LNCS 547*, Springer-Verlag, 1991, pp. 17-38.
11. M. Matsui, "Linear cryptanalysis method for DES cipher", Advances in Cryptology - Eurocrypt'93, LNCS 765, Springer-Verlag, 1993, pp. 386-397.
12. M. Matsui, "New Structure of Block Ciphers with Provable Security against Differential and Linear Cryptalaysis", *Fast Software Encryption, LNCS 1039*, Springer-Verlag, 1996, pp. 205-218.
13. M. Matsui, "New Block Encryption Algorithm MISTY", *Fast Software Encryption'97, LNCS 1267*, Springer-Verlag, 1997, pp. 54-68.
14. National Institute of Standards and Technology, "Skipjack and KEA Algorithm Specifications", http://csrc.nist.gov/encryption/skipjack-jea.htm, 1998.
15. K. Nyberg, "Linear Approximation of Block Ciphers", Advances in Cryptology - Eurocrypt'94, LNCS 950, Springer-Verlag, 1994, pp. 439-444.
16. K. Nyberg and L. R. Knudsen, "Provable Security against Differential Cryptanalysis", *Journal of Cryptology*, 1995, No. 8, (1), pp. 27-37.
17. T. Shimoyama, S. Moriai, and T. Kaneko, "Improving the higher order differential attack and cryptanalysis of the KN Cipher", *Information Security Workshop, Proceedings*, 1997, pp. 1-8.
18. M. Sugita, "Higher order differential attack of block ciphers MISTY1,2", *Technical Report of IEICE, ISEC 98-4*, 1998, pp. 31-40.

On the Strength of KASUMI
without FL Functions
Against Higher Order Differential Attack

Hidema TANAKA, Chikashi ISHII, Toshinobu KANEKO

Science University of TOKYO
2641 Yamazaki Noda CHIBA, 278-8510, JAPAN
tel:+81-471-24-1501 fax:+81-471-25-8651
email:{tanaka,isii,kaneko}@kaneko01.ee.noda.sut.ac.jp

Abstract. The encryption algorithm KASUMI is referred to MISTY, proposed by Matsui, is a provably secure against Linear cryptoanalysis and Differential attack. We attacked KASUMI without FL functions by using Higher Order Differential Attack. The necessary order of Higher Order Differential Attack depends on the degree of F function and it is determined by the chosen plaintext. We found effective chosen plaintext which enables the attack to 4 round KASUMI without FL functions. As the result, we can attack it using 2nd order differentials. This attack needs about 1,416 chosen plaintexts.

1 Introduction

Within the security architecture of the 3GPP system is based on the KASUMI. KASUMI is a block cipher that produce a 64[bit] output from a 64[bit] input under the control of a 128[bit] key. It has DES-like structure with 8 round f function named FO. KASUMI is based on MISTY1 proposed by Matsui which is probably secure against Linear cryptoanalysis and Differential cryptoanalysis. From this fact, the main part of security of KASUMI will be guaranteed by FO function. Thus we estimate the strength of FO function against Higher Order Differential Attack in this paper.

Higher Order Differential Attack is a chosen plaintext attack which uses the fact that the value of higher order differential of the output does not depend on the key. The sufficient order for the attack affects the necessary number of chosen plaintexts and computational cost. So the attacker needs to search for the effective choice of plaintexts.

In this paper, we show that 4 round KASUMI without FL functions can be attackable using effective 2nd order differentials. This attack needs 1,416 chosen plaintexts and $2^{22.11}$ computational cost.

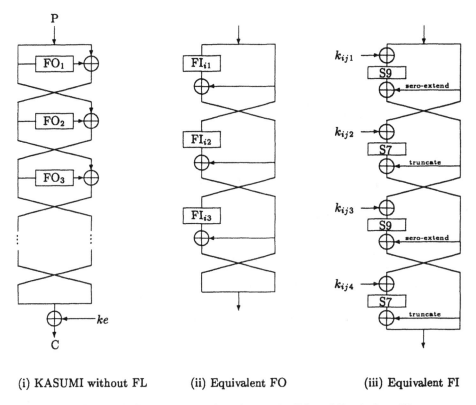

(i) KASUMI without FL   (ii) Equivalent FO   (iii) Equivalent FI

Fig. 1. KASUMI without FL, Equivalent FO and Equivalent FI

2 Modified KASUMI

KASUMI is referred to MISTY1. The differences between them are as follows.

- The design of S-boxes, S7 and S9.
- The design of FL function.
- The number of S-box (S7) in a FI function.

Since the main part of security of KASUMI will be guaranteed by FO function, we omit FL function in this paper. And to simplify the Attack equation, we deduce the equivalent FO function and FI function. Figure 1 shows them. FO function is consisted of 3 round FI functions and FI function is consisted of 2 kinds of S-boxes called S7 and S9. S7 is 7[bit] table and the algebraic degree is 3. S9 is 9[bit] table and the degree is 2. We denote FO function in i-th round as FO_i, j-th FI function in FO_i as FI_{ij}, and equivalent sub-key as k_{ijk} for k-th S-box.

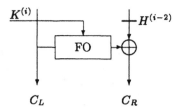

Fig. 2. Last round of r round KASUMI without FL

3 Higher Order Differential Attack

3.1 Higher Order Differential

Let $F(X; K)$ be a function : $GF(2)^n \times GF(2)^s \mapsto GF(2)^m$.

$$Y = F(X; K), \quad (X \in GF(2)^n, Y \in GF(2)^m, K \in GF(2)^s) \tag{1}$$

Let $(A_1, A_2, \ldots, A_i)$ be a set of linear independent vectors in $GF(2)^n$ and $V^{(i)}$ be a subspace spanned by the set. We define $\Delta^{(i)} F(X; K)$ as the i-th order differential of $F(X; K)$ with respect to X as follows.

$$\Delta^{(i)} F(X; K) = \bigoplus_{A \in V^{(i)}} F(X + A; K) \tag{2}$$

If $\deg_X F(X; K) = N$, we have the following properties.

$$\deg_X \{F(X; K)\} = N \Rightarrow \Delta^{(N+1)} F(X; K) = 0 \tag{3}$$

3.2 Attacking Equation

Figure 2 shows the last round of i round KASUMI. $H^{(i-2)}(P)$ which is the output from FO_{i-2} can be calculated by cipher text $C_L(P)$, $C_R(P) \in GF(2)^{32}$ and sub-key $K^{(i)}$, as follows.

$$H^{(i-2)}(P) = FO(C_L(P); K^{(i)}) + C_R(P) \tag{4}$$

If $\deg_X H^{(i-2)}(P) = N$, following equation holds.

$$\Delta^{(N+1)} H^{(i-2)}(P) = 0 \tag{5}$$

where $F(\cdot)$ denotes the function $GF(2)^{32} \times GF(2)^{75 \times (i-2)} \mapsto GF(2)^{32}$. $K^{(1\cdots(i-2))}$ denotes the set of keys for previous $(i - 2)$ rounds.

From equations (4) and (5), we can derive the following equation.

$$\bigoplus_{A \in V^{(N+1)}} \{FO(C_L(P + A); K^{(i)}) + C_R(P + A)\} = 0 \tag{6}$$

If the right hand of this equation can be determined for some analytical method, we can use this equation (6) as the Attack equation for $K^{(i)}$.

4 Analysis of Modified KASUMI

4.1 Effective chosen plaintext

The order of Higher Order Differential Attack affects the necessary number of chosen plaintexts and computational cost. The attacker needs to search for the minimum one. The order of output is determined by the choice of variable bits.

Let $P = (P_L, P_R)$ $P_L, P_R \in GF(2)^{32}$ be the plaintext. Then $H^{(2)}$ can be calculated as follows.

$$FO_2(FO_1(P_L; K_1) \oplus P_R; K_2) \oplus P_L = H^{(2)} \qquad (7)$$

P_L passes 2 FO functions. On the other hand, P_R passes only 1. So we fix P_L in the following to slow increase of the degree of output.

4.2 Formal analysis of degree

In this section, we show the formal analysis of degree. Figure 3 shows the increase of degree in FI_{21} and FO_2. The symbol $< a|b >$ denotes that the degree of left block is a and the right block is b. This figure shows that the degree of output from FI_{21} and FI_{22} is $< 9|4 >$ and the degree of output from FI_{23} is $< 54|36 >$.

In the same way, the degree of output from FO_3 will be $< 324|216 >$. As mentioned before, we fix half of 64[bit] plaintext, so we can not calculate larger than 32nd order differential. Thus we can not calculate the higher order differential of $H^{(2)}$ in the simple way.

4.3 Computational analysis of degree

As mentioned before, the degree of output depends on the choice of variable bits. We made brute force search for the effective choice of variable bits in P_R for 2nd ~ 7th order differential by the computer simulations.

Due to the limited space, we show a part of result of $H^{(2)}$ in Table 1. The symbol A^{aL} and A^{aR} denote left a[bit] of sub-block A and right a[bit] of sub-block A. The symbol A_m^{aL} denotes m-th bit of A^{aL} and A_m^{aR} denotes m-th bit of A^{aR} $(m = 0 \sim (a-1))$.

We have not found the effective choice for $H^{(3)}$ yet. This is our future work.

5 Attack of KASUMI

From the computer simulations, we conclude that 4round KASUMI without FL functions can be attackable using 2nd order differential. For 2nd order differential, following holds.

$$\Delta^{(2)} H^{16L} = 0 \qquad (8)$$

From equation (6), we can derive following Attack equation.

$$\bigoplus_{A \in V^{(1)}} FO(C_L(P); K_4) \oplus C_R(P) = \Delta^{(2)} H^{16L}(P) = 0 \qquad (9)$$

n	Choice of variable bits	Output sub-blocks or bits hold $\Delta^{(n)} = 0$
2	$P^{16L}_{(1)}, P^{16R}_{(1)}$ $P^{16L}_{(1)}, P^{16R}_{(2)}$ ⋮ $P^{16L}_{(16)}, P^{16R}_{(16)}$	H^{16L}
3	$P^{16R}_{(1,3)}, P^{16L}_{(1)}$ $P^{16R}_{(1,3)}, P^{16L}_{(2)}$ ⋮ $P^{16R}_{(1,3)}, P^{16L}_{(16)}$	$H^{16L}, H^{16R}_{(1,5,7,8,9)}$
4	$P^{16R}_{(1,2)}, P^{16L}_{(1)}, P^{16R}_{(3)}$ $P^{16R}_{(1,2)}, P^{16L}_{(1)}, P^{16R}_{(4)}$ ⋮ $P^{16R}_{(1,2)}, P^{16L}_{(16)}, P^{16R}_{(16)}$	$H^{16L}, H^{16R}_{(1\sim9)}$
5	$P^{16R}_{(1,2,3)}, P^{16L}_{(1)}, P^{16R}_{(4)}$ $P^{16R}_{(1,2,3)}, P^{16L}_{(1)}, P^{16R}_{(5)}$ ⋮ $P^{16R}_{(1,2,3)}, P^{16L}_{(16)}, P^{16R}_{(16)}$	$H^{16L}, H^{16R}_{(1\sim9)}$
6	$P^{16R}_{(1,3,4)}, P^{16R}_{(8)}, P^{16R}_{(9)}, P^{16R}_{(10)}$ $P^{16R}_{(1,3,4)}, P^{16R}_{(8)}, P^{16R}_{(9)}, P^{16R}_{(11)}$ ⋮ $P^{16R}_{(1,3,4)}, P^{16R}_{(14)}, P^{16R}_{(15)}, P^{16R}_{(16)}$	$H^{(2)}$
7	$P^{16R}_{(1,2,3,4)}, P^{16R}_{(8)}, P^{16R}_{(9)}, P^{16R}_{(10)}$ $P^{16R}_{(1,2,3,4)}, P^{16R}_{(8)}, P^{16R}_{(9)}, P^{16R}_{(11)}$ ⋮ $P^{16R}_{(1,2,3,4)}, P^{16R}_{(14)}, P^{16R}_{(15)}, P^{16R}_{(16)}$	$H^{(2)}$

Table 1. Result of computer simulation for $H^{(2)}$

where

$$\mathcal{K}_4 = (k_{411}, k_{412}, k_{413}, k_{421}, k_{422}, k_{423})$$
$$k_{411}, k_{413}, k_{421}, k_{423} \in \mathrm{GF}(2)^9 \qquad k_{412}, k_{422} \in \mathrm{GF}(2)^7$$

5.1 Preparation

There are 50[bit] unknowns in the Attack equation (9). We can not determine them by brute force search. In this paper, we adapt the algebraic method to solve the equation. The attacker solves an Attack equation using known cipher texts and plaintexts. Let Z be the output from FO_4.

$$Z = (Z^{L7L}, Z^{L9R}, Z^{R7L}, Z^{R9R}) \tag{10}$$

We can calculate Z^{L9R} which has the smallest degree among Z as follows.

$$Z^{L9R} = S9(k_{413} \oplus C_3^L \oplus S9(k_{411} \oplus C_1^L | C_2^L))$$

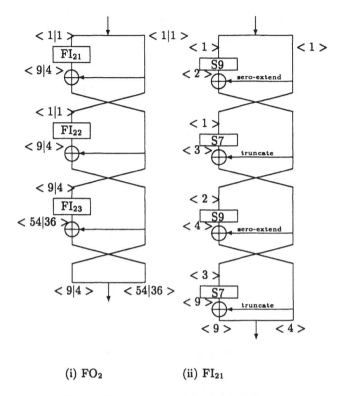

(i) FO$_2$ (ii) FI$_{21}$

Fig. 3. Increasemnet of degree in FO$_2$

$$\oplus S9(k_{411} \oplus C_1^L | C_2^L) \oplus S7(k_{412} \oplus C_3^L)$$
$$\oplus S9(k_{423} \oplus C_3^R \oplus S9(k_{421} \oplus C_1^R | C_2^R))$$
$$\oplus S9(k_{421} \oplus C_1^R | C_2^R) \oplus S7(k_{422} \oplus C_3^R)$$
$$\oplus C_2^R | C_3^L \oplus ke_2^R | ke_3^L \tag{11}$$

We regard all the variable term with respect to $\mathcal{K}$ as the independent variables. The highest term of Z^{L9R} is $S9(k_{413} \oplus C_3^L \oplus S9(k_{411} \oplus C_1^L | C_2^L))$. So the degree of Z^{L9R} with respect to C is 4. Since $\Delta^{(2)} H^{(2)} = 0$ holds, we have following.

$$\Delta^{(2)} \{ Z^{L9R} \oplus C_R(P) \} = 0 \tag{12}$$

Note that the appropriate bits of $C_R(P)$ is selected for this equation. By solving this equation, we can determine k_{411}, k_{412}, k_{413}, k_{421}, k_{422}, k_{423}. From equation (9), we have following.

$$\bigoplus_{P \in V^{(2)}} Z^{L9R}(C_L(P); \mathcal{K}_4) \oplus C_R(P) = 0 \tag{13}$$

We can transform this equation as follows.

$$\bigoplus_{P \in V^{(2)}} Z^{L9R}(C_L(P); \mathcal{K}_4) = \bigoplus_{P \in V^{(2)}} C_R(P)$$

$$\bigoplus_{P \in V^{(2)}_{\backslash\{0\}}} Z^{L9R}(C_L(P) \oplus C_L(0); \mathcal{K}_4) = \bigoplus_{P \in V^{(2)}} C_R(P)$$

$$\bigoplus_{P \in V^{(2)}_{\backslash\{0\}}} \Delta^{(1)} Z^{L9R}(C_L(P); \mathcal{K}_4) = \bigoplus_{P \in V^{(2)}} C_R(P)$$

wfere $V^{(2)}_{\backslash\{0\}}$ denotes that $V^{(2)}$ except All-zero.

By this transformation, the degree of left hand side of the equation with respect to $\mathcal{K}$ becomes 3. Attacker can calculate the constant value of right hand side of the equation. As the result, we have following Attack equation.

$$\bigoplus_{P \in V^{(2)}_{\backslash\{0\}}} \Delta^{(1)} Z^{L9R}(C_L(P); \mathcal{K}_4) = \text{const} \tag{14}$$

5.2 Estimation of necessary number of chosen plaintexts

The term $S9(k_{411} \oplus C_1^L | C_2^L)$ has 9[bit] unknown sub-key whose degree is 2. So there are ${}_9C_2 + {}_9C_1 = 45$ unknowns. The $S7(k_{412} \oplus C_3^L)$ has 7[bit] unknown sub-key whose degree is 3. So there are ${}_7C_3 + {}_7C_2 + {}_7C_1 = 63$ unknowns. The term $S9(k_{413} \oplus C_3^L \oplus S9(k_{411} \oplus C_1^L | C_2^L))$ has ${}_9C_2 + {}_9C_1 + 9 = 54$ unknowns whose degree is 2. So there are ${}_{54}C_2 + {}_{54}C_1 = 1,485$ unknowns. In the same way we calculate number of unknowns in the term $S9(k_{421} \oplus C_1^R | C_2^R)$, $S7(k_{422} \oplus C_3^R)$, $S9(k_{423} \oplus C_3^R \oplus S9(k_{421} \oplus C_1^R | C_2^R))$. As the result, we found there are 3,186 unknowns in the Attack equation.

We can deduce 9 linear equations from one Attack equation because Attack equation is the vector equation on $GF(2)^9$. To solve the equation, we need $3,186/9 = 354$ different 2nd order differentials. Thus we need $2^2 \times 354 = 1,416$ chosen plaintexts to attack 4 round KASUMI without FL functions.

6 Conclusion

We showed that 4 round KASUMI cipher without FL functions is attackable by Higher Order Differential Attack using 2nd order differentials. We adapt the algebraic method to solve Attack equation. By this attack, we will be able to determine 6 sub-keys in 4th round using 1,416 chosen plaintexts. And we estimate that necessary computational cost is $2^{22.11}$ times FO function operations. The algebraic method will be $2^{29.89}$ times faster than brute force search which needs 2^{52} times FO function operations.

References

1. "KASUMI", available at
   http://www.etsi.org/dvbandca/3GPP/3gppspecs.htm
2. Matui, "Block Encryption Algorithm MISTY", ISEC96-11(1996-07)
3. Shimoyama, Moriai, Kaneko, "Improving the Higher Order Differential Attack and Cryptanalysis of the $\mathcal{KN}$ Cipher", ISEC97-29(1997-09)
4. Moriai, Shimoyama, Kaneko, "Higher Order Attack of a CAST Cipher (II)", SCIS98-12E
5. Jakobsen, Knudsen, "The Interpolation Attack on Block Cipher", FSE-4th International Workshop, LNCS.1008
6. Nyberg, Knudsen, "Provable Security against Differential Cryptanalsis", Jounal of Cryptology, Vol.8-no.1 (1995)
7. Shimoyama, Moriai, Kaneko, "Improving the Higher Order Differential Attack and Cryptanalysis of the $\mathcal{KN}$ Cipher", 1997 Information Security Workshop
8. Matsui, "New Structure of Block Ciphers with Provable Security against Differential and Linear cryptanalysis", FSE-3rd International Workshop, LNCS.1039
9. Moriai, Shimoyama, Kaneko, "Higher Order Attack of a CAST Cipher", FSE-4th International Workshop, LNCS.1372

On MISTY1 Higher Order Differential Cryptanalysis

Steve Babbage[1] and Laurent Frisch[2]

[1] Vodafone Ltd
The Courtyard, 2-4 London Road,
Newbury RG14 1JX, ENGLAND
steve.babbage@vf.vodafone.co.uk
[2] France Télécom Recherche & Développement,DTL/SSR
38-40, avenue du General Leclerc
92794 Issy les Moulineaux Cedex 9, FRANCE
laurent.frisch@francetelecom.fr

Abstract. MISTY1 is a block cipher whose design relies on an assertion of provable security against linear and differential cryptanalysis. Yet, a simplified and round reduced version of MISTY1 that does not alter the security provability can be attacked with higher order differential cryptanalysis. We managed to explain this attack by deriving the attacking property from the choice of an atomic component of the algorithm, namely one of the two MISTY1 S-boxes. This allowed us to classify the good and the bad S-boxes built with the same principles and to show that none of the S-boxes with optimal linear and differential properties has an optimal behaviour with respect to higher order differential cryptanalysis.

1 Introduction

The encryption algorithm MISTY1 is a block cipher designed by Matsui in 1996 (see [5]) with the achievements of provable security and reasonable speed in both software and hardware. It is built on a Feistel-scheme, iterating on 8 rounds a function named FO. The original built also uses in each round a small and fast linear function named FL. As the design with FO is proved secure against differential and linear cryptanalysis, the purpose of adding FL functions is to prevent attacks other than differential and linear, but with no proved security. Section 2 gives the scheme of MISTY1.

Restricting themselves to the "provable secure part" of MISTY1, [9] and [8] showed that a reduced version of MISTY1 with only 5 rounds and with no FL functions is attackable by higher order differential cryptanalysis. The authors gained hope in such an attack by noticing that the algebraic degree of the overall cipher does not grow very fast in the first 3 rounds. Indeed, the scheme uses two S-boxes which themselves have rather low algebraic degree, well suited for hardware implementation. The attack will be explained in Section 3.

Increasing the algebraic degree of S-boxes would prevent such an attack. Yet, [9] and [8] do not explain the origin of the attacking property. For instance, it could not be said that increasing the algebraic degree of the S-boxes was the only way to change the degree of the overall cipher and prevent higher order differential attacks. We managed to derive the higher order differential property used by the attackers from a property on S_7, an S-box permutation on 7 bits, which will provide a link between the chosen exponent and the attack. Moreover, we were able to classify the exponents in S_7 that would allow a higher order differential attack on the weaker version of MISTY1, and to compare them to the good exponents for provable security against differential and linear cryptanalysis. We explain how in Section 4.

2 Presentation of MISTY1

This Section recalls the construction of the block-cipher MISTY1 as defined in [5].

MISTY1 is designed around small block schemes used recursively. There are three levels of these schemes. The small one, with 16-bit input, is called FI (figure 2), and is built with three calls to two permutation S-boxes, S_7 and S_9 with respective inputs on 7 and 9 bits. The intermediate scheme, with 32-bit input, is called FO (figure 2) and makes three calls to FI. The main scheme MISTY1 is a block cipher with 64-bit input, whose number of rounds should be a multiple of 4. Each two rounds (figure 1) is made of two calls to FO and two calls to a small linear function FL (figure 2). The proof of security of such a scheme against differential and linear cryptanalysis uses generalisations of techniques first introduced in [7] (resistance against differential attacks) and in [6] (resistance against linear attacks).

The key schedule produces three series of keys (i referring to the round number) : $KL_{ij}(j = 1, 2)$ used in FL_i, $KO_{ij}(j = 1 \ldots 4)$ used in FO_i, and $KI_{ijk}(j = 1 \ldots 4, k = 1, 2)$ used in FI_{ij}.

There is an equivalent key schedule that makes things a little bit easier to see on MISTY1's schemes (see figure 3). KO_i and KI_{ij} series can be joined in a single series $K_{ijk}(j = 1 \ldots 3, k = 1 \ldots 3)$ used in FI_{ij}, and a single key word $K^{\oplus}$ XORed with the algorithm's output word. A XOR with K_{ijk} is made just before entering an S-box, i.e. before entering the only non-linear parts of the cipher.

The two S-boxes are chosen in order to have good differential and linear properties [1], to be fast to compute in hardware, and to have the highest reasonable algebraic degree. Here are some basic definitions :

Definition 1 ($GF(2^m)$). $GF(2^m)$ *denotes the unique finite field with 2^m elements. We choose on $GF(2^m)$ a polynomial basis $(X^i)_{0 \le i \le m-1}$, thanks to which we can assimilate $GF(2^m)$ to $\{0, 1\}^m$. For any function $GF(2^m) \rightarrow GF(2^n)$, $x \mapsto y(x)$, we write $\deg_x y$ its algebraic degree in x ($0 \le \deg_x y \le m$). We will use the fact that $\deg_x x^e = |e|$ for an exponent e that is coprime with the order of the multiplicative group and $|e|$ the Hamming weight of e [2] .*

---

[1] This property is to have a minimal average differential and linear probability. See [4].

[2] We obviously have $deg_x x^e \le |e|$. For a complete proof, see for example[2].

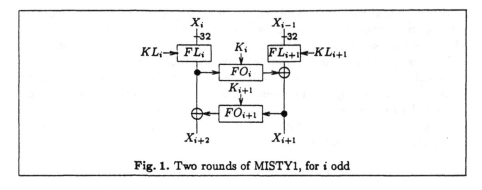

Fig. 1. Two rounds of MISTY1, for i odd

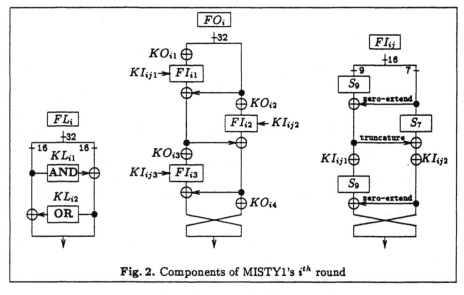

Fig. 2. Components of MISTY1's i^{th} round

Both S-boxes are chosen among permutations in $GF(2^m)$ having the form $x \mapsto L(x^e)$ [3] and a linear permutation L. Choosing the degree in order to keep a reasonably short hardware computation time, the exponent is of Hamming weight 3 for S_7 and 2 for S_9. Explicitly, it is $e_3 = 81$ for S_7.

3 Higher order differential attack on a reduced version of MISTY1

3.1 Basic definitions, notations, and facts

Definition 2 (k^{th}-order differentials). *Let $F : GF(2^m) \to GF(2^n)$ be a function. We call the kth-order differential of F according to a k dimensional vector*

---

[3] Note that for such a permutation, one necessarily has $\gcd(2^m - 1, e) = 1$.

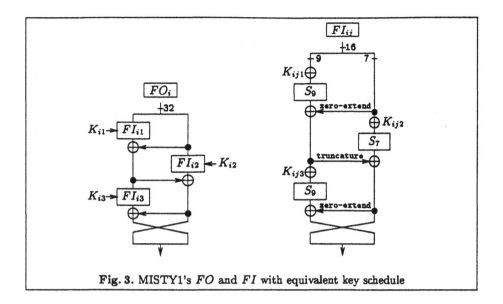

Fig. 3. MISTY1's *FO* and *FI* with equivalent key schedule

space $\mathcal{V} \subset GF(2^m)$ the mapping :

$$\Delta_{\mathcal{V}} F : GF(2^m) \to GF(2^n)$$
$$x \mapsto \Delta_{\mathcal{V}} F(x) = \bigoplus_{v \in \mathcal{V}} F(x \oplus v)$$

We denote by $[F]_d$ the sum of the terms in the algebraic normal form of F whose degrees are at least d. For example, $[x_0 x_2 x_4 x_5 + x_0 x_1 x_2 + x_1 x_2 x_3 + x_0 x_1 + 1]_3 = x_0 x_2 x_4 x_5 + x_0 x_1 x_2 + x_1 x_2 x_3$.

Fact 1 We have the following properties for all $x, x' \in GF(2^m)$, $x' \oplus x \in \mathcal{V}$:

$$\Delta_{\mathcal{V}} F(x') = \Delta_{\mathcal{V}} F(x) \tag{1}$$
$$= \Delta_{\mathcal{V}} [F]_{dim\ \mathcal{V}}(x) \tag{2}$$

Using (1) when $\mathcal{V}$ is $GF(2^m)$, we can write $\Delta_{\mathcal{V}} F$ to refer to $\Delta_{\mathcal{V}} F(0)$, or even $\Delta_v F$ (with variable v instead of vector space $\mathcal{V}$) when no mistake is possible.

Definition 3 (Bit manipulations). If E belongs to $\{0,1\}^m$, we will write $(E : n)$ to refer to the n-bit word deduced from E either by extension on the left with $n - m$ zero bits if $n \geq m$ or by truncature on the left if $n < m$. For instance, $0 : 32$ will mean a 32-bit zero word. E^L, E^{L9}, E^R, and E^{R7} will respectively refer to : the left half (more significant bits), the 9 left most bits, the right half (less significant bits), and the 7 right most bits of E. $E_1 || E_2$ will be the concatenation of E_1 and E_2. We will let $rev(E)$ be the mirror-reversed bits of E. We denote $\cap$ the bitwise AND and $\cup$ the bitwise OR. $E \lll k$ denotes the cyclic left shift of k bits of E ; in other words, it is E^{2^k} where E is seen as an element of $\mathbb{Z}/m\mathbb{Z}$.

3.2 Description of the attack

Definition 4 (M1'). *We name M1' the reduced version of MISTY1 that has 5 rounds and no FL-function. As in the presentation of MISTY1 above, we let X_0 be the right 32-bit word entering M1', X_1 the left 32-bit word, and (X_{i+1}, X_i) the intermediate value in M1' after i rounds. We also let $x = \sum_{i=0}^{6} x_i X^i$ be the 7 right bits of X_0 ($x = X_0^{R7}$), and $y(X_1, X_0, K)$ (or y) be the 7 left most bits of FO_3's output word, while (X_1, X_0) is the entering word of M1' and K is the key. We call "constant" and denote cst everything that does not depend on $x = X_0^{R7}$ (i.e. whose degree in x is zero).*

With these notations, we can easily check on M1' the following property :

$$\Delta_x y = C \tag{3}$$

to be understood as :

$$\bigoplus_{x \in GF(2^7)} y(X_0^{L25}\|x, X_1, K) = C \tag{4}$$

C being a constant independant of K and of the input bits (X_1, X_0^{L25}). (3) says that C is the 7th order differential of the function $(X_0, X_1, K) \mapsto y$ with respect to the subspace described by x. In other words, this means that the highest degree monomial in x coefficients [4] in the expression of the y bits as boolean functions of x's 7 bits are independant of all other parameters K, X_1, and X_0^{L25}. Property (3) will turn out to come from a property of S_7 (see next Section) ; thus, the constant C will strongly depend on the choice of S_7. C is actually very easy to compute, provided that we have an implementation of S_7, with formula (4). With the S_7 chosen in [5], the constant turns out to be 1101101_2.

This property can be used to perform a chosen plaintext attack of M1' in the following way. In M1', the plaintext is (X_1, X_0), and the ciphertext is (X_5, X_6). X_5 and X_6 are functions of the plaintext (X_1, X_0^{L25}, x). Considering the last round, we have $X_4(x) = X_6(x) \oplus FO(X_5(x), K_5)$. We can compute the 7th order differential with respect to x of the 7 left most bits in this expression. As X_4^{L7} is the sum of y and of $(x \oplus \text{cst})$, we have $\Delta_x X_4^{L7} = \Delta_x y$. With (3), we obtain the following attacking equation :

$$\bigoplus_{x \in GF(2^7)} (X_6^{L7}(X_1, X_0^{L25}, x) \oplus FO^{L7}(X_5(X_1, X_0^{L25}, x); K_5)) = C \tag{5}$$

We do not need to compute the whole FO_5 function, but just the 7 left most bits. This means that we only need to compute FI_{51}^{L7} and FI_{52}^{L7}. To compute these, we just go through S_9 once and S_7 once. When we sum in (5), some monomials of key bits vanish. With one chosen plaintext, we get one equation on monomials on subkey bits. The subkeys involved are K_{511} and K_{521}, each

---

[4] the highest degree monomial is $x_0 x_1 x_2 x_3 x_4 x_5 x_6$.

one with degree 1 in 9 bits, and K_{512} and K_{522}, each one with degree 2 in 7 bits. One can linearize terms of degree two. Finally, the attack can be performed by choosing enough set of 2^7 plaintexts associated to various X_1, X_0^{L25} values in order to get enough such linearized equations (see [9]), or simply by exhaustive search on those subkeys.

4 Where does this higher order differential property come from ?

In this section, we derives property (3) from another property on S_7. This deduction starts with analytic details on M1' scheme in order to have an explicit expression of y (subsection 4.1). The analysis of the highest degree monomial of y gives us a sufficient property on S_7 for property (3) to hold. (subsection 4.2). This allows us to characterize the exponents of S_7-like S-boxes with respect to property (3) (subsection 4.3). Finally, we comment the choice of the exponent e_3 in S_7 (subsection 4.4).

4.1 Low-degree property on 3 rounds

We can summarize the phenomenon as follows : the 7 left bits y of FO_3's output appear to come from the 7 right bits x of X_0. x's bits having gone through few S-boxes, the degree of most intermediate bits involved in computation of y in the x bits is pretty low.

To help the reader, we have tried to summarize the steps on figures 4 and 5.

1st **round.** X_1 goes through FO_1 and is added to X_0 : the intermediate value after one round is ($X_2 = X_0 \oplus \text{cst}$, $X_1 = \text{cst}$).

2nd **round.** X_2 goes through FO_2 and is added to X_1. X_2 is a 32-bit word, whose constant left 16-bit word X_2^L goes through FI_{21}, which gives another constant. X_2^R is added onto this constant. The result $\text{cst} \oplus X_2^R$ on one hand is added to the output α of FI_{22}, and on the other hand goes through FI_{23} to give β. We have $\alpha = FI_{22}(X_0^R \oplus \text{cst})$ and $\beta = FI_{23}(X_0^R \oplus \text{cst})$ (see figure 4).

Let us consider the degree (in x) going out of FO_2. If we call λ the right 16-bit word of FO_2's output, we can see that λ's degree in x is only 2. Indeed, $\lambda = \beta \oplus \alpha \oplus X_0^R \oplus \text{cst}$. Moreover, $FI_{ij}(X_0^R \oplus \text{cst})$ will raise an expression of degree in x equal to 3, and with coefficients of degree 3 independant of constants [5]. So that we have $[\alpha]_3 = [\beta]_3$, and then $[\lambda]_3 = [\alpha]_3 \oplus [\beta]_3 = 0$. Thus, $\deg(\lambda) \leq 2$.

---

[5] This is easy to check on the design of an FI-box : the variable are concentrated in the 7 right most bits of the 16-bit input, so that they just go through at most one S-box. S_9 has degree 2 and S_7 has degree 3, which means that $\deg(S_9(P)) \leq 2.\deg(P)$ and $\deg(S_7(P)) \leq 3.\deg(P)$ for any polynomial P. For $P = x \oplus \text{cst}$, ie $\deg(P) = 1$, $S_9(P)$ and $S_7(P)$ have respective degree 2 and 3, with highest degree coefficients independant of constants, because highest degree monomials of the output come from highest degree monomials of the input.

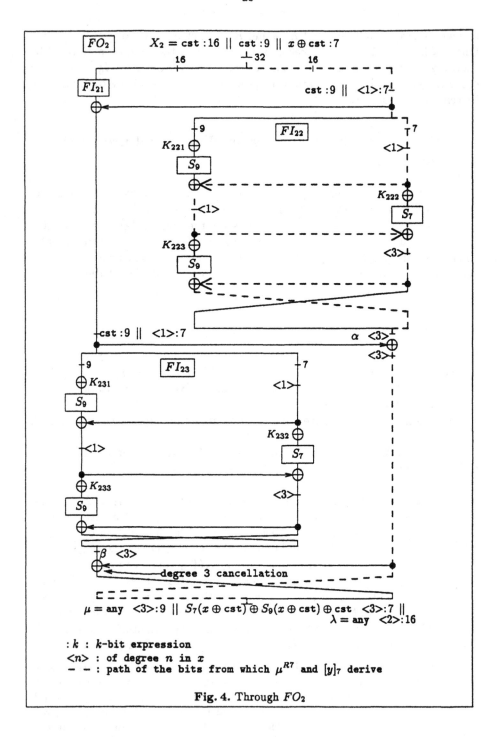

Fig. 4. Through FO_2

Now, let us consider μ, the left 16-bit word of FO_2's output. We have $\mu = \alpha \oplus X_0^R \oplus \text{cst}$, and $\deg(\mu) = 3$ for the same reason as above. We will actually need to be more precise. We will need the expression of μ^{R9}. This is easy to derive from FI design (we omit to write any 9-bit to 7-bit truncature and any 7-bit to 9-bit zero-extension) :

$$
\begin{aligned}
\mu^{R9} &= (x \oplus \text{cst}) \oplus FI_{22}^{R9}(x \oplus \text{cst}) \\
&= (x \oplus \text{cst}) \oplus (S_9(x \oplus \text{cst}) \oplus S_7(x \oplus \text{cst}) \oplus x \oplus \text{cst}) \\
&= S_7(x \oplus \text{cst}) \oplus S_9(x \oplus \text{cst}) \oplus \text{cst} \quad (6)
\end{aligned}
$$

Note that, by truncature, this is also the expression of μ^{R7}.

To summarize the output X_3 of the 2nd round, we have, from left to right, 9 bits of degree 3 (μ^{L9}), 7 bits of degree 3 (μ^{R7} in (6)), and 16 bits of degree 2 (λ).

3rd **round.** Looking at the 7th degree monomial at the output of FO_3, we will establish the following result :

$$
[y]_7 = [S_7(\mu^{R7} \oplus \text{cst})]_7 \quad (7)
$$

As for the 2nd round, we need to follow thoroughly the design of FO_3 to derive this equation (see figure 5). As input of FI_{31}, we have the 16-bit word $\mu = \mu^{L9}||\mu^{R7}$, whose degree is 3. The 7 left most bits going out of FI_{31} come from the expression $S_7(\mu^{R7} \oplus \text{cst}) \oplus S_9(\mu^{L9} \oplus \text{cst}) \oplus \mu^{R7} \oplus \text{cst}$. Considering the 7th degree part of the terms, we obviously have $[\mu^{R7}]_7 = [\text{cst}]_7 = 0$. As $\deg_x \mu^{L9} = 3$, we also have $\deg_x S_9(\mu^{L9} \oplus \text{cst}) = 6$, so that $[S_9(\mu^{L9} \oplus \text{cst})]_7 = 0$, and finally we see that the 7 left most bits of the output of FI_{31} have 7th degree monomials that are $[S_7(\mu^{R7} \oplus \text{cst})]_7$.

Following the intermediate values in FO_3, we then see that onto this output is added the right 16-bit part λ of FO_3's input, whose degree is 2. The 7th degree monomial of the 7 left most bits stays unchanged. It is then added to the output of FI_{32} with input of degree 2, before reaching y. The 7 left most bits of the output of FI_{32} actually have a degree equal to 6 : as above, we call λ FI_{32}'s input with $\deg_x \lambda = 2$, and we see that the 7 left most bits of the output come from the expression $S_7(\lambda^{R7} \oplus \text{cst}) \oplus S_9(\lambda^{L9} \oplus \text{cst}) \oplus \lambda^{R7} \oplus \text{cst}$, whose highest degree monomials come from $S_7(\lambda^{R7})$, so that this highest degree is 6. Thus, It has no contribution to the 7th degree of the 7 left most bits. This shows that equation (7) holds.

As a conclusion, we derive from (6) and (7) the expression of $[y]_7$ according to x and constants :

$$
[y]_7 = [S_7(S_7(x \oplus \text{cst}) \oplus S_9(x \oplus \text{cst}) \oplus \text{cst})]_7
$$

Note that $[y]_7 = (\Delta_x y).(x_0 x_1 x_2 x_3 x_4 x_5 x_6)$, and then, with properties (1) and (2), we can make a change of variable that leads to a monomial F of degree 7 (and giving a name to constants) :

$$
[y]_7 = F(x) = [S_7(S_7(x) \oplus S_9(x \oplus c) \oplus c')]_7 \quad (8)
$$

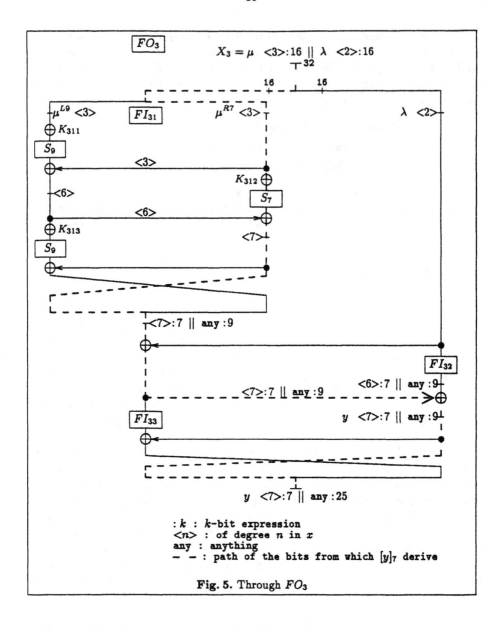

Fig. 5. Through FO_3

We should notice that there is no such basic equation that holds when we put in the FL functions. Indeed, as they mix bits, even though linearly, they make the highest degree terms spread and interfere with each others. So that things do not go as well for the attacker as, for example, on figure (4).

4.2 Analysis of the 7th degree monomial

If we call $t(x) = S_7(x) \oplus S_9(x \oplus c) \oplus c'$ the argument of S_7 in (8), S_7 being a polynomial function of x of degree 3, the possible ways of obtaining a degree 7 in x involve products of 3 monomial of $t(x)$ of degree 1, 2, or 3 in x, but no constant term. Thus, if we replace $t(x)$ by $t'(x) = [S_7(x)]_{\deg_x \geq 1} \oplus [S_9(x \oplus c)]_{\deg_x \geq 1}$, we still have $F(x) = [S_7(t'(x))]_7$. We can say more : to obtain a degree 7, we need to pick in $t'(x)$ 3 terms whose sum of degrees in x is greater than 7. But this is not precise enough to raise easy conclusions.

It is better to consider the degrees in (x, c) [6] and not just x. We classify the possible combinations of $t'(x)$ monomials that give a degree 7 in x by their degree in (x, c). $S_7(x)$ has just monomials in x and none in c. $S_9(x \oplus c)$ has monomials of degree 2, 1, and 0 in (x, c), so that the possibilities are :

- $\deg_x = 7$ and $\deg_{x,c} = 9$: these monomials would only come from a product of 3 monomials of degree 3 in (x, c), in which there is some monomials in c. But this can never happen, because any degree 3 come from the term $S_7(x)$ in $t'(x)$, and this term does not have any monomial in c;
- $\deg_x = 7$ and $\deg_{x,c} = 8$: these monomials come from a product of two monomials of degree 3 (*a priori* in (x, c), but actually in x because they come from $S_7(x)$) and from one monomial of degree 2 with a degree 1 in c (it comes from $S_9(x \oplus c)$). So that this part of $F(x)$ can be written $[S_7(t''(x))]_7$ with $t''(x) = [S_7(x)]_3 \oplus [S_9(x \oplus c)]_{(1,1)}$, where $[f]_{(1,1)}$ means monomials of degree 2 that are degree 1 in x and 1 in c;
- $\deg_x = 7$ and $\deg_{x,c} = 7$: this raises a monomial in $x_0 x_1 x_2 x_3 x_4 x_5 x_6$ with a coefficient C independant of c.

The conclusion is that $F(x) = [S_7(t''(x))]_7 \oplus C.x_0 x_1 x_2 x_3 x_4 x_5 x_6$ where $t''(x) = [S_7(x)]_3 \oplus [S_9(x \oplus c)]_{(1,1)}$ and where the c-dependant part of $F(x)$ is a sum of products of two monomials of $[S_7(x)]_3$ and one monomial of $[S_9(x \oplus c)]_{(1,1)}$.

The fact that makes property (3) hold with the C above, is that $[S_7(t''(x))]_7 = 0$. This property is not trivial, and comes straight from the following fact :

Fact 2 *The product of two output bits of $S_7(x)$ has 6th degree terms that always zeroize.*

Fact 2 is a sufficient property on S_7 for property (3) on M'1 to hold. Taking into account the fact that there exists an e_3 of Hamming weight 3 and a linear permutation L such that $S_7(x) = L(x^{e_3})$, the former property can be restated as follow [7] :

$$\forall L_1, L_2, \ \forall L, \ \deg_x L_1(L(x^{e_3}).L_2(L(x^{e_3})) < 6 \quad \text{where } |e_3| = 3 \qquad (9)$$

---

[6] this is the degree of the expression seen as a polynomial in the bits of x and of c.

[7] The property turns out to be true for the chosen S_7 but the equivalence that will be staten in fact 3 holds for any L and any e_3 of Hamming weight 3

or, by changing L_i into $L_i \circ L$:

$$\forall L_1, L_2, \quad \deg_x L_1(t).L_2(t) < 6 \quad \text{where } t = x^{e_3} \text{ and } |e_3| = 3 \qquad (10)$$

where L_1 and L_2 are two linear maps $GF(2^7) \to GF(2)$, and $L \in GL(GF(2^7))$. It is equivalent to prove it for (L_1, L_2) describing a basis, i.e. :

$$\forall i, j, \quad \deg_x t_i.t_j < 6 \quad \text{where } t = x^{e_3} \text{ and } |e_3| = 3 \qquad (11)$$

This property is true on $deg_x t_i t_j$ for any i, j if and only if it is true on $deg_x t^{e_2} = deg_x x^{e_2 e_3}$ for any e_2 with Hamming weight 2. The "only if" part is easy [8]. So that property (11) is equivalent to :

$$\forall e_2, \ |e_2| = 2, \quad \deg_x t^{e_2 e_3} < 6 \quad \text{where } |e_3| = 3 \qquad (12)$$

Fact 2 states that the above property holds for the exponent e_3 used in MISTY1. In the subsequent sections we will show that the same property necessarly holds for any possible exponent e_3 that satisfies the other properties required for provable security of MISTY1.

4.3 Characterizations of S_7 exponents

The property that allows the higher order differential attack is expressed in terms of the S-box S_7 in (9) and in terms of the exponent chosen in S_7 in (12). Property (12) turns out to be very useful to find a characterization of exponents with respect to the higher order differential attack.

First, we need to put the stress on the fact that, given an S-box of the form $x \mapsto L(x^e)$, there is not just one exponent e that can be taken but many of them.

Definition 5 (Set of S-box exponents). *On $GF(2^m)$, we call set of S-box exponents and we write $\mathcal{E}_m$ the set of exponents e such that $x \mapsto x^e$ is one to one. This is the set of integers between 1 and $2^m - 1$ that are coprime with $2^m - 1$. Thus, $\mathcal{E}_m$ is the subset of invertible elements of $\mathbb{Z}/(2^m - 1)\mathbb{Z}$. For example, $\mathcal{E}_7 = [1, 126]$.*

Definition 6 (S-box equivalent exponents). *Let $\sim$ be the equivalence relation defined on $\mathcal{E}_m$ by : $e_1 \sim e_2$ if there exist two linear permutations L_1 and L_2 such that $\forall x \in GF(2^m)$, $L_1(x^{e_1}) = L_2(x^{e_2})$. We say that e_1 and e_2 are S-box equivalent exponents. We write $EQU_m(e)$ the equivalence class of e (i.e. $EQU_m(e) \in \mathcal{E}_m/\sim$).*

The linear permutations of the form $x \mapsto x^e$ are the cyclic shift maps $x \mapsto x^{2^k}$. Indeed, we have $deg_x x^e = 1$ because it is linear and $deg_x x^e = |e|$ as in definition 1. Thus, e is an exponent of Hamming weight 1 and such an exponent is 2^k for some $k \in \mathbb{Z}/m\mathbb{Z}$.

---

[8] We have checked (but not proved) that : $\max_{\{i,j\} \subset [0;6]} deg_x t_i t_j = \max_{|e_2|=2} deg_x t^{e_2}$ for t such that $t = x^e$ and $|e| \le 3$.

The relation $e_1 \sim e_2$ is equivalent to :

$$\exists M \in GL(GF(2^m)) \; \forall x \in GF(2^m) \; x^{e_1/e_2 \pmod{2^m-1}} = M(x)$$

which means, with the previous remark, that $x \mapsto x^{e_1/e_2}$ is a cyclic shift map $x \mapsto x^{2^k}$ for some $k \in \mathbb{Z}/m\mathbb{Z}$. Thus, $e_1 \sim e_2$ is equivalent to :

$$\exists k \in \mathbb{Z}/m\mathbb{Z}, \; e_1 = 2^k e_2 \pmod{2^m - 1} \qquad (13)$$

Generally, the cardinal of $\mathrm{EQU}_m(e)$ depends on e, particularly on the bit periodicity of e. By definition, the bit periodicity of an m-bit word w is the smallest positive integer T such that $w \lll T = w$. We write its period $T(w)$. It is obvious that $T(w)$ divides m. With (13), we have : $\#\mathrm{EQU}_m(e) = T(e)$.

If m is prime, exponents $e \in \mathcal{E}_m$ have a period equal to m. In the case of S_7 where $m = 7$, each class of S-box equivalent exponents has exactly 7 elements. Permutations on $GF(2^7)$ of the form $L(x^e)$ are defined modulo S-box equivalence classes, which means for example that a given S_7 can be expressed with 7 couples (L, e).

We also notice that power maps with S-box equivalent exponents have exactly the same differential and linear properties, since they only differ by a linear permutation.

Definition 7 (Complementary Cyclic Shift Property). *We say that a n-bit integer e has the n-bit complementary cyclic shift property (n-bit CCS property) if there exists an integer k modulo n such that $e \cap (e \lll k) = 0$. Equivalently stated, e has the n-bit CCS property if there is an exponent e_2 of Hamming weight 2 such that the product ee_2 has Hamming weight $2|e|$.*

Definition 8 (CCS equivalent exponents). *Let $\mathcal{R}$ be the equivalence relation defined on $\mathcal{E}_m$ by : $e_1 \mathcal{R} e_2$ if $\exists k \in \mathbb{Z}/m\mathbb{Z}, \; e_1 = 2^k e_2 \pmod{2^m - 1}$ or $e_1 = 2^k \mathrm{rev}(e_2) \pmod{2^m - 1}$. We say that e_1 and e_2 are complementary cyclic shift (CCS) equivalent exponents. We write $CCS_m(e)$ the equivalence class of e (i.e. $CCS_m(e) \in \mathcal{E}_m/\mathcal{R}$).*

This equivalence relation $\mathcal{R}$ naturally comes from the definition of the CCS property, because if e has this property, so have $e \lll k$ ($\forall k$) and $\mathrm{rev}(e \lll k)$. In other words, if e has the CCS property, all the exponents in $CCS_m(e)$ have it. We talk about a CCS-positive class if it has ; a CCS-negative class if it has not.

The link between S-box equivalent exponents and CCS equivalent is easy to make with property (13) and definition 8. We have :

$$\forall e \in \mathcal{E}_m, \; CCS_m(e) = \mathrm{EQU}_m(e) \cup \mathrm{EQU}_m(\mathrm{rev}(e)) \qquad (14)$$

To be more precise, if $e \sim \mathrm{rev}(e)$, $CCS_m(e) = \mathrm{EQU}_m(e)$ and $\#CCS_m(e) = T(e)$. If $e \nsim \mathrm{rev}(e)$, $CCS_m(e) = \mathrm{EQU}_m(e) \oplus \mathrm{EQU}_m(\mathrm{rev}(e))$ and $\#CCS_m(e) = 2.T(e)$ (since $T(\mathrm{rev}(e)) = T(e)$).

Given an S-box of the form $L(x^e)$, we can define its S-box class as the S-box class of e ($\text{EQU}_m(x \mapsto L(x^e)) = \text{EQU}_m(e)$) and its CCS class as the CCS class of e ($\text{CCS}_m(x \mapsto L(x^e)) = \text{CCS}_m(e)$). $\text{EQU}_m(x \mapsto L(x^e))$ is well-defined by definition of $\sim$, and $\text{CCS}_m(x \mapsto L(x^e))$ is well-defined because $\text{EQU}_m(e) \subset \text{CCS}_m(e)$. We say that such an S-box is CCS-positive or CCS-negative according to its CCS class.

Here is why the CCS property is interesting.

Fact 3 *The following properties are equivalent.*

- *(12) holds.*
- *S_7 is CCS-negative.*
- *for any e_2 of Hamming weight 2, the product $e_2 e_3$ has a Hamming weight strictly lower than 6.*

If they are true, 7th order differential attack on M1' is possible. With the choice of S_7 in MISTY1, they are true.

This fact comes straight from property (12), from definitions 7 and 8, and from $deg_x x^{e_2 e_3} = |e_2 e_3|$. To see that this holds for MISTY1, one can easily check (9) with L and e_3 chosen such that $S_7(x) = L(x^{e_3})$.

This allows us to classify the exponents with respect to their belonging to a positive or to a negative class. The 7th order differential property (3) will hold, and the attack will be possible, as soon as the exponent has the CCS property.

4.4 On the choice of S_7 exponent

After a short study of all exponents of Hamming weight 3 in $\mathcal{E}_7$, we saw that there are 4 CCS classes, 3 of them being positive. The negative class has 14 elements, which means that it is made by 2 S-box equivalent exponents classes. It also necessarily contains e_3. We give the two subclasses explicitly :

$$\text{EQU}_7(81) = \{81, 35, 70, 13, 26, 52, 104\} \tag{15}$$

$$\text{EQU}_7(\text{rev}(81)) = \{69, 11, 22, 44, 88, 49, 98\} \tag{16}$$

The 3 other classes are positive ; they are generated (for example) by 7 ($7 \cap 7 < \ll 3 = 0$), 19 ($19 \cap 19 \lll 2 = 0$), 21 ($21 \cap 21 \lll 1 = 0$). Each of these CCS classes are also S-box equivalent exponents classes, with 7 elements. As they all are CCS-positive, the use of them in S_7 would prevent the 7th order differential attack on M1'.

We have computed the average differential probability DP^{S_7} and average linear probability LP^{S_7} (see [5]) for functions $x \mapsto x^{e_3'}$ with e_3' in each of these 3 classes. They are always worse than the original S_7 box, with a CCS-negative class.

Fact 4 *There is no Hamming weight 3 optimal exponent e_3 with respect to linear and differential properties and to higher order differential properties. None of the S-boxes with optimal linear and differential properties has an optimal behaviour with respect to higher oreder differential cryptanalysis.*

Remark 1. MISTY1's S_7 has the S-box class $EQU_7(13)$. $13 = 2^{2k} - 2^k + 1$ is a Kasami exponent with $k = 2$. The other S-box class that is CCS negative is $EQU_7(11)$. $11 = 2^{(m-1)/2} + 3$ is a Welsh exponent with $m = 7$ (proved in [3]). These two classes are proved to be optimal with respect to differential and linear properties.

The proven optimality of these two classes with respect to differential and linear properties are linked in our case. Indeed, the optimality of a permutation f remains on the inverse permutation f^{-1}. On power maps $x \mapsto x^{e_3}$ with $e_3 \in CCS_7(11)$, we have a property that makes e_3^{-1} and $\text{rev}(e_3)$ S-box equivalent exponents. This property is that e_3 and $\text{rev}(e_3)$ are CCS equivalent but not S-box equivalent. This implies that we can find k such that $e_3.(2^k\text{rev}(e_3))$ is of Hamming weight 6. Thus, $e_3.(2^k\text{rev}(e_3))$ is $2^7 - 2$ up to a cyclic shift. As $x^{2^7-2} = 1$ in $GF(2^7)$, we see that the power maps $x \mapsto x^{e_3^{-1}}$ and $x \mapsto x^{\text{rev}(e_3)}$ are the same up to a linear permutation. Then, if $x \mapsto x^{e_3}$ is optimal with respect to differential and linear properties, so is $x \mapsto x^{\text{rev}(e_3)}$. This proves that all exponents in $CCS_7(11)$ are equivalent with respect to differential, linear, and (9) properties.

5 Conclusion

We showed that the 7th order differential property on MISTY1 with no FL function and on 5 rounds comes from the choice of the S_7 box. It is actually not possible to find an S_7 box coming from a mapping $x \mapsto x^{e_3}$ with an exponent e_3 of Hamming weight 3, that is at the same time an optimum from the points of view of average differential/linear probabilities and of the 7th order differential property. It may be possible to prevent this higher order property without greatly altering the scheme, and still keeping the provable security properties. For instance, the S-boxes could be chosen with higher weight exponents (although that might increase the hardware implementation complexity), or the number of rounds in the FI scheme could be increased [9] (although there is a 2nd order differential attack on KASUMI with no FL function and on 4 rounds ; see [10]).

6 Acknowledgements

The second author would like to thank Henri Gilbert and Marine Minier for guiding, advicing, supporting, and printing support.

---

[9] In the block cipher KASUMI, which was developed from MISTY1 for use in the 3GPP confidentiality and integrity algorithms [1], the number of rounds in the FI function was increased from 3 to 4.

References

1. *Specification of the 3GPP Confidentiality and Integrity Algorithms, Document 2 : KASUMI Specification*, available for download at http://www.etsui.org/dvbandca/3GPP/3gppspecs.htm
2. Carlet, *Codes de Reed-Muller, Codes de Kerdock et de Preparata*, Thesis, Publication of LITP, Institut Blaise Pascal, Université Paris 6, 90.59, 1990
3. Hans Dobbertin, *Almost Perfect Nonlinear Power Functions on $GF(2^n)$: The Welch Case*, IEEE Transactions on Information Theory, Vol.45, No 4, May 1999
4. Mitsuru Matsui, *New Structure of Block Ciphers with Provable Security against Differential and Linear Cryptanalysis*, Proceedings of the Third International Workshop of Fast Software Encryption, Lecture Notes in Computer Science 1039, Springer Verlag, 1996
5. Mitsuru Matsui, *New Block Encryption Algorithm MISTY*, Proceedings of the Fourth International Workshop of Fast Software Encryption, Lecture Notes in Computer Science 1039, Springer Verlag, 1997
6. Kaisa Nyberg, *Linear Approximation of Block Ciphers*, Advances in Cryptology - Eurocrypt'94, Lecture Notes in Computer Science 950, Springer Verlag, 1994
7. Kaisa Nyberg, Lars Knudsen, *Provable Security against Diffrential Cryptanalysis*, Journal of Cryptology, Vol.8, no.1, 1995
8. Makoto Sugita, *Higher Order Differential Attack of Block Cipher MISTY1,2*, technical report of IEICE, ISEC98-1
9. Hidema Tanaka, Kazuyuki Hisamatsu, Toshinobu Kaneko, *Strength of MISTY1 without FL function for Higher Order Differential Attack*, SCI/ISAS99
10. Hidema Tanaka, Chikashi Ishii, Toshinobu Kaneko, *On the strength of KASUMI without FL functions against Higher Order Differential Attack*, To appear in : Proceedings of ICISC 2000, Lecture Notes in Computer Science, Springer Verlag

Difference Distribution Attack on DONUT and Improved DONUT*

Dong Hyeon Cheon*, Seok Hie Hong*, Sang Jin Lee*,
Sung Jae Lee**, Kyung Hwan Park**, and Seon Hee Yoon**

*Center for Information and Security Technologies(CIST),
Korea University, Seoul, 136-701, Korea
{dhcheon, hsh}@cist.korea.ac.kr, sangjin@tiger.korea.ac.kr
**Korea Information Security Agency(KISA),
5th FL., Dong-A Tower, 1321-6, Seocho-Dong, Seocho-Gu, Seoul, 137-070, Korea
{sjlee, khpark, shyoon, hslee}@kisa.or.kr

Abstract. Vaudenay[12] proposed a new way of protecting block ciphers against classes of attacks, which was based on the notion of decorrelation. He also suggested two block cipher families COCONUT and PEANUT. Wagner[14] suggested a new differential-style attack called boomerang attack and cryptanalyzed COCONUT'98. Cheon[5] suggested a new block cipher DONUT which was made by two pairwise perfect decorrelation modules and is secure against boomerang attack. In this paper we suggest an attack called difference distribution attack on DONUT. We also suggest an improved DONUT which is secure against difference distribution attack.

Key words : Decorrelation, DONUT, Differential Cryptanalysis(DC), Linear Cryptanalysis(LC), Difference Distribution Attack(DDA).

1 Introduction

Vaudenay[12] proposed a new way of protecting block ciphers against differential cryptanalysis(DC)[2,3] and linear cryptanalysis(LC)[9] which was based on the notion of decorrelation. This notion is similar to that of universal functions which was introduced by Carter and Wegman[4,15]. Vaudenay also suggested two block cipher families COCONUT(Cipher Organized with Cute Operations and NUT) and PEANUT(Pretty Encryption algorithm with NUT). COCONUT family used a pairwise perfect decorrelation module and PEANUT family used a partial decorrelation module.

COCONUT'98 cipher is a product cipher $C_3 \circ C_2 \circ C_1$, where C_1 and C_3 are 4-round Feistel ciphers and C_2 is a pairwise perfect decorrelation module. Wagner[14] suggested a new differential-style attack called boomerang attack and cryptanalysed COCONUT'98. Cheon[5] suggested a new block cipher called

* This work is supported by Korea Information Security Agency(KISA) grant 2000-S-078.

DONUT(Double Operations with NUT) which was made by two pairwise perfect decorrelation modules. DONUT is secure against boomerang attack. In this paper we will suggest an attack called difference distribution attack(DDA) and apply this attack to DONUT. We will also suggest an improved DONUT which is secure against difference distribution attack.

This paper is organized as follows. In section 2, we recall the basic definitions used in the decorrelation theory and present the previous results of decorrelation theory. In section 3, we describe the structure of DONUT and in section 4, we introduce the difference distribution attack and apply this attack to DONUT. In section 5, we suggest an improved DONUT and in section 6, we estimate the security of improved DONUT against DC, LC, boomerang attack. Finally in section 7, we conclude the paper.

2 Preliminaries

In this section, we recall the basic definitions used in the decorrelation theory and briefly present the previous result[12, 13].

Definition 1. *Given a random function F from a given set M_1 to a given set M_2 and an integer d, we define the "d-wise distribution matrix" $[F]^d$ of F as a $M_1^d \times M_2^d$-matrix where the (x, y)-entry of $[F]^d$ corresponding to the multipoints $x = (x_1, \cdots, x_d) \in M_1^d$ and $y = (y_1, \cdots, y_d) \in M_2^d$ is defined as the probability that we have $F(x_i) = y_i$ for $i = 1, \cdots, d$.*

Each row of the d-wise distribution matrix corresponds to the distribution of the d-tuple $(F(x_1), \cdots, F(x_d))$ where $(x_1, \cdots, x_d)$ corresponds to the index of the row.

Definition 2. *Given two random functions F and G from a given set M_1 to a given set M_2, an integer d and a distance D over the matrix space $R^{M_1^d \times M_2^d}$, we define the "d-wise decorrelation D-distance between F and G" as being the distance*

$$DecF_D^d(F, G) = D([F]^d, [G]^d).$$

We also define the "d-wise decorrelation D-bias of function F" as being the distance

$$DecF_D^d(F) = D([F]^d, [F^*]^d)$$

where F^* is a uniformly distributed random function from M_1 to M_2. Similarly, for $M_1 = M_2$, if C is a random permutation over M_1 we define the "d-wise decorrelation D-bias of permutation C" as being the distance

$$DecP_D^d(C) = D([C]^d, [C^*]^d)$$

where C^* is a uniformly distributed random permutation over M_1.

In the above definition, C^* is called the Perfect Cipher. If a cipher C has zero d-wise decorrelation bias, we call C a perfectly decorrelated cipher. When message space $M = \{0,1\}^m$ has a field structure, we can construct pairwise perfectly perfectly decorrelated ciphers on M by $C(y) = A \cdot y + B$ where key $K = (A, B)$ is uniformly distributed on $M^\times \times M$. This pairwise perfect decorrelation module is used for COCONUT family.

COCONUT is a family of ciphers parameterized by $(m, p(x))$, where m is the block size and $p(x)$ is an irreducible polynomial of degree m in $GF(2)[x]$. A COCONUT cipher is a product cipher $C_3 \circ C_2 \circ C_1$, where C_1 and C_3 are any (possibly weak) ciphers, and C_2 is defined as follows:

$$C_2(y) = A \cdot y + B \bmod p(x),$$

where A, B and y are polynomials of degree at most $m - 1$ in $GF(2)[x]$. The polynomials A and B are secret and act as round keys. Since the COCONUT family has pairwise perfect decorrelation, the ciphers are secure against the basic differential and basic linear cryptanalysis[12].

COCONUT'98 is a member of COCONUT family parameterized by $(64, x^{64} + x^{11} + x^2 + x + 1)$ and uses 4-round Feistel structures for C_1 and C_3, respectively. Wagner[14] cryptanalysed COCONUT'98 using boomerang attack, which exploits that high probability differentials exist for both C_1 and C_3.

3   Structure of DONUT

Frame Structure DONUT transforms a 128-bit plaintext block into a 128-bit ciphertext block. DONUT uses variable key length and consists of 4 rounds. The first round and the fourth round consist of pairwise perfect decorrelation modules $A_1 \cdot y + B_1$ and $A_2 \cdot y + B_2$ where A_1, B_1, A_2, B_2 are 128-bit subkeys. The inner 2-round transformation consists of two Feistel permutations and each round uses six 32-bit subkeys(see Fig.1).

Round Function F The round function F of DONUT consists of three G functions. A 64-bit input of F is split into two 32-bit words and 3-round transformation is followed with inner function G. Fig.2 shows the structure of F function.

inner function G The function G is a key-dependent permutation on 32-bit words with two 32-bit subkeys. The function G which we call the SDS function consists of 5 layers as follows:

1. The first key addition layer.
2. The first substitution layer.
3. The diffusion layer.
4. The second key addition layer.
5. The second substitution layer.

Fig.3 shows the structure of G function.

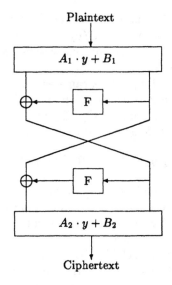

Fig. 1. Frame Structure of DONUT

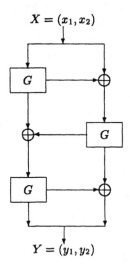

Fig. 2. structure of F function

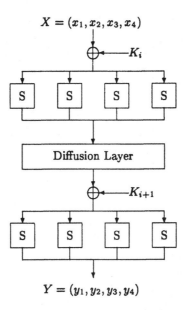

$$X = (x_1, x_2, x_3, x_4)$$

$$Y = (y_1, y_2, y_3, y_4)$$

Fig. 3. structure of G function

Substitution Layer We use the same S-box which is 8-bit input/output permutation as the substitution layer. The S-box is constructed by the function of the form $a \cdot x^{-1} \oplus b$, where $a = 0xa5, b = 0x37 \in GF(2^8)$. The Galois field $GF(2^8)$ is defined by the irreducible polynomial $x^8 + x^4 + x^3 + x^2 + 1$(hex : 0x11d). In the $GF(2^8)$, the x^{-1} has a good resistance against differential and linear attacks. The purpose of using affine transform is preventing from two fixed points such as zero to zero and one to one in the function x^{-1}.

Diffusion Layer The diffusion layer is performed by the 4×4 circulant-matrix D.

$$D = \begin{pmatrix} 01\ 06\ 07\ 02 \\ 06\ 07\ 02\ 01 \\ 07\ 02\ 01\ 06 \\ 02\ 01\ 06\ 07 \end{pmatrix}$$

Let $X = (x_3, x_2, x_1, x_0) = \sum_{i=0}^{3} x_i 2^{8i}$ be the input of the diffusion layer and $Y = (y_3, y_2, y_1, y_0) = \sum_{i=0}^{3} y_i 2^{8i}$ be the output of the diffusion layer. Then we have the followings:

$$\begin{pmatrix} y_0 \\ y_1 \\ y_2 \\ y_3 \end{pmatrix} = \begin{pmatrix} 01\ 06\ 07\ 02 \\ 06\ 07\ 02\ 01 \\ 07\ 02\ 01\ 06 \\ 02\ 01\ 06\ 07 \end{pmatrix} \begin{pmatrix} x_0 \\ x_1 \\ x_2 \\ x_3 \end{pmatrix} = \begin{pmatrix} 01 \cdot x_0 \oplus 06 \cdot x_1 \oplus 07 \cdot x_2 \oplus 02 \cdot x_3 \\ 06 \cdot x_0 \oplus 07 \cdot x_1 \oplus 02 \cdot x_2 \oplus 01 \cdot x_3 \\ 07 \cdot x_0 \oplus 02 \cdot x_1 \oplus 01 \cdot x_2 \oplus 06 \cdot x_3 \\ 02 \cdot x_0 \oplus 01 \cdot x_1 \oplus 06 \cdot x_2 \oplus 07 \cdot x_3 \end{pmatrix}$$

Key Scheduling DONUT has variable key length. Since two decorrelation modules need four 128-bit subkeys and every G function needs two 32-bit(4 bytes) subkeys, we need to generate twenty eight 32-bit subkeys. Our design strategy of key schedule is to prevent someone from finding some round keys from another round keys.

In the following let $G(X, Y, Z)$ be a G-function with input X, the first addition key Y, and the second addition key Z. The notation $<< n$ means n-bit left shift and $<<< n (>>> n)$ means n-bit left(right) rotation. Let $b(> 16)$ be the byte number of key length and $uk[0], uk[1], \cdots, uk[b-1]$ be a user-supplied key. Then the following is the key schedule of DONUT:

- Input : $uk[0], uk[1], \cdots, uk[b-1]$.
- Output : $k[0], k[1] \cdots, k[27]$.
  1. for($i = 0; i < 112; i + +$) $L[i] = uk[i\%b]$;
  2. for($i = 5; i < 109; i + +$) $L[i] = S[L[i]] \oplus S[L[i-5]] \oplus S[L[i+3]]$;
  3. $X[0] = 0x9e3779b9$;
  4. $Y[0] = 0xb7e15163$;
  5. for($i = 0; i < 28; i + +$) do the followings:
     (a) $T[i] = L[4i] \mid (L[4i+1] << 8) \mid (L[4i+2] << 16) \mid (L[4i+3] << 24)$;
     (b) $X[i+1] = G(X[i], (T[i] >>> 7), (T[i] <<< 5))$;
     (c) $Y[i+1] = G(Y[i], (T[i] <<< 13), (T[i] >>> 9))$;
     (d) $K[i] = X[i+1] \oplus Y[i+1] \oplus T[i]$;

4 Difference Distribution Attack

In this section we suggest an attack called difference distribution attack(DDA). This attack is a chosen plaintext attack and uses the distribution of output differences when the input difference is fixed. In Fig.4, if the input difference of DONUT is fixed, then the output differences are non-uniform with probability 2^{-16}. When this occurs, we can say that the input difference of second decorrelation module is of the form $(\alpha, 0)$.

In the G function, since XOR distribution table of S-box is 4-uniform, all components of XOR distribution table is bounded by 4. So the probability of all $a \xrightarrow{S} b$ is bounded by 2^{-6} where a and b are 8-bit difference and the probability of $c \xrightarrow{G} d$ is bounded by $(2^{-6})^4 = 2^{-24}$ where c and d are 16-bit difference. So the probability of $\alpha \xrightarrow{F} \beta$ is bounded by $(2^{-24})^2 = 2^{-48}$. This means that if we consider 2^{16} input differences, then we can find the input difference such that $\alpha \xrightarrow{F} \beta$. Consider 2^{64} input pairs with above input difference. Since minimum non-zero probability of $\alpha \xrightarrow{F} \beta$ is 2^{-63} and we consider 2^{64} input pairs, at least two pairs have the same output difference.

The attack starts as follows. First, choose an input difference Δ and choose 2^{64} plaintext pairs with difference Δ. Obtain the corresponding ciphertext pairs and determine whether the same output differences exist. This occurs with probability 2^{-16}. So if we consider 2^{16} input differences, then we can obtain an input difference Δ such that the output differences have the same value, say Δ'. When

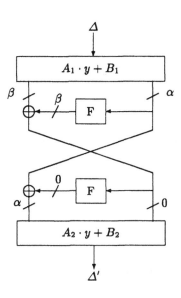

Fig. 4. Difference of Frame Structure

this occurs, we can say that the input difference of the last decorrelation module is of the form $(\alpha, 0)$. For all $(\alpha, 0)$, we compute the corresponding key A_2 such that $(\alpha, 0) \xrightarrow{A_2 \cdot y + B_2} \Delta'$ and store them. Second, as above, choose another input difference such that the output differences have the same value, say Δ''. For all $(\alpha, 0)$, compute A_2 such that $(\alpha, 0) \xrightarrow{A_2 \cdot y + B_2} \Delta''$. For the same $(\alpha, 0)$, the right key has the same A_2 value. This attack needs 2^{64} storage and $2 \times 2^{16} \times 2^{64} = 2^{81}$ chosen plaintext pairs.

5 Structure of Improved DONUT

In this section we describe the structure of improved DONUT. It has two differences between improved DONUT and original DONUT.

Frame Structure Improved DONUT consists of 5 rounds. The first round and the fifth round consist of pairwise perfect decorrelation modules $A_1 \cdot y + B_1$ and $A_2 \cdot y + B_2$. The inner three rounds consist of three round Feistel permutations. Fig.5 shows the frame structure of improved DONUT. The inner functions F and G are the same as DONUT.

Plaintext

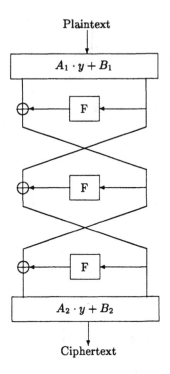

Ciphertext

Fig. 5. Frame Structure of Improved DONUT

Key Scheduling Since improved DONUT needs four 128-bit subkeys and every G function needs two 32-bit(4 bytes) subkeys, we need to generate thirty four 32-bit subkeys. The following is the key schedule of improved DONUT:

- Input : $uk[0], uk[1], \cdots, uk[b-1]$.
- Output : $k[0], k[1] \cdots, k[33]$.
    1. for($i = 0; i < 136; i++$) $L[i] = uk[i\%b]$;
    2. for($i = 5; i < 133; i++$) $L[i] = S[L[i]] \oplus S[L[i-5]] \oplus S[L[i+3]]$;
    3. $X[0] = 0x9e3779b9$;
    4. $Y[0] = 0xb7e15163$;
    5. for($i = 0; i < 34; i++$) do the followings:
        (a) $T[i] = L[4i] \mid (L[4i+1] << 8) \mid (L[4i+2] << 16) \mid (L[4i+3] << 24)$;
        (b) $X[i+1] = G(X[i], (T[i] >>> 7), (T[i] <<< 5))$;
        (c) $Y[i+1] = G(Y[i], (T[i] <<< 13), (T[i] >>> 9))$;
        (d) $K[i] = X[i+1] \oplus Y[i+1] \oplus T[i]$;

6 Security of Improved DONUT

6.1 Resistance against DC and LC

In this section, we estimate the security of improved DONUT against DC and LC.

Definition 3. *For any given* $\Delta x, \Delta y, \Gamma x, \Gamma y \in \mathbb{Z}_2^m$, *the differential and linear probability of each S-box are defined by*

$$DP^S(\Delta x \rightarrow \Delta y) = \frac{\#\{x \in \mathbb{Z}_2^m | S(x) \oplus S(x \oplus \Delta x) = \Delta y\}}{2^m}$$

and

$$LP^S(\Gamma x \rightarrow \Gamma y) = \left(\frac{\#\{x \in \mathbb{Z}_2^m | \Gamma x \cdot x = \Gamma y \cdot S(x)\}}{2^{m-1}} - 1 \right)^2$$

where $\Gamma x \cdot x$ *denotes the parity of bitwise XOR of* Γx *and* x.

Definition 4. *The maximal differential and linear probability of S-box are defined by*

$$DP^S_{max} = \max_{\Delta x \neq 0, \Delta y} DP^S(\Delta x \rightarrow \Delta y)$$

and

$$LP^S_{max} = \max_{\Gamma x, \Gamma y \neq 0} LP^S(\Gamma x \rightarrow \Gamma y),$$

respectively.

Since S-box in the G function of improved DONUT is 8×8 and differentially and linearly 4-uniform, $DP^S_{max} = 2^{-6}$ and $LP^S_{max} = 2^{-6}$. Not only the S-box but also the diffusion is an important factor to give the provable security of the SDS function.

Now the minimum number of differentially and linearly active S-boxes of the SDS function are defined by

$$n_d(D) = \min_{\Delta x \neq 0} \{Hw(\Delta x) + Hw(\Delta y)\}$$

and

$$n_l(D) = \min_{\Gamma y \neq 0} \{Hw(\Gamma x) + Hw(\Gamma y)\},$$

respectively[11].

Theorem 1. *[6] Let* M *be the* $n \times n$ *matrix representing a diffusion layer* D. *Then* $n_d(D) = n + 1$ *if and only if the rank of each* $k \times k$ *submatrix of* M *is* k *for all* $1 \leq k \leq n$.

Since we construct the 4×4 matrix D which satisfies the theorem 1, the diffusion layer used in improved DONUT is maximal with $n_d(D) = 5$ and $n_l(D) = 5$. Therefore, we can give the provable security of the function G with the following two theorems.

Theorem 2. *[6] Assume that the round keys, which are XORed to the input data at each round, are independent and uniformly distributed. If $n_d(D) = n+1$, then each differential probability of SDS function is bounded by $(DP_{max}^S)^n$.*

Theorem 3. *[6] Assume that the round keys, which are XORed to the input data at each round, are independent and uniformly distributed. If $n_l(D) = n+1$, then each linear probability of SDS function is bounded by $(LP_{max}^S)^n$.*

If we assume that the round keys of improved DONUT are independent and uniformly distributed, then we can obtain that the maximal differential and linear probability of G function are bounded by 2^{-30}. The F function consists of 3-round Skipjack-like structure and the maximal differential and linear probability of F function are bounded by $(2^{-30})^2 = 2^{-60}$. Aoki and Ohta[1] showed that the differential probability of 3-round Feistel structure is bounded by DP_{max}^2 when the maximum differential probability of inner function is bound by DP_{max}. So we can conclude that the inner 3-round Feistel structure of improved DONUT has the maximal differential and linear probability bounded by $((2^{-30})^2)^2 = 2^{-120}$.

Let $M = M_0^2$ be a message space where $M_0 = \{0,1\}^{64}$ has a group structure. Since improved DONUT uses three round Feistel permutations as round functions and decorrelation module $A_1 \cdot y + B_1$ is very key dependent[7], the maximum probability of improved DONUT is 2^{-120}. This case occurs only when the input difference of second round F is zero. For a given nonzero input difference of decorrelation module, the probability that the input difference of second round F is zero is 2^{-64}. As a point of view of attacker, he must find the characteristic with probability higher than 2^{-128} in order to attack improved DONUT. But this occurs with probability 2^{-64} and though he can find the characteristic with probability 2^{-120}, he must find the key A_1 and A_2 with computational complexity $2 \cdot 2^{-128}$ field multiplication, because he cannot know the input difference of inner round Feistel permutations. Therefore we can expect to obtain a higher resistance against differential and linear cryptanalysis with only five rounds.

6.2 Resistance against Boomerang Attack

In this section we introduce the boomerang attack and show that improved DONUT is secure against boomerang attack.

Boomerang attack was introduced by D. Wagner[14]. It is a differential attack that attempts to generate a quartet structure at an intermediate value halfway through the cipher. If the best characteristic for half of the rounds of the cipher has probability q, then the boomerang attack can be used in a successful attack needing $O(q^{-4})$ chosen texts. The attacker considers four plaintexts P, P', Q, Q', along with their respective ciphertexts C, C', D, D'. Let $E(\cdot)$ represent the encryption operation, and decompose the cipher into $E = E_1 \circ E_0$, where E_0 represents the first half of the cipher and E_1 represents the last half. We will use two differential characteristics, $\triangle \to \triangle^*$ for E_0, as well as $\nabla \to \nabla^*$ for E_1^{-1}.

The attacker wants to cover the pair P, P' with the characteristic for E_0, and to cover the pairs P, Q and P', Q' with the characteristic for E_1^{-1}. Then the pair Q, Q' is perfectly set up to use the characteristic $\triangle^* \to \triangle$ for E_0^{-1} as follows:

$$E_0(Q) \oplus E_0(Q') = E_0(P) \oplus E_0(P') \oplus E_0(P) \oplus E_0(Q) \oplus E_0(P') \oplus E_0(Q')$$
$$= E_0(P) \oplus E_0(P') \oplus E_1^{-1}(C) \oplus E_1^{-1}(D) \oplus E_1^{-1}(C') \oplus E_1^{-1}(D')$$
$$= \triangle^* \oplus \nabla^* \oplus \nabla^*$$
$$= \triangle^*.$$

We define a right quartet as one where all four characteristics hold simultaneously. The only remaining issue is how to choose the texts so they have the right differences. We can get this as follows. First, we generate $P' = P \oplus \triangle$, and get the encryptions C, C' of P, P' with two chosen-plaintext queries. Then we generate D, D' as $D = C \oplus \nabla$ and $D' = C' \oplus \nabla$. Finally we decrypt D, D' to obtain the plaintexts Q, Q' with two adaptive chosen-ciphertext queries.

Let M_0 be the first pairwise perfect decorrelation module of improved DONUT and let M_1 be the second one. Let Ψ_0 be the first two rounds Feistel permutations of improved DONUT and let Ψ_1 be the last one round Feistel permutation of improved DONUT. Take $E_0 = \Psi_0 \circ M_0$ and $E_1 = M_1 \circ \Psi_1$. First, Ψ_0 has characteristic with probability bounded by 2^{-60}. So Ψ_0 has no good characteristic for boomerang attack. Second, Ψ_1^{-1} has characteristics with probability 1 as follows:

$$(0, c) \rightarrow (c, 0)$$

where c is a nonzero 64-bit value and 0 is a 64-bit zero value. But we can get a characteristic $(a, b) \rightarrow (0, c)$ through M_1^{-1} with probability 2^{-64} because we cannot know the subkey of M_1. So we cannot apply the boomerang attack to improved DONUT and improved DONUT is secure against boomerang attack. M_0 plays an important role when the number of rounds of Ψ_0 and Ψ_1 are exchanged.

7 Conclusion

In this paper we suggested the difference distribution attack(DDA) and applied this attack to DONUT. We also suggested an improved DONUT which consists of five rounds, which uses two pairwise perfect decorrelation modules and 3 rounds Feistel permutations. We showed that improved DONUT was secure against conventional DC, LC, boomerang attack.

References

1. K. Aoki and K. Ohta, *Stict evaluation for the maximum average of differential probability and the maximem average of linear probability*, IEICE Transcations fundamentals of Elections, Communications and Computer Sciences, No.1, 2–8, 1997.
2. E. Biham and A. Shamir, *Differential cryptanalysis of DES-like cryptosystems*, Advances in Cryptology - CRYPTO'90, Lecture Notes in Computer Science, Vol. 537, Springer-Verlag, 2–21, 1991.
3. E. Biham and A. Shamir, *Differential cryptanalysis of DES-like cryptosystems*, Journal of Cryptology, Vol. 4, 3–72, 1991.
4. L. Carter and M. Wegman, *Universal classes of hash functions*, Journal of Computer and System Science, Vol. 18, 143–154, 1979.

5. D. H. Cheon, S. J. Lee, J. I. Lim, and S. J. Lee, *New Block Cipher DONUT Using Pairwise Perfect Decorrelation*, Advances in Cryptology - INDOCRYPT'2000, 2000, to appear.

6. S. H. Hong, S. J. Lee, J. I. Lim, J. C. Sung, D. H. Cheon, and I. H. Cho, *The provable security against differential and linear cryptanalysis for the SPN structure*, Fast Software Encryption Workshop 2000, 2000, to appear.

7. L. R. Knudsen and V. Rijmen, *On the Decorrelated Fast Cipher (DFC) and Its Theory*, Fast Software Encryption Workshop 99, Lecture Notes in Computer Science, Vol. 1636, Springer-Verlag, 137–151, 1999.

8. M. Luby, C. Rackoff, *How to construct pseudorandom permutations from pseudorandom functions*, SIAM Journal of Computing, Vol. 17, 373–386, 1988.

9. M. Matsui, *Linear cryptanalysis method for DES cipher*, Advances in Cryptology - EUROCRYPT'93, Lecture Notes in Computer Science, Vol. 765, Springer-Verlag, 386–397, 1994.

10. M. Naor, S. Reingold, *On the Construction of Pseudo-Random Permutations: Luby-Rackoff Revised*, Journal of Cryptology, Vol. 12, 29–66, 1999.

11. J. Daemen and V. Rijmen, *The Rijndael block cipher*, AES proposal, 1998.

12. S. Vaudenay, *Provable Security for Block Ciphers by Decorrelation*, STACS'98, Lecture Notes in Computer Science, Vol. 1373, Springer-Verlag, 1998.

13. S. Vaudenay, *Feistel Ciphers with L_2-Decorrelation*, SAC'98, Springer-Verlag, 1998.

14. D. Wagner, *The boomerang attack*, Fast Software Encryption Workshop 99, Lecture Notes in Computer Science, Vol. 1636, Springer-Verlag, 156–170, 1999.

15. M. N. Wegman and J. L. Carter, *New hash functions and their use in authentication and set equality*, Journal of Computer and System Science, Vol. 22, 265–279, 1981.

New Results on Correlation Immunity

Yuliang Zheng[1] and Xian-Mo Zhang[2]

[1] Monash University, Frankston, Melbourne, VIC 3199, Australia
yuliang.zheng@monash.edu.au, www.netcomp.monash.edu.au/links/
[2] The University of Wollongong, Wollongong, NSW 2522, Australia
xianmo@cs.uow.edu.au

Abstract. The absolute indicator for GAC forecasts the overall avalanche characteristics of a cryptographic Boolean function. From a security point of view, it is desirable that the absolute indicator of a function takes as small a value as possible. The first contribution of this paper is to prove a tight lower bound on the absolute indicator of an mth-order correlation immune function with n variables, and to show that a function achieves the lower bound if and only if it is affine. The absolute indicator for GAC achieves the upper bound when the underlying function has a non-zero linear structure. Our second contribution is about a relationship between correlation immunity and non-zero linear structures. The third contribution of this paper is to address an open problem related to the upper bound on the nonlinearity of a correlation immune function. More specifically, we prove that given any odd mth-order correlation immune function f with n variables, the nonlinearity of f, denoted by N_f, must satisfy $N_f \leq 2^{n-1} - 2^{m+1}$ for $\frac{1}{2}n - 1 \leq m < 0.6n - 0.4$ or f has a non-zero linear structure. This extends a known result that is stated for $0.6n - 0.4 \leq m \leq n - 2$.

Key Words:

Correlation Immunity, Absolute Indicator, Nonlinearity, Linear Structures, Stream Ciphers

1 Introduction

Correlation immunity has long been recognized as one of the critical indicators of nonlinear combining functions of shift registers in stream generators (see [10]). A high correlation immunity is generally a very desirable property, in view of various successful correlation attacks against a number of stream ciphers (see for instance [5]).

Another class of cryptanalytic attacks against stream ciphers, called best approximation attacks, were advocated in [3]. Success of these attacks in breaking a stream cipher is made possible by exploiting the low nonlinearity of functions employed by the cipher, and it highlights the significance of nonlinearity in the analysis and design of encryption algorithms.

However it should be pointed out that correlation immunity is not harmonious with some other cryptographic requirements. In particular, high correlation immunity may introduce weaknesses in terms of a low algebraic degree, a small avalanche degree and a low nonlinearity and so on. This can be seen, for instance, from recent work in [14, 15].

GAC is a nonlinearity indicator introduced in [11] to study the global or overall avalanche characteristics of a cryptographic function. Two different indicators were proposed to measure numerically the GAC of a functions, namely, the sum-of-squares indicator and the absolute indicator. A small value for the absolute indicator of a function is generally more desirable.

In the first part of this paper we show that functions with a high order correlation immunity necessarily has weaknesses in its avalanche characteristics. More specifically, we prove that if f is a balanced mth-order correlation immune function with n variables, then the absolute indicator for GAC of f, denoted by Δ_f, satisfies $\Delta_f \geq 2^m \sum_{i=0}^{+\infty} 2^{i(m-n)}$. For an unbalanced function f, we show that $\Delta_f \geq 2^{m-1} \sum_{i=0}^{+\infty} 2^{i(m-1-n)}$. We further investigate the tightness of the lower bounds and identify a necessary and sufficient condition on when the two lower bounds are achieved.

When $\Delta_f = 2^n$, f must have a non-zero linear structure, which is considered cryptographically undesirable. In the second part of this paper, we employ correlation immunity to characterize Boolean functions having non-zero linear structures.

Recently, Zheng and Zhang [14] have proved that if f is an mth-order correlation immune function f with n variables, then its nonlinearity satisfies $N_f \leq 2^{n-1} - 2^{m+1}$, when $0.6n - 0.4 \leq m \leq n - 2$, regardless of the balance of the function. Note that the inequality $N_f \leq 2^{n-1} - 2^{m+1}$ does not hold for $m = n - 1$. Fortunately, this is a trivial case, as an $(n-1)$th-order correlation immune function f with n variables must be affine. In the same paper, Zheng and Zhang have also shown that the equality holds if and only if f is a plateaued function. The authors leave as an open problem for the case of $\frac{1}{2}n - 1 \leq m < 0.6n - 0.4$. This open problem is addressed in the third part of this paper. In particular, we prove that the inequality $N_f \leq 2^{n-1} - 2^{m+1}$ does hold for odd m with $\frac{1}{2}n - 1 \leq m < 0.6n - 0.4$ otherwise f has a non-zero linear structure. This brings us a step closer to finally solving the open problem.

2 Boolean Functions

We consider functions from V_n to $GF(2)$ (or simply functions on V_n), where V_n is the vector space of n tuples of elements from $GF(2)$. The *truth table* of a function f on V_n is a $(0, 1)$-sequence defined by $(f(\alpha_0), f(\alpha_1), \ldots, f(\alpha_{2^n - 1}))$, and the *sequence* of f is a $(1, -1)$-sequence defined by $((-1)^{f(\alpha_0)}, (-1)^{f(\alpha_1)}, \ldots, (-1)^{f(\alpha_{2^n - 1})})$, where $\alpha_0 = (0, \ldots, 0, 0)$, $\alpha_1 = (0, \ldots, 0, 1)$, $\ldots$, $\alpha_{2^{n-1} - 1} = (1, \ldots, 1, 1)$. The *matrix* of f is a $(1, -1)$-matrix of order 2^n defined by $M = ((-1)^{f(\alpha_i \oplus \alpha_j)})$ where $\oplus$ denotes the addition in $GF(2)$.

Given two sequences $\tilde{a} = (a_1, \cdots, a_m)$ and $\tilde{b} = (b_1, \cdots, b_m)$, their *component-wise product* is defined by $\tilde{a} * \tilde{b} = (a_1 b_1, \cdots, a_m b_m)$. In particular, if $m = 2^n$ and $\tilde{a}, \tilde{b}$ are the sequences of functions f and g on V_n respectively, then $\tilde{a} * \tilde{b}$ is the sequence of $f \oplus g$ where $\oplus$ denotes the addition in $GF(2)$.

Let $\tilde{a} = (a_1, \cdots, a_m)$ and $\tilde{b} = (b_1, \cdots, b_m)$ be two sequences or vectors, the *scalar product* of $\tilde{a}$ and $\tilde{b}$, denoted by $\langle \tilde{a}, \tilde{b} \rangle$, is defined as the sum of the component-wise multiplications. In particular, when $\tilde{a}$ and $\tilde{b}$ are from V_m, $\langle \tilde{a}, \tilde{b} \rangle = a_1 b_1 \oplus \cdots \oplus a_m b_m$, where the addition and multiplication are over $GF(2)$, and when $\tilde{a}$ and $\tilde{b}$ are $(1, -1)$-sequences, $\langle \tilde{a}, \tilde{b} \rangle = \sum_{i=1}^m a_i b_i$, where the addition and multiplication are over the reals.

An *affine* function f on V_n is a function that takes the form of $f(x_1, \ldots, x_n) = a_1 x_1 \oplus \cdots \oplus a_n x_n \oplus c$, where $a_j, c \in GF(2)$, $j = 1, 2, \ldots, n$. Furthermore f is called a *linear* function if $c = 0$.

A $(1, -1)$-matrix N of order n is called a *Hadamard* matrix if $NN^T = nI_n$, where N^T is the transpose of N and I_n is the identity matrix of order n. A Sylvester-Hadamard matrix of order 2^n, denoted by H_n, is generated by the following recursive relation

$$H_0 = 1, \ H_n = \begin{bmatrix} H_{n-1} & H_{n-1} \\ H_{n-1} & -H_{n-1} \end{bmatrix}, \ n = 1, 2, \ldots.$$

Obviously H_n is symmetric. Let $\ell_i, 0 \leq i \leq 2^n - 1$, be the i row of H_n. It is known that ℓ_i is the sequence of a linear function $\varphi_i(x)$ defined by the scalar product $\varphi_i(x) = \langle \alpha_i, x \rangle$, where α_i is the ith vector in V_n according to the ascending alphabetical order.

The *Hamming weight* of a $(0, 1)$-sequence ξ, denoted by $HW(\xi)$, is the number of ones in the sequence. Given two functions f and g on V_n, the *Hamming distance* $d(f, g)$ between them is defined as the Hamming weight of the truth table of $f(x) \oplus g(x)$, where $x = (x_1, \ldots, x_n)$.

A function f is said to be *balanced* if its truth table contains an equal number of ones and zeros.

3 Cryptographic Criteria of Boolean Functions

The following criteria for cryptographic Boolean functions are often considered: balance, nonlinearity, avalanche criterion, correlation immunity, algebraic degree and non-zero linear structures. In this paper we focus mainly on nonlinearity and correlation immunity.

The so-called Parseval's equation (Page 416 [6]) is a useful tool in this work: Let f be a function on V_n and ξ denote the sequence of f. Then $\sum_{i=0}^{2^n-1} \langle \xi, \ell_i \rangle^2 = 2^{2n}$ where ℓ_i is the ith row of H_n, $i = 0, 1, \ldots, 2^n - 1$.

The *nonlinearity* of a function f on V_n, denoted by N_f, is the minimal Hamming distance between f and all affine functions on V_n, i.e.,

$$N_f = \min_{i=1,2,\ldots,2^{n+1}} d(f, \varphi_i)$$

where φ_1, φ_2, ..., φ_{2^n+1} are all the affine functions on V_n. High nonlinearity is useful in resisting a linear attack and a best approximation attack. The following characterization of nonlinearity will be useful (for a proof see for instance [7]).

Lemma 1. *The nonlinearity of f on V_n can be expressed by*

$$N_f = 2^{n-1} - \frac{1}{2}\max\{|\langle \xi, \ell_i \rangle|, 0 \leq i \leq 2^n - 1\}$$

where ξ is the sequence of f and ℓ_0, ..., ℓ_{2^n-1} are the rows of H_n, namely, the sequences of linear functions on V_n.

From Lemma 1 and Parseval's equation, it is easy to verify that $N_f \leq 2^{n-1} - 2^{\frac{1}{2}n-1}$ for any function f on V_n. If $N_f = 2^{n-1} - 2^{\frac{1}{2}n-1}$, then f is called a *bent function* [8]. It is known that a bent function on V_n exists only when n is even.

Let f be a function on V_n. For a vector $\alpha \in V_n$, denote by $\xi(\alpha)$ the sequence of $f(x \oplus \alpha)$. Thus $\xi(0)$ is the sequence of f itself and $\xi(0) * \xi(\alpha)$ is the sequence of $f(x) \oplus f(x \oplus \alpha)$. Set $\Delta_f(\alpha) = \langle \xi(0), \xi(\alpha) \rangle$, the scalar product of $\xi(0)$ and $\xi(\alpha)$. $\Delta(\alpha)$ is called the auto-correlation of f with a shift α. We omit the subscript of $\Delta_f(\alpha)$ if no confusion occurs. Obviously, $\Delta(\alpha) = 0$ if and only if $f(x) \oplus f(x \oplus \alpha)$ is balanced, i.e., f satisfies the avalanche criterion with respect to α. In the case that f does not satisfy the avalanche criterion with respect to a vector α, it may be desirable for $f(x) \oplus f(x \oplus \alpha)$ to be almost balanced. That is, one may require $|\Delta(\alpha)|$ to be a small value. In an extreme case, $\alpha \in V_n$ is called a *linear structure* of f if $|\Delta(\alpha)| = 2^n$ (i.e., $f(x) \oplus f(x \oplus \alpha)$ is a constant). For any function f, $\Delta(\alpha_0) = 2^n$, where α_0 is the zero vector on V_n. It is easy to verify that the set of all linear structures of a function f form a linear subspace of V_n, whose dimension is called the *linearity of f*, denoted by L_f. A non-zero linear structure is cryptographically undesirable hence we should avoid non-zero linear structures in the design of cryptographic functions as possible as we can. It is also well-known that if f has non-zero linear structures, then there exists a nonsingular $n \times n$ matrix B over $GF(2)$ such that $f(xB) = g(y) \oplus \psi(z)$, where $x = (y, z)$, $y \in V_p$, $z \in V_q$, g is a function on V_p that has no non-zero linear structures, and ψ is a linear function on V_q.

The concept of correlation immune functions was introduced by Siegenthaler [10]. Xiao and Massey gave an equivalent definition [1, 4]: A function f on V_n is called a *mth-order correlation immune function* if

$$\sum_{x \in V_n} f(x)(-1)^{\langle \beta, x \rangle} = 0$$

for all $\beta \in V_n$ with $1 \leq HW(\beta) \leq m$, where in the the sum, $f(x)$ and $\langle \beta, x \rangle$ are regarded as real-valued functions. From the first equality in Section 4.2 of [1], a correlation immune function can also be equivalently restated as follows: Let f be a function on V_n and let ξ be its sequence. Then f is called a *mth-order correlation immune function* if $\langle \xi, \ell \rangle = 0$ for every ℓ, where ℓ is the sequence of a linear function $\varphi(x) = \langle \alpha, x \rangle$ on V_n constrained by $1 \leq HW(\alpha) \leq m$. In fact, $\langle \xi, \ell_i \rangle = 0$, where ℓ_i is the ith row of H_n, if and only if $f(x) \oplus \langle \alpha_i, x \rangle$ is balanced, where α_i is the binary representation of an integer i, $0 \leq i \leq 2^n - 1$. Correlation immune functions are used in the design of running-key generators in stream ciphers to resist a correlation attack and the design of hash functions. Relevant discussions on correlation immune functions, more generally on resilient functions, can be found in [12].

4　A Tight Lower Bound on the Absolute Indicators of Correlation Immune Functions

Let f be a function on V_n and ξ denote the sequence of f. We introduce two new notations:

1. Set $\Im_f = \{i \mid \langle \xi, \ell_i \rangle \neq 0, \ 0 \leq i \leq 2^n - 1\}$ where ℓ_i is the ith row of H_n,
2. set $\Im_f^* = \{\alpha_i \mid \langle \xi, \ell_{\alpha_i} \rangle \neq 0, \ 0 \leq i \leq 2^n - 1\}$ where α_i is the binary representation of an integer i, $0 \leq i \leq 2^n - 1$ and ℓ_{α_i} is identified with ℓ_i.

$\Im_f^*$ is essentially the same as $\Im_f$ with the only difference being that its elements are represented by a binary vector in V_n. We will simply write $\Im_f$ as $\Im$ and $\Im_f^*$ as $\Im^*$ when no confusion arises. It is easy to verify that $\#\Im_f$ and $\#\Im_f^*$ are invariant under any nonsingular linear transformation on the variables of the function f. $\#\Im_f$ ($\#\Im_f^*$) together with the distribution of $\Im_f$ ($\Im_f^*$) determines the correlation immunity and other cryptographic properties of a function.

Lemma 2. *Let f be a function on V_n, β be a vector in V_n and B be a nonsingular $n \times n$ matrix over $GF(2)$. Then the following statements hold:*

(i) Set $g(x) = f(xB \oplus \beta)$. Then $\#\Im_g^* = \#\Im_f^*$.
(ii) Set $g(x) = f(x \oplus \beta)$. Then $\Im_g^* = \Im_f^*$.
(iii) Set $g(x) = f(xB)$. Then $\Im_g^* = \Im_f^* B^T$ where $XB^T = \{\alpha B^T | \alpha \in X\}$.
(iv) Set $g(x) = f(x) \oplus \varphi(x)$, where $\varphi(x) = \langle \beta, x \rangle$. Then $\Im_g^* = \beta \oplus \Im_f^*$ where $X = \{\beta \oplus \gamma | \gamma \in X\}$.

Proof. Since (ii), (iii) and (iv) together imply (i), we prove (ii), (iii) and (iv) only.

(ii) $\alpha \in \Im_g^* \iff g(x) \oplus \langle \alpha, x \rangle$ is unbalanced, i.e., $f(x \oplus \beta) \oplus \langle \alpha, x \rangle$ is unbalanced $\iff f(x \oplus \beta) \oplus \langle \alpha, x \oplus \beta \rangle$ is unbalanced $\iff f(u) \oplus \langle \alpha, u \rangle$ is unbalanced where $u = x \oplus \beta \iff \alpha \in \Im_f^*$. This proves $\Im_g^* = \Im_f^*$.

(iii) $\alpha \in \Im_g^* \iff g(x) \oplus \langle \alpha, x \rangle$ is unbalanced, i.e., $f(xB) \oplus \langle \alpha, x \rangle$ is unbalanced $\iff f(u) \oplus \langle \alpha, uB^{-1} \rangle$ is unbalanced where $xB = u$. Note that $\langle \alpha, uB^{-1} \rangle = (uB^{-1})\alpha^T = u(B^{-1}\alpha^T) = (B^{-1}\alpha^T)^T u^T = \alpha(B^T)^{-1}u^T = \langle \alpha(B^T)^{-1}, u \rangle$. Therefore $f(u) \oplus \langle \alpha, uB^{-1} \rangle$ is unbalanced $\iff f(u) \oplus \langle \alpha(B^T)^{-1}, u \rangle$ is unbalanced $\iff \alpha(B^T)^{-1} \in \Im_f^* \iff \alpha \in \Im_f^* B^T$. This proves $\Im_g^* = \Im_f^* B^T$.

(iv) $\alpha \in \Im_g^* \iff g(x) \oplus \langle \alpha, x \rangle$ is unbalanced, i.e., $f(x) \oplus \langle \beta, x \rangle \oplus \langle \alpha, x \rangle$ is unbalanced $\iff f(x) \oplus \langle \beta \oplus \alpha, x \rangle$ is unbalanced $\iff \beta \oplus \alpha \in \Im_f^* \iff \alpha \in \beta \oplus \Im_f^*$. This proves $\Im_g^* = \beta \oplus \Im_f^*$.

The following definition is from [11].

Definition 1. *For a function f on V_n, the* absolute indicator *for GAC of f is defined as*

$$\Delta_f = \max_{\alpha \in V_n, \alpha \neq 0} |\Delta(\alpha)|$$

Obviously $\Delta_f = 2^n$ if and only if f has a non-zero linear structure, while $\Delta_f = 0$ if and only if f is bent. Since balanced functions are not bent, we have $\Delta_f > 0$ where f is balanced. In designing cryptographic algorithms, we are concerned with a balanced nonlinear function f that shows a small Δ_f, as was discussed in [11] where it was argued that a smaller Δ_f is cryptographically more desirable. This section shows that a high order of correlation immunity may result in weaknesses in avalanche characteristics.

The following lemma is the re-statement of a relation proved in Section 2 of [2].

Lemma 3. *For every function f on V_n, we have*

$$(\Delta(\alpha_0), \Delta(\alpha_1), \ldots, \Delta(\alpha_{2^n-1}))H_n = (\langle \xi, \ell_0 \rangle^2, \langle \xi, \ell_1 \rangle^2, \ldots, \langle \xi, \ell_{2^n-1} \rangle^2).$$

where ξ denotes the sequence of f and ℓ_i is the ith row of H_n, $i = 0, 1, \ldots, 2^n-1$.

From [14], we have the following statement.

Lemma 4. *Consider a function f on V_n. Let $\xi = (a_0, a_1, \ldots, a_{2^n-1})$, where $a_j = \pm 1$ denote the sequence of f and ℓ_i denote the ith row of H_n, $i = 0, 1, \ldots, 2^n - 1$. Let p be an integer with $1 \leq p \leq n-1$. Write $\xi = (\xi_0, \xi_1, \ldots, \xi_{2^p-1})$ where each ξ_i is of length 2^{n-p}. Let e_i denote the ith row of H_{n-p}, $i = 0, 1, \ldots, 2^{n-p} - 1$.*

$$2^p(\langle \xi_0, e_j \rangle, \langle \xi_1, e_j \rangle, \langle \xi_2, e_j \rangle, \ldots, \langle \xi_{2^p-1}, e_j \rangle)$$
$$= (\langle \xi, \ell_j \rangle, \langle \xi, \ell_{j+2^{n-p}} \rangle, \langle \xi, \ell_{j+2 \cdot 2^{n-p}} \rangle, \ldots, \langle \xi, \ell_{j+(2^p-1)2^{n-p}} \rangle)H_p$$

where $j = 0, 1, \ldots, 2^{n-p} - 1$.

The following lemma is useful in proving one of our main theorems.

Lemma 5. *Let $(k_0, k_1, \ldots, k_{2^n-1})H_n = (r_0, r_1, \ldots, r_{2^n-1})$, where $k_0 = 0$ and each k_j and each r_j are both real numbers. Then*

$$\max\{|r_1|, \ldots, |r_{2^n-1}|\} \geq \max\{|k_1|, \ldots, |k_{2^n-1}|\}$$

Proof. Without loss of generality, we assume that $|k_{2^n-1}| = \max\{|k_1|, \ldots, |k_{2^n-1}|\}$. Let $H_n = [P \ Q]$ where both P and Q are $2^n \times 2^{n-1}$ matrices. Hence we have $(k_0, k_1, \ldots, k_{2^n-1})Q = (r_{2^{n-1}}, r_{2^{n-1}+1}, \ldots, r_{2^n-1})$. Let e_0 denote the all-one sequence of length 2^{n-1}. It is obvious that

$$(k_0, k_1, \ldots, k_{2^n-1})Qe_0^T = (r_{2^{n-1}}, r_{2^{n-1}+1}, \ldots, r_{2^n-1})e_0^T \quad (1)$$

Note that $Q = \begin{bmatrix} H_{n-1} \\ -H_{n-1} \end{bmatrix}$ and hence we have $Qe_0^T = 2^{n-1}(b_0, b_1, \ldots, b_{2^n-1})^T$ where $(b_0, b_1, \ldots, b_{2^n-1})$ satisfies $b_0 = 1$, $b_{2^{n-1}} = -1$ and other $b_j = 0$. Due to (1), we have $2^{n-1}(k_0 - k_{2^{n-1}}) = \sum_{j=2^{n-1}}^{2^n-1} r_j$, where $k_0 = 0$. This proves that there exits some i_0, $2^{n-1} \leq i_0 \leq 2^n - 1$, such that $|r_{i_0}| \geq |k_{2^{n-1}}|$. Thus the lemma holds. $\square$

We notice that $\max\{|r_0|, |r_1|, \ldots, |r_{2^n-1}|\} \geq \max\{|k_0|, |k_1|, \ldots, |k_{2^n-1}|\}$ is still true. However this inequality is less useful in this paper as $\Delta(\alpha_0) = 2^n$ holds for every function on V_n, and we are concerned with Δ_f where $\Delta_f = \max_{\alpha \in V_n, \alpha \neq 0} |\Delta(\alpha)|$.

Theorem 1. *Let f be a function on V_n. Then the following statements hold:*

(i) *If there exist an m-dimensional linear subspace W, $1 \leq m \leq n - 1$, and a vector α^* in V_n such that $\Im_f^* \cap (\alpha^* \oplus W) = \emptyset$ where $\emptyset$ denotes the empty set, then*

$$\Delta_f \geq 2^m \sum_{i=0}^{+\infty} 2^{i(m-n)} \tag{2}$$

(ii) *Under the assumption of (i), the following three statements are equivalent:*
   (a) $\Delta_f = 2^m \sum_{i=0}^{+\infty} 2^{i(m-n)}$,
   (b) $m = n - 1$,
   (c) f has a non-zero linear structure.

Proof. First we prove (i). Due to Lemma 2, we can assume, without loss of generality, that $\alpha^* = \alpha_0$, where α_0 denotes the zero vector in V_n, and $W = \{\alpha_0, \alpha_1, \ldots, \alpha_{2^m-1}\}$. Let ξ denote the sequence of f and ℓ_i be the ith row of H_n, $i = 0, 1, \ldots, 2^n - 1$. Since $\Im_f^* \cap W = \emptyset$, we have $\langle \xi, \ell_i \rangle = 0$, $i = 0, 1, \ldots, 2^m - 1$. Due to Lemma 3, we have

$$(\Delta(\alpha_0), \Delta(\alpha_1), \ldots, \Delta(\alpha_{2^n-1}))H_n = (0, \ldots, 0, \langle \xi, \ell_{2^m} \rangle^2, \ldots, \langle \xi, \ell_{2^n-1} \rangle^2)$$

or

$$(0, \ldots, 0, \langle \xi, \ell_{2^m} \rangle^2, \ldots, \langle \xi, \ell_{2^n-1} \rangle^2)H_n = 2^n(\Delta(\alpha_0), \Delta(\alpha_1), \ldots, \Delta(\alpha_{2^n-1})) \tag{3}$$

Applying Lemma (4) (with $p = n - m$ and $j = 0$) to Equation (3), we obtain

$$(0, \sum_{j=2^m}^{2 \cdot 2^m - 1} \langle \xi, \ell_j \rangle^2, \ldots, \sum_{j=2^n-2^m}^{2^n-1} \langle \xi, \ell_j \rangle^2)H_{n-m}$$
$$= 2^n(\Delta(\alpha_0), \Delta(\alpha_{2^m}), \Delta(\alpha_{2 \cdot 2^m}), \ldots, \Delta(\alpha_{(2^{n-m}-1) \cdot 2^m})) \tag{4}$$

Applying Parseval's equation to f, we have $\sum_{i=1}^{2^{n-m}-1} \sum_{j=i \cdot 2^m}^{(i+1) \cdot 2^m - 1} \langle \xi, \ell_j \rangle^2 = 2^{2n}$. It is easy to see that there exists some i_0, $1 \leq i_0 \leq 2^{n-m} - 1$, such that

$$\sum_{j=i_0 \cdot 2^m}^{(i_0+1) \cdot 2^m - 1} \langle \xi, \ell_j \rangle^2 \geq \frac{2^{2n}}{2^{n-m} - 1} = 2^{n+m} \sum_{i=0}^{+\infty} 2^{i(m-n)}$$

Applying Lemma 5 to (4), we conclude that there exists some j_0, $1 \leq j_0 \leq 2^{n-m} - 1$, such that

$$2^n |\Delta(\alpha_{j_0 \cdot 2^m})| \geq 2^{n+m} \sum_{i=0}^{+\infty} 2^{i(m-n)}$$

This proves that $\Delta_f \geq 2^m \sum_{i=0}^{+\infty} 2^{i(m-n)}$ and hence (i) holds. Next we prove (ii).

First we prove (a) $\Longleftrightarrow$ (b). Assume that (a) holds, i.e., $\Delta_f = 2^m \sum_{i=0}^{+\infty} 2^{i(m-n)}$, or equivalently, $\Delta_f = 2^m \sum_{i=0}^{+\infty} 2^{i(m-n)} = \frac{2^n}{2^{n-m}-1}$. Therefore $\frac{2^n}{2^{n-m}-1}$ must be an integer. Since 2^n is not divisible by $2^{n-m}-1$ if $n-m \geq 2$, we conclude that $m = n-1$, i.e., (b) holds. Conversely, assume that (b) holds, i.e., $m = n-1$. In this case, by using (i) of the theorem, we have $\Delta_f \geq 2^m \sum_{i=0}^{+\infty} 2^{i(m-n)} = \frac{2^n}{2^{n-m}-1} = 2^n$. Hence $\Delta_f = 2^n$, i.e., (a) holds.

We now prove (b) $\Longleftrightarrow$ (c). Assume that (b) holds, i.e., $m = n-1$. In this case, by using (i) of the theorem, we have $\Delta_f \geq 2^m \sum_{i=0}^{+\infty} 2^{i(m-n)} = \frac{2^n}{2^{n-m}-1} = 2^n$. Hence $\Delta_f = 2^n$. This means that f has a non-zero linear structure and hence (c) holds. Conversely, assume that (c) holds, i.e., f has a non-zero linear structure. Due to Lemma 2, without loss of generality, assume that α_{2^n-1} is a non-zero linear structure. Hence we can write f as $f(x) = cx_1 \oplus g(y)$ where g is a function on V_{n-1}, $x = (x_1, \ldots, x_n)$, $y = (x_2, \ldots, x_n)$ and c is a constant in $GF(2)$. Once again, due to Lemma 2, without loss of generality, assume that $c = 0$. Let η denote the sequence of g. Then the sequence ξ of f can be denoted as $\xi = (\eta, \eta)$. It is easy to verify that $\langle \xi, l_i \rangle = 0$ where l_i is the ith row of H_n, $i = 0, 1, \ldots, 2^{n-1}-1$. This proves that $\mathfrak{S}_f^* \cap W = \emptyset$ where W is specialized as an $(n-1)$-dimensional subspace, that is, $W = \{\alpha_0, \alpha_1, \ldots, 2^{n-1}-1\}$. This proves that $m = n-1$ and hence (b) holds. $\qquad\square$

From the definition of correlation immune functions [1, 4], if f is a balanced mth-order correlation immune functions, then $m \leq n-1$, and a function on V_n is $(n-1)$th-order correlation immune if and only if $f(x) = x_1 \oplus \cdots \oplus x_n \oplus c$ where $x = (x_1, \ldots, x_n)$ and c is a constant in $GF(2)$. Using Theorem 1, we obtain

Theorem 2. *Let f be a balanced mth-order correlation immune function on V_n $(1 \leq m \leq n-1)$. Then*

$$\Delta_f \geq 2^m \sum_{i=0}^{+\infty} 2^{i(m-n)}$$

where the equality holds if and only if $f(x) = x_1 \oplus \cdots \oplus x_n \oplus c$ where $x = (x_1, \ldots, x_n)$ and c is a constant in $GF(2)$.

Let f be a function on V_n whose sequence is ξ. Assume that f satisfies $\langle \xi, l_i \rangle = 0$ for every $i = 1, \ldots, 2^n - 1$, or equivalently, $f(x) \oplus \langle \alpha, x \rangle$ is balanced for every non-zero vector in V_n. It is easy to verify that f must be a constant in $GF(2)$. For this reason, we define the zero function on V_n and the non-zero constant function on V_n as an nth-order correlation immune function on V_n.

Theorem 3. *Let f be an unbalanced mth-order correlation immune function on V_n $(2 \leq m \leq n)$. Then*

$$\Delta_f \geq 2^{m-1} \sum_{i=0}^{+\infty} 2^{i(m-1-n)}$$

where the equality holds if and only if f is a constant. (Note that an nth-order correlation immune function is defined as a constant).

Proof. Let $\beta \in V_n$ and $HW(\beta) = m$. Set $\psi(x) = \langle \beta, x \rangle$ and $g = f \oplus \psi$. It is easy to see that g is a balanced $(m-1)$th-order correlation immune function on V_n. Due to Theorem 2, the statement holds. □

Theorems 2 and 3 indicate that correlation immunity is not harmonious with avalanche characteristics.

5 A Relationship between Correlation Immunity and Linear Structures

In this section, we consider the case when the absolute indicator for SAC achieves the maximum value i.e., $\Delta_f = 2^n$.

Theorem 4. *Let f be a function on V_n. If there exist a p-dimensional linear subspace W with $1 \le p \le n-1$ and a vector α in V_n such that $\Im_f^* \subseteq \alpha \oplus W$ if and only if f has a non-zero linear structure.*

Proof. We first prove the necessity. Since the existence of non-zero linear structures is invariant under a nonsingular linear transformation on the variables, without loss of generality, we can assume $W = \{(a_1, \ldots, a_p, 0, \ldots, 0) | (a_1, \ldots, a_p, 0, \ldots, 0) \in V_n\}$. In other words, $W = \{\alpha_0, \alpha_{2^{n-p}}, \alpha_{2 \cdot 2^{n-p}}, \ldots, \alpha_{(2^p-1) \cdot 2^{n-p}}\}$, where each $\alpha_j \in V_n$ and α_j is the binary representation of an integer j. Let $W^* = \{(0, \ldots, 0, c_1, \ldots, c_{n-p}) | (0, \ldots, 0, c_1, \ldots, c_{n-p}) \in V_n\}$. In other words, $W^* = \{\alpha_0, \alpha_1, \ldots, \alpha_{2^{n-p}-1}\}$, where each $\alpha_j \in V_n$ is the binary representation of an integer j, $j = 0, 1, \ldots, 2^{n-p}$, and

$$V_n = (\alpha_0 \oplus W) \cup (\alpha_1 \oplus W) \cup \cdots \cup (\alpha_{2^{n-s}-1} \oplus W)$$

where $(\alpha_j \oplus W) \cap (\alpha_i \oplus W) = \emptyset$ whenever $j \ne i$.

Since $\Im_f^* \subseteq \alpha_{j_0} \oplus W$ for some j_0, $0 \le j_0 \le 2^{n-p}-1$, $\langle \xi, \ell_i \rangle = 0$ if $\alpha_i \in \alpha_j \oplus W$ with $j \ne j_0$, where α_i is the representation of an integer i. Note that $\alpha_i \in \alpha_j \oplus W$ if and only if $i \in \{j, j + 2^{n-p}, \ldots, j + (2^p - 1)2^{n-p}\}$. By using Lemma 4, we have

$$(\langle \xi_0, e_i \rangle), \langle \xi_1, e_i \rangle), \ldots, \langle \xi_{2^p-1}, e_i \rangle) H_p$$
$$= (\langle \xi, \ell_i \rangle, \langle \xi, \ell_{i+2^{n-p}} \rangle, \ldots, \langle \xi, \ell_{i+(2^p-1)2^{n-p}} \rangle) = (0, 0, \ldots, 0)$$

whenever $i \ne j_0$. Therefore

$$(\langle \xi_0, e_i \rangle, \langle \xi_1, e_i \rangle, \ldots, \langle \xi_{2^p-1}, e_i \rangle) = (0, 0, \ldots, 0) \tag{5}$$

whenever $i \ne j_0$. Since $\langle \xi_0, e_i \rangle = 0$, whenever $i \ne j_0$, we conclude $\xi_0 = b_0 e_{j_0}$ where $b_0 = \pm 1$. Similarly $\xi_1 = b_1 e_{j_0}$ where $b_1 = \pm 1$, ..., $\xi_{2^p-1} = b_{2^p-1} e_{j_0}$ where $b_{2^p-1} = \pm 1$. Therefore the sequence of f, ξ, satisfies

$$\xi = (b_0 e_{j_0}, b_1 e_{j_0}, \ldots, b_{2^p-1} e_{j_0}) \tag{6}$$

Since e_{j_0} is a row of H_{n-p}, e_{j_0} is the sequence of a linear function on V_{n-p}, denoted by ψ. Let $(b_0, b_1, \ldots, b_{2^p-1})$ be the sequence of a function on V_p, denoted by g. Due to (6), f can be expressed as $f(x) = g(y) \oplus \psi(z)$ where $x = (y, z)$, $y \in V_p$, $z \in V_{n-p}$. This proves that f has a non-zero linear structure.

Conversely, assume that f has a non-zero linear structure. Then f is equivalent to $g(x) = cx_1 \oplus h(y)$ under a nonsingular linear transformation on the variables, where h is a function on V_{n-1}, $x = (x_1, \ldots, x_n)$ and $y = (x_2, \ldots, x_n)$. Without loss of generality, assume that $c = 0$. Let ξ' denote the sequence of g and η denote the sequence of h. Then $\xi' = (\eta, \eta)$. Obviously, if ℓ_i satisfies $\ell_i = (e, -e)$, where ℓ_i denotes the ith row of H_n and e is a row of H_{n-1}, we have $\langle \xi', \ell_i \rangle = 0$. Therefore if $\langle \xi', \ell_j \rangle \neq 0$ then ℓ_j must take the form of $\ell_j = (e, e)$. Due to the structure of H_n, j satisfies $0 \leq j \leq 2^{n-1} - 1$. This proves that $\Im_g \subseteq \{0, 1, \ldots, 2^{n-1} - 1\}$, equivalently, $\Im_g^* \subseteq W = \{\alpha_0, \alpha_1, \ldots, \alpha_{2^{n-1}-1}\}$, where W obviously is an $(n-1)$-dimensional subspace of V_n. Since the linearity is invariant under any nonsingular linear transformation on the variables, we have the same conclusion on $\Im_f^*$. Thus we have proved the sufficiency. □

Theorem 4 can be viewed as a way of characterizing Boolean functions having non-zero linear structures by the use of correlation immunity. This result will be used in the next section.

6 A New Result on Upper Bound on Nonlinearity of Correlation Immune Functions

6.1 Previously Known Results

Recently Zheng and Zhang proved that when $0.6n - 0.4 \leq m \leq n - 2$, the nonlinearity N_f of an mth-order correlation immune function f with n variables satisfies the condition of $N_f \leq 2^{n-1} - 2^{m+1}$. In the same paper they also showed that if a correlation immune function achieves the maximum nonlinearity for such a function, then it is a *plateaued function*.

The concept of plateaued functions was introduced in [13]. Let f be a function on V_n and ξ denote the sequence of f. If there exists an even number $r, 0 \leq r \leq n$, such that $\#\Im = 2^r$ and each $\langle \xi, \ell_j \rangle^2$ takes the value of 2^{2n-r} or 0 only, where ℓ_j denotes the jth row of H_n, $j = 0, 1, \ldots, 2^n - 1$, then f is called a *rth-order plateaued function* on V_n. f is also simply called a *plateaued function* on V_n if we ignore the particular order r. Some facts about plateaued functions follow: if f is a rth-order plateaued function, then r must be even; f is an nth-order plateaued function if and only if f is bent; and f is a 0th-order plateaued function if and only if f is affine. Plateaued functions are interesting as they have a number of cryptographically useful properties [13]. For instance: $\sum_{j=0}^{2^n-1} \Delta^2(\alpha_j) \geq \frac{2^{3n}}{\#\Im}$ where the equality holds if and only if f is a plateaued function.

We now introduce a main result in [14].

Theorem 5. *Let f be an mth-order correlation immune function on V_n. If m and n satisfy the condition of $0.6n - 0.4 \leq m \leq n - 2$, then $N_f \leq 2^{n-1} - 2^{m+1}$,* *where the equality holds if and only if f is also a $2(n-m-2)$th-order plateaued function.*

Note that Theorem 5 is an improvement on Sarkar and Maitra's upper bound [9], $N_f \leq 2^{n-1} - 2^m$ when $m > \frac{1}{2}n - 1$.

The following result was given by Sarkar and Maitra [9].

Theorem 6. *Let f be an mth-order correlation immune function on V_n, where $m \leq n-2$. Then $\langle \xi, \ell \rangle \equiv 0 \pmod{2^{m+1}}$ where ℓ is any row of H_n. In particular, if f is balanced mth-order correlation immune, then $\langle \xi, \ell \rangle \equiv 0 \pmod{2^{m+2}}$.*

The following two Lemmas can be found from [14].

6.2 A New Result

Lemma 4 can be generalized. Let f be a function on V_n and W be a p-dimensional subspace of V_n. Let $U = \{0, \alpha_{2^{n-p}}, \alpha_{2 \cdot 2^{n-p}}, \ldots, \alpha_{(2^p-1)2^{n-p}}\}$. Since both W and U are p-dimensional subspaces of V_n, we can find an $n \times n$ matrix B over $GF(2)$ satisfying $WB^T = U$, where $WB^T = \{\alpha B^T | \alpha \in W\}$. Set $x = uB$ and $g(u) = f(uB)$. Consider $f(x) \oplus \langle \alpha, x \rangle$ where $\alpha \in W$. Note that $\langle \alpha, x \rangle = x\alpha^T = uB\alpha^T = u(\alpha B^T)^T = \langle \alpha B^T, u \rangle$. Therefore $f(x) \oplus \langle \alpha, x \rangle = g(u) \oplus \langle \alpha B^T, u \rangle$ where $\alpha \in W$ and $\alpha B^T \in U$. Let η denote the sequence of g. Equivalently, we have $\langle \xi, \ell_j \rangle = \langle \eta, \ell_i \rangle$ where j is the binary representation of $\alpha \in W$, and i is the binary representation of $\alpha B^T \in U$.

Define a permutation π on $\{0, 1, \ldots, 2^n - 1\}$ as follows: $\pi(j) = i$ if $\alpha_j B^T = \alpha_i$, where i and j are the the binary representations of α_i and α_j respectively. Therefore

$$\langle \xi, \ell_j \rangle = \langle \eta, \ell_{\pi(j)} \rangle \text{ or } \langle \xi, \ell_{\pi^{-1}(j)} \rangle = \langle \eta, \ell_j \rangle \tag{7}$$

Rewrite $\eta = (\eta_0, \eta_1, \ldots, \eta_{2^p-1})$ where each η_i is of length 2^{n-p}. Applying Lemma 4 to the function g and the subspace U, we have

$$2^p(\langle \eta_0, e_j \rangle, \langle \eta_1, e_j \rangle, \langle \eta_2, e_j \rangle, \ldots, \langle \eta_{2^p-1}, e_j \rangle)$$
$$= (\langle \eta, \ell_j \rangle, \langle \eta, \ell_{j+2^{n-p}} \rangle, \ldots, \langle \eta, \ell_{j+(2^p-1)2^{n-p}} \rangle)H_p$$

where e_j denotes the jth row of H_p, $j = 0, 1, \ldots, 2^{n-p} - 1$. Due to (7), we obtain

$$2^p(\langle \eta_0, e_j \rangle, \langle \eta_1, e_j \rangle, \langle \eta_2, e_j \rangle, \ldots, \langle \eta_{2^p-1}, e_j \rangle)$$
$$= (\langle \xi, \ell_{\pi^{-1}(j)} \rangle, \langle \xi, \ell_{\pi^{-1}(j+2^{n-p})} \rangle, \ldots, \langle \xi, \ell_{\pi^{-1}(j+(2^p-1)2^{n-p})} \rangle)H_p \tag{8}$$

where e_j denotes the jth row of H_p, $j = 0, 1, \ldots, 2^{n-p} - 1$.

Lemma 6. *Let f be a function on V_n and ξ denote the sequence of f. Let q be an odd number with $1 \leq q \leq n - 2$, such that*

$$\langle \xi, \ell_j \rangle = 0 \text{ for all } j \text{ such that } HW(\alpha_j) \leq q \text{ and } HW(\alpha_j) \text{ is odd}$$

where $\alpha_j \in V_n$ is the binary representation of integer j. Then $\langle \xi, \ell_j \rangle \equiv 0 \pmod{2^{q+2}}$ holds for all j with $HW(\alpha_j) = q + 2$ where $\alpha_j \in V_n$ is the binary representation of an integer j.

Proof. Let $U = \{0, \alpha_{2^{n-q-1}}, \alpha_{2 \cdot 2^{n-q-1}}, \ldots, \alpha_{(2^{q+1}-1)2^{n-q-1}}\}$. Obviously U can be rewritten as $U = \{(a_1, a_2, \ldots, a_{q+1}, 0, \ldots, 0) | (a_1, a_2 \ldots, a_{q+1}, 0, \ldots, 0) \in V_n\}$.

Set

$$W = \{(a_1, a_2, \ldots, a_{q+2}, 0, \ldots, 0) | (a_1, a_2 \ldots, a_{q+2}, 0, \ldots, 0) \in V_n,$$
$$HW(a_1, a_2, \ldots, a_{q+2}) \text{ is even}\}$$

Since both U and W are $(q+1)$-dimensional subspaces of V_n, there exists an $n \times n$ matrix B over $GF(2)$ satisfying

(i) $WB^T = U$, where $WB^T = \{\alpha B^T | \alpha \in W\}$, in particular, we require
$$\alpha_{(2^{q+1}-1)2^{n-q-1}} B^T = \alpha_{(2^{q+1}-1)2^{n-q-1}},$$
(ii) $\alpha_j B^T = \alpha_j, j = 1, \ldots, 2^{n-q-1} - 1$.

Set $x = uB$ and $g(u) = f(uB)$. Let $\eta = (\eta_0, \eta_1, \ldots, \eta_{2^{n-q-1}-1})$ denote the sequence of g, where each η_i is of length 2^{n-q-1}. Obviously $HW(\alpha)$ is even for any $\alpha \in W$, i.e., $HW(\alpha)$ takes the values, $0, 2, 4, \ldots, q-1, q+1$. Therefore $HW(\alpha_{2^{n-q-2}} \oplus \alpha)$ must be odd, i.e., $HW(\alpha)$ takes the values, $1, 3, 5, \ldots, q, q+2$. Note that $\alpha_{(2^{q+1}-1)2^{n-q-1}} = (1, \ldots, 1, 0, \ldots, 0)$ and $HW(\alpha_{2^n - 2^{n-q-1}}) = q+1$. Obviously, $\alpha_{2^{n-q-2}} \oplus \alpha_{(2^{q+1}-1)2^{n-q-1}} = (1, \ldots, 1, 1, 0, \ldots, 0) = \alpha_{(2^{q+2}-1)2^{n-q-2}}$. Note that $HW(\alpha_{(2^{q+2}-1)2^{n-q-2}}) = q+2$, and for any other $\alpha \in W$ with $\alpha \neq \alpha_{(2^{q+1}-1)2^{n-q-1}}$, we have $1 \leq HW(\alpha_{2^{n-q-2}} \oplus \alpha) \leq q$. Due to the property of f, $\langle \xi, \ell_j \rangle = 0$ for all j, where j is the integer representation of $\alpha_{2^{n-q-2}} \oplus \alpha$, if $\alpha \in W$ and $\alpha \neq \alpha_{(2^{q+1}-1)2^{n-q-1}}$, From the properties of B, $(\alpha_{2^{n-q-2}} \oplus \alpha_j)B^T = \alpha_{2^{n-q-2}} \oplus \alpha_j B^T$ for all $\alpha_j \in W$. In particular, $(\alpha_{2^{n-q-2}} \oplus \alpha_{(2^{q+1}-1)2^{n-q-1}})B^T = \alpha_{2^{n-q-2}} \oplus \alpha_{(2^{q+1}-1)2^{n-q-1}}$.

Using (8) with $j = 1$, we have

$$2^{q+1}(\langle \eta_0, e_1 \rangle, \langle \eta_1, e_1 \rangle, \langle \eta_2, e_1 \rangle, \ldots, \langle \eta_{2^{q+1}-1}, e_1 \rangle)$$
$$= (0, \ldots, 0, \langle \xi, \ell_{(2^{q+1}-1)2^{n-q-1}} \rangle)H_{q+1} \tag{9}$$

Since $2^{n-q-1} \geq 2$, $\langle \eta_{2^{q+1}-1}, e_1 \rangle$ is even. Comparing the rightmost term in both sides of (9), we conclude that $\langle \xi, \ell_{(2^{q+1}-1)2^{n-q-1}} \rangle \equiv 0 \pmod{2^{q+2}}$. By the same reasoning, we can prove that $\langle \xi, \ell_j \rangle \equiv 0 \pmod{2^{q+2}}$ holds for all j with $HW(\alpha_j) = q+2$. □

Lemma 7. *Let f be a function on V_n and ξ denote the sequence of f. Let q be an odd number with $1 \leq q \leq n-2$, such that*

$$\langle \xi, \ell_j \rangle = 0 \text{ for all } j \text{ such that } HW(\alpha_j) \text{ is odd and } HW(\alpha_j) \leq q$$

where $\alpha_j \in V_n$ is the binary representation of integer j. Then either there exists some j_0 such that $|\langle \xi, \ell_{j_0} \rangle| \geq 2^{q+2}$, or $\langle \xi, \ell_j \rangle = 0$ for all j where $HW(\alpha_j)$ is odd.

Proof. By using Lemma 6, $\langle \xi, \ell_j \rangle \equiv 0 \pmod{2^{q+2}}$ holds for all j with $HW(\alpha_j) = q+2$ where $\alpha_j \in V_n$. There exist two cases to be considered.

Case 1: there exists some j_0 with $HW(\alpha_{j_0}) = q+2$ satisfying $\langle \xi, \ell_{j_0} \rangle \neq 0$. In this case we have $|\langle \xi, \ell_{j_0} \rangle| \geq 2^{q+2}$. Thus the lemma holds.

Case 2: $\langle \xi, \ell_j \rangle = 0$ holds for all j with $HW(\alpha_j) = q + 2$ where $\alpha_j \in V_n$. In this case, we conclude that $\langle \xi, \ell_j \rangle = 0$ holds for all j such that $HW(\alpha_j) \leq q + 2$ and $HW(\alpha_j)$ is odd.

Once again we use Lemma 6. There exist two cases to be considered.

Case 2.1: we have an integer $t > 1$ such that $\langle \xi, \ell_j \rangle = 0$ for all j where $HW(\alpha_j) \leq q + 2(t - 1)$ and $HW(\alpha_j)$ is odd, and there also exists j_0 with $HW(\alpha_{j_0}) = q + 2t$ satisfying $\langle \xi, \ell_{j_0} \rangle \neq 0$. By using Lemma 6, we can conclude that $|\langle \xi, \ell_{j_0} \rangle| \geq 2^{q+2t}$. Thus the lemma holds in Case 2.1.

Case 2.2: $\langle \xi, \ell_j \rangle = 0$ for all j where $HW(\alpha_j)$ is odd. Clearly the lemma holds.

□

Applying Lemma 7, we can extend Theorem 5 in the following way.

Theorem 7. *Let f be an odd mth-order correlation immune function on V_n. Then either $N_f \leq 2^{n-1} - 2^{m+1}$ holds for $\frac{1}{2}n - 1 \leq m < 0.6n - 0.4$ or f has a non-zero linear structure.*

Proof. If f is balanced, $N_f \leq 2^{n-1} - 2^{m+1}$ holds due to Theorem [9]. Thus we only need to consider the unbalanced case. From Lemma 7, there there two cases to be considered. Case 1: there exists some j_0 such that $|\langle \xi, \ell_{j_0} \rangle| \geq 2^{m+2}$. In this case, we have proved the theorem by using Lemma 1. Case 2: $\langle \xi, \ell_j \rangle = 0$ for all j where $HW(\alpha_j)$ is odd. Set $W = \{\alpha | \alpha \in V_n, HW(\alpha) \text{ is even}\}$. Thus W is an $(n-1)$-dimensional subspace of V_n. From the property of f, obviously, $\mathfrak{S}^* \subseteq W$. From Theorem 4, f has a non-zero liear structure. □

Note that the nonlinearity of any Boolean function on V_n is upper-bounded by $2^{n-1} - 2^{\frac{1}{2}n-1}$. For $m \leq \frac{1}{2}n - 2$, we have $2^{n-1} - 2^{\frac{1}{2}n-1} \leq 2^{n-1} - 2^{m+1}$. Hence the inequality $N_f \leq 2^{n-1} - 2^{m+1}$ is trivial when $m \leq \frac{1}{2}n - 2$, although it still holds. For this reason, we require that $m \geq \frac{1}{2}n - 1$ in Theorem 7.

Theorem 7 represents an extension of Theorem 5. The latter is stated for the case of $0.6n - 0.4 \leq m \leq n - 2$.

7 Conclusion Remarks

This paper includes three main results. (1) We have presented a tight lower bound on the absolute indicator for GAC of an mth-order correlation immune function on V_n, and proved that a correlation immune function achieves the low bound for the absolute indicator if and only if it is affine. (2) We have established a relationship between correlation immunity and non-zero linear structures. (3) We have shown that given an odd mth-order correlation immune function f on V_n, the nonlinearity N_f of f satisfies $N_f \leq 2^{n-1} - 2^{m+1}$ for $\frac{1}{2}n - 1 \leq m < 0.6n - 0.4$ otherwise f has a non-zero linear structure. This is an extension of a known result that holds for $0.6n - 0.4 \leq m \leq n - 2$. It would be interesting to known whether or not Theorem 7 can be extended to the case of an even m.

Some observations on upper bounds on nonlinearity for a "small" m were made by Sarkar and Maitra in [9]. For instance, they showed that $N_f \leq 2^{n-1} -$

$2^{\frac{1}{2}n-1} - 2^m$ when n is even and $m \le \frac{1}{2}n - 1$, and $N_f \le 2^{n-1} - 2^{\frac{1}{2}n-1} - 2^{m+1}$ when f is balanced, n is even and $m \le \frac{1}{2}n - 1$. It is not clear whether these bounds are tight.

Acknowledgment

The second author was supported by a Queen Elizabeth II Fellowship (227 23 1002). Both authors would like to thank Yuriy Tarannikov and Subhamoy Maitra for pointing out an error in an earlier version.

References

1. P. Camion, C. Carlet, P. Charpin, and N. Sendrier. On correlation-immune functions. In *Advances in Cryptology - CRYPTO'91*, volume 576 of *Lecture Notes in Computer Science*, pages 87–100. Springer-Verlag, Berlin, Heidelberg, New York, 1991.
2. Claude Carlet. Partially-bent functions. *Designs, Codes and Cryptography*, 3:135–145, 1993.
3. C. Ding, G. Xiao, and W. Shan. *The Stability Theory of Stream Ciphers*, volume 561 of *Lecture Notes in Computer Science*. Springer-Verlag, Berlin, Heidelberg, New York, 1991.
4. Xiao Guo-Zhen and J. L. Massey. A spectral characterization of correlation-immune combining functions. *IEEE Transactions on Information Theory*, 34(3):569–571, 1988.
5. M. Hermelin and K. Nyberg. Correlation properties of the bluetooth combiner generator. In *The 2nd International Conference on Information Security and Cryptology (ICISC'99), Seoul, Korea*, volume 1787 of *Lecture Notes in Computer Science*, pages 17–29. Springer-Verlag, Berlin, Heidelberg, New York, 1999.
6. F. J. MacWilliams and N. J. A. Sloane. *The Theory of Error-Correcting Codes*. North-Holland, Amsterdam, New York, Oxford, 1978.
7. W. Meier and O. Staffelbach. Nonlinearity criteria for cryptographic functions. In *Advances in Cryptology - EUROCRYPT'89*, volume 434 of *Lecture Notes in Computer Science*, pages 549–562. Springer-Verlag, Berlin, Heidelberg, New York, 1990.
8. O. S. Rothaus. On "bent" functions. *Journal of Combinatorial Theory*, Ser. A, 20:300–305, 1976.
9. P. Sarkar and S. Maitra. Nonlinearity bounds and constructions of resilient boolean functions. In *Advances in Cryptology - CRYPTO2000*, volume 1880 of *Lecture Notes in Computer Science*, pages 515–532. Springer-Verlag, Berlin, Heidelberg, New York, 2000.
10. T. Siegenthaler. Correlation-immunity of nonlinear combining functions for cryptographic applications. *IEEE Transactions on Information Theory*, IT-30 No. 5:776–779, 1984.
11. X. M. Zhang and Y. Zheng. GAC — the criterion for global avalanche characteristics of cryptographic functions. *Journal of Universal Computer Science*, 1(5):316–333, 1995. (http://www.jucs.org/).
12. X. M. Zhang and Y. Zheng. Cryptographically resilient functions. *IEEE Transactions on Information Theory*, 43(5):1740–1747, 1997.

13. Y. Zheng and X. M. Zhang. Plateaued functions. In *Advances in Cryptology - ICICS'99*, volume 1726 of *Lecture Notes in Computer Science*, pages 284–300. Springer-Verlag, Berlin, Heidelberg, New York, 1999.

14. Y. Zheng and X. M. Zhang. Improved upper bound on the nonlinearity of high order correlation immune functions. In *Selected Areas in Cryptography, 7th Annual International Workshop, SAC2000*, volume xxxx of *Lecture Notes in Computer Science*, pages xxx–xxx. Springer-Verlag, Berlin, Heidelberg, New York, 2000. (in Pre-Proceedings pages 258-269).

15. Y. Zheng and X. M. Zhang. On relationships among avalanche, nonlinearity and correlation immunity. In *Advances in Cryptology - ASIACRYPT2000*, volume 1976 of *Lecture Notes in Computer Science*, pages 470–482. Springer-Verlag, Berlin, Heidelberg, New York, 2000.

Elliptic Curves and Resilient Functions

Jung Hee Cheon[1] and Seongtaek Chee[2]

[1] Mathematics Department, Brown University, USA and Securepia, Korea
jhcheon@math.brown.edu,
[2] Section 8110, NSRI, 161 Kajong-Dong, Yusong-Gu, Taejon, 305-345, Korea
chee@etri.re.kr

Abstract. In this paper, we propose a novel relationship between the correlation of two polynomial-type Boolean functions and the order of an associated algebraic curve. By this relationship, we propose a method to generate a resilient(correlation immune and balanced) function from a cubic polynomial. Since our resilient function is derived from a polynomial over a finite field, its nonlinearity is much easier to control. Moreover we can construct a resilient function with multi-bit outputs. We present several examples of a resilient function with 2 outputs.

1 Introduction

A correlation immune function has been an active area since T. Siegentalar firstly introduced its concept[10], because most stream ciphers with nonlinear filter function or nonlinear combination of LFSR's are vulnerable to correlation attack[8]. P. Camion *et al.*[1] presented a method for construction balanced correlation immune function. J. Sebbery *et al.*[9] discussed the nonlinearity and propagation characteristic of such functions. Later, S. Chee *et al.*[3] verified the relationship between correlation immunity and nonlinearity. However, since all of this methods construct each Boolean function independently, it seems to be very difficult to construct a correlation immune vector Boolean function. Recently, a vector Boolean function which is bent[5] or which has small correlation[12] was introduced, but none of which provides a result on correlation immune functions.

In this paper, we propose a novel relationship between the correlation of two polynomial-type Boolean functions and the order of an associated algebraic curve. As a result, we show that several component functions of a cubic polynomial are bent (or semi-bent) for even (or odd, resp.) n. Also, we analyze correlation between a cubic function and affine functions and propose a method to generate a resilient(correlation immune and balanced) function from a cubic polynomial. While every previous method generates a resilient function systematically, we firstly derive a resilient function from a polynomial over a finite field. Hence we can easily construct a resilient function with multi-bit outputs and its nonlinearity is much easier to control. Our vector resilient functions can be used in designing a stream cipher with several output bits.

In section 2, we gives basic definitions on cryptographic properties of Boolean functions. In section 3, we prove the main theorem. In section 4, we analyze correlation properties of cubic polynomials over $\mathbb{F}_{2^n}$. In section 5, we propose a new method to generate a resilient function and gives several examples. In section 6, we define a vector resilient function and propose a method to generate it. Also, we present several examples of a resilient function with two outputs. In section 7. we conclude this paper.

2 Basic Definitions

Let $\mathbb{Z}_2$ be the finite field with two elements and $\mathbb{Z}_2^n$ the n-dimensional vector space over the field $\mathbb{Z}_2$. A Boolean function on $\mathbb{Z}_2^n$ is a function whose input is binary n-tuples $x = (x_1, x_2, \cdots, x_n)$ and takes the values 0 and 1. The set of all Boolean functions defined over $\mathbb{Z}_2^n$ is denoted by $\mathcal{B}_n$. In some cases it will be more convenient to work with the function $(-1)^{f(x)} = 1 - 2f(x)$ that takes the value $\{1, -1\}$.

For $a = (a_1, a_2, \cdots, a_n)$, $b = (b_1, b_2, \cdots, b_n)$ in $\mathbb{Z}_2^n$, $a \cdot b = a_1 b_1 \oplus \cdots \oplus a_n b_n$ is the inner product of two vectors. The Hamming weight of an element a is the number of components equal to 1 and denoted by $\mathrm{wt}(a)$. The Hamming weight of a function f is the number of function values equal to 1. The distance $d(f, g)$ between two functions f and g is the number of function values in which they differ:

$$d(f, g) = \mathrm{wt}(f \oplus g) = \#\{x | f(x) \neq g(x)\}.$$

A function is said to be linear if there is $w \in \mathbb{Z}_2^n$ such that it can be written as $l_w(x) = w \cdot x = a_1 x_1 \oplus a_2 x_2 \oplus \cdots \oplus a_n x_n$. The set of all linear functions on $\mathbb{Z}_2^n$ is denoted by $\mathcal{L}_n$. A function f is said to be affine if there is $w \in \mathbb{Z}_2^n$ and $c \in \mathbb{Z}_2$ such that $f(x) = l_w(x) \oplus c$, and the set of all affine function on $\mathbb{Z}_2^n$ is denoted by Γ_n.

Now we introduce some basic definitions on properties of Boolean functions.

Definition 1. $f \in \mathcal{B}_n$ is balanced if $\#\{x \in \mathbb{Z}_2^n | f(x) = 0\} = \#\{x \in \mathbb{Z}_2^n | f(x) = 1\}$.

Definition 2. The correlation value between the function f and g in $\mathcal{B}_n$ is defined by

$$c(f, g) = 1 - \frac{d(f, g)}{2^{n-1}}.$$

If $c(f, l_w) = 0$ for all $w \in \mathbb{Z}_2^n$ with $1 \leq \mathrm{wt}(w) \leq k$, we say that f is k-th order correlation immune. If a balanced function is k-th order correlation immune, it is called a k-th order resilient function.

From the definition, we know that $-1 \leq c(f, g) \leq 1$. In particular, $c(f, f) = 1$, $c(f, f \oplus 1) = -1$ and $c(f, g) = 0$ if and only if $d(f, g) = 2^{n-1}$.

Definition 3. For any Boolean function f over $\mathbb{Z}_2^n$, we define the Walsh-Hadamard transformation $\hat{f} : \mathbb{Z}_2^n \to \mathbb{R}$ as follows:

$$\hat{f}(\mathbf{w}) = \sum_{\mathbf{x} \in \mathbb{Z}_2^n} f(\mathbf{x}) \cdot (-1)^{\mathbf{w} \cdot \mathbf{x}}, \quad w \in \mathbb{Z}_2^n.$$

When we apply Walsh-Hadamard transformation to $\chi_f = (-1)^f$, we have

$$\widehat{\chi_f}(w) = 2^n c(f, w \cdot x), \tag{1}$$

which has the following properties.

Lemma 1. Let $f \in \mathcal{B}_n$ and $l_v = v \cdot x$.

1. f is balanced if and only if $\widehat{\chi_f}(0) = 0$.
2. f is a k-th order correlation immune function if and only if $\widehat{\chi_f}(w) = 0$ for all $w \in \mathbb{Z}_2^n$ with $\mathrm{wt}(w) \leq k$.
3. $\hat{\chi}_{f+1}(w) = -\widehat{\chi_f}(w)$, $\widehat{\chi}_{f+l_v}(w) = \widehat{\chi_f}(w + v)$.

Now we introduce a bent and semi-bent function. A bent function is a Boolean function whose correlation value is unique up to sign, and a semi-bent function is a balanced Boolean function whose correlation value is at most 2 up to sign. The exact definition is as follows:

Definition 4. Let $f \in \mathcal{B}_n$.

1. f is called a bent function if for any $w \in \mathbb{Z}_2^n$

$$|\widehat{\chi_f}(\mathbf{w})| = 2^{\frac{n}{2}}.$$

2. f is called a semi-bent function[2] if $\widehat{\chi_f}(0) = 0$, and $|\widehat{\chi_f}(\mathbf{w})| = 0$ or $2^{\lfloor \frac{n}{2} \rfloor + 1}$, where $\lfloor m \rfloor$ is the greatest integer not greater than m.

Observe that a bent function exists only when n is even, and f is a bent function if and only if $c(f, l_w) = 2^{-n/2}$ for any $w \in \mathbb{Z}_2^n$.

Consider a vector Boolean function $F : \mathbb{Z}_2^n \to \mathbb{Z}_2^m$. Let $b = (b_1, b_2, \cdots, b_m) \in \mathbb{Z}_2^m$ be a nonzero element. We denote by $b \cdot F$ the Boolean function on $\mathbb{Z}_2^n$, which is the linear combination $b_1 f_1 \oplus b_2 f_2 \oplus \cdots \oplus b_n f_n$ of the components functions $f_1, f_2, \cdots, f_m$ of F on $\mathbb{Z}_2^n$.

3 Main Theorem

Let a be an element of a finite field $\mathbb{F}_{2^n}$, F a polynomial over $\mathbb{F}_{2^n}$, $B = \{\beta_1, \cdots, \beta_n\}$ a basis of $\mathbb{F}_{2^n}$ over $\mathbb{F}_2$. Then we can embed all elements of $\mathbb{F}_{2^n}$ to $\mathbb{Z}_2^n$. That is, we have a natural isomorphism depending on a basis B of $\mathbb{F}_{2^n}$:

$$\phi_B : \mathbb{F}_{2^n} \to \mathbb{Z}_2^n, \quad \sum_{i=1}^{n} a_i \beta_i \mapsto (a_1, a_2, \cdots, a_n), \quad a_i \in \mathbb{Z}_2.$$

Throughout this paper, we consider an element of $\mathbb{F}_{2^n}$ as an element of $\mathbb{Z}_2^n$, and vice versa without notifying, unless confused.

Now we can define a Boolean function $a \cdot F$ as follows:

$$a \cdot F : \mathbb{Z}_2^n \to \mathbb{Z}_2, \quad x \mapsto a \cdot F(x),$$

where $a \cdot F(x)$ is the inner product of two binary vectors, one of which is obtained by expressing a by the basis B and another is obtained by expressing $F(x)$ by its dual basis $\hat{B}$. Note that $a \cdot F$ does not depend on the choice of a basis B since $a \cdot F(x) = Tr[aF(x)]$ [4].

Throughout this paper, we use the following notation:

$$Tr[\cdot] = Tr_{\mathbb{F}_{2^n}/\mathbb{F}_2}[\cdot], \quad Te[\cdot] = Tr_{\mathbb{F}_{2^n}/\mathbb{F}_4}[\cdot].$$

Before presenting a main theorem, we introduce an useful fact to prove the main theorem.

$$x^2 + ax + b = 0 \text{ has a root in } \mathbb{F}_{2^n} \text{ if and only if } Tr[b/a^2] = 0. \tag{2}$$

Theorem 1. *Let $F(x), G(x)$ be polynomials over $\mathbb{F}_{2^n}$, and $a, b \in \mathbb{F}_{2^n}$. Consider an algebraic curve $C : y^2 + y = aF(x) + bG(x)$ over $\mathbb{F}_{2^n}$. Then we have*

$$c(a \cdot F, b \cdot G) = \frac{\#C(\mathbb{F}_{2^n})}{2^n} - 1,$$

where $\#C(\mathbb{F}_{2^n})$ is the number of $\mathbb{F}_{2^n}$-rational points of the curve C in the affine plane.

Proof. Take a basis B of $\mathbb{F}_{2^n}$ over $\mathbb{F}_2$. If we represent a, b and $F(x), G(x)$ by the basis B and its dual basis $\tilde{B}$ respectively, we have

$$a \cdot F(x) = Tr[aF(x)], \quad b \cdot G(x) = Tr[bG(x)].$$

Hence we have

$$\begin{aligned} d(a \cdot F, b \cdot G) &= \#\{x \in \mathbb{Z}_2^n | Tr[aF(x)] \neq Tr[bG(x)]\} \\ &= \#\{x \in \mathbb{Z}_2^n | Tr[aF(x) + bG(x)] \neq 0\}. \end{aligned}$$

On the other hand, since it has no multiple root, the equation of y, $y^2 + y = aF(x) + bG(x)$, has two roots if and only if $Tr[aF(x) + bG(x)] = 0$ by (2). Hence we have

$$\#C(\mathbb{F}_{2^n}) = 2(2^n - d(a \cdot F, b \cdot G))$$

so

$$c(a \cdot F, b \cdot G) = 1 - \frac{d(a \cdot F, b \cdot G)}{2^{n-1}} = \frac{\#C}{2^n} - 1.$$

Using Theorem 1, we can derive easily the following corollary.

Corollary 1. *Let $F(x), G(x)$ be polynomials over $\mathbb{F}_{2^n}$ and $a, b \in \mathbb{F}_{2^n}$. Then we have*

1. *$c(a \cdot F, b \cdot G) = 0$ if and only if an algebraic curve $y^2 + y = aF(x) + bG(x)$ has 2^n $\mathbb{F}_{2^n}$-rational points on the affine plane.*

2. *F is k-th order correlation immune if and only if an algebraic curve $y^2 + y = aF(x) + wx$ has 2^n $\mathbb{F}_{2^n}$-rational points on the affine plane for each $w \in \mathbb{Z}_2^n$ with $1 \leq wt(w) \leq k$. (Note that the imbedding of w to $\mathbb{F}_{2^n}$ changes by a basis.)*

By the above corollary, if we find some family of algebraic curves which has 2^n $\mathbb{F}_{2^n}$-rational points, we may obtain two functions without correlation.

4 Correlation Properties of Cubic Polynomials

Let $F(x)$ be a monic polynomial of degree 3 over $\mathbb{F}_{2^n}$ and $a \in \mathbb{F}_{2^n}$. In this section, we investigate correlation properties of a Boolean function $a \cdot F$. By Theorem 1, the correlation of $a \cdot F$ and l_w is determined by the order of an elliptic curve $y^2 + y = aF(x) + wx$. In fact, $a \cdot F$ is k-th order correlation immune if and only if $y^2 + y = aF(x) + wx$ has order $2^n + 1$ (including the infinity point) over $\mathbb{F}_{2^n}$ for all $w \in \mathbb{Z}_2^n$ with $1 \le \text{wt}(w) \le k$.

Now we introduce a lemma which is useful to prove the results in this section.

Lemma 2. *[6] Consider a quartic equation*

$$x^4 + ax + b = 0, \quad a, b \in \mathbb{F}_{2^n}, \quad a \ne 0. \tag{3}$$

- *If n is odd, then (3) has either no solution or exactly two solutions.*
- *If n is even and a is not a cube, then (3) has exactly one solution.*
- *If n is even and a is a cube, then (3) has four solutions if $Te[b/a^{4/3}] = 0$, and no solutions if $Te[b/a^{4/3}] \ne 0$.*

4.1 The case of odd n

First of all, we introduce a lemma for orders of supersingular elliptic curves over $\mathbb{F}_{2^n}$ for odd n.

Lemma 3. *[6] For odd n, any supersingular elliptic curve over $\mathbb{F}_{2^n}$ is isomorphic to one of following three curves.*

- $y^2 + y = x^3$ *whose order is* $2^n + 1$.
- $y^2 + y = x^3 + x$ *whose order is* $2^n + 1 \pm 2^{(n+1)/2}$.
- $y^2 + y = x^3 + x + 1$ *whose order is* $2^n + 1 \pm 2^{(n+1)/2}$.

Observe that if n is odd, $x^3 - a$ has a root in $\mathbb{F}_{2^n}$ for any $a \in \mathbb{F}_{2^n}$ because $\gcd(2^n - 1, 3) = 1$.

Theorem 2. *Let n be an odd integer, $a, v \in \mathbb{F}_{2^n}$, and $F(x) = x^3 + vx \in \mathbb{F}_{2^n}[x]$. Assume that $x, w \in \mathbb{Z}_2^n$ are embedded into $\mathbb{F}_{2^n}$ by a basis and its dual basis, respectively. If we let $\sqrt[3]{a}$ is a root of $x^3 - a$ in $\mathbb{F}_{2^n}$, we have*

$$\widehat{\chi_{a \cdot F}}(w) = \begin{cases} 0 & \text{if } w/\sqrt[3]{a} = s^4 + s \text{ for some } s \in \mathbb{F}_{2^n} \\ \pm 2^{\frac{n+1}{2}} & \text{otherwise} \end{cases}.$$

Proof. Consider an elliptic curve $E : y^2 + y = a(x^3 + vx) + wx$. Note that E is isomorphic to $E_1 : y^2 + yx = x^3 + (\frac{av+w}{\sqrt[3]{a}})x$ by a linear transformation $(x, y) \mapsto (\sqrt[3]{a}x, y)$. Moreover E_1 is isomorphic to $y^2 + y = x^3$ if and only if the following two equations are solvable simultaneously in $\mathbb{F}_{2^n}$ [6].

$$s^4 + s + \frac{(av + w)}{\sqrt[3]{a}} = 0, \tag{4}$$

$$t^2 + t + s^6 = 0. \tag{5}$$

Observe that (5) is equivalent to $Tr[s^6] = 0$ by (2), and so $Tr[s^3] = 0$. If s_0 is a root of (4), then $s_0 + 1$ is also a root of (4). If (4) has a root one of roots of (4) satisfies $Tr[s^3] = 0$ since $Tr[(s_0 + 1)^3] = Tr[s_0^3] + 1$. Hence we see that E_1 is isomorphic to $y^2 + y = x^3$ if and only if (4) has a root in $\mathbb{F}_{2^n}$. In this case, we have $c(a \cdot F, l_w) = 0$ by Theorem 1 because $y^2 + y = x^3$ has order $2^n + 1$.

On the other hand, if (4) has no root in $\mathbb{F}_{2^n}$, then E_1 should be isomorphic to either $y^2 + y = x^3 + x$ or $y^2 + y = x^3 + x$. In any cases, E_1 has order $2^n + 1 \pm 2^{(n+1)/2}$. Hence we have $c(a \cdot F, l_w) = \pm 2^{(1-n)/2}$ by Theorem 1.

If we use $c(a \cdot F, l_w) = \widehat{\chi_{a \cdot F}}(w)/2^n$, we obtain the theorem.

Note that $a \cdot F$ is a semi-bent function for any basis B of $\mathbb{F}_{2^n}$. From Lemma 2, we know that $s^4 + s + w/\sqrt[3]{a}$ has exactly 0 or 2 different roots. Hence given $a \in \mathbb{F}_{2^n}$, the number of w such that $s^4 + s + w/\sqrt[3]{a}$ has a root in $\mathbb{F}_{2^n}$ is exactly $2^n/2 = 2^{n-1}$. Hence $a \cdot F$ has the correlation value 0 for a half of all $w \in \mathbb{Z}_2^n$.

4.2 The case of even n

We have the following lemma for orders of supersingular elliptic curves over $\mathbf{F}_{2^n}$ for even n.

Lemma 4. *[6] Let $\alpha, \gamma, \delta \in \mathbf{F}_{2^n}$, $\gamma \notin \mathbf{F}_{2^n}^3$, and $Te[\delta] \neq 0$. For even n, any supersingular elliptic curve over $\mathbf{F}_{2^n}$ is isomorphic to one of following three curves.*

- $y^2 + y = x^3 + \delta x$ whose order is $2^n + 1$.
- $y^2 + y = x^3 + \alpha$ whose order is $2^n + 1 \pm 2^{n/2+1}$.
- $y^2 + \gamma y = x^3 + \alpha$ whose order is $2^n + 1 \pm 2^{n/2}$.

Theorem 3. *Let n be an even integer, $a, v \in \mathbf{F}_{2^n}$, and $F(x) = x^3 + vx \in \mathbf{F}_{2^n}[x]$. Assume that $x, w \in \mathbf{Z}_2^n$ are embedded into $\mathbf{F}_{2^n}$ by a basis and its dual basis, respectively.*

1. *If $x^3 - a$ has a root, say $\sqrt[3]{a}$, in $\mathbf{F}_{2^n}$, then we have*

$$\widehat{\chi_{a \cdot F}}(w) = \begin{cases} 0 & \text{if } Te[\frac{va+w}{\sqrt[3]{a}}] \neq 0 \\ \pm 2^{\frac{n}{2}+1} & \text{if } Te[\frac{va+w}{\sqrt[3]{a}}] = 0 \end{cases}.$$

2. *If $a \notin \mathbf{F}_{2^n}^3$, then we have*

$$\widehat{\chi_{a \cdot F}}(w) = \pm 2^{\frac{n}{2}} \quad \text{for all } w.$$

Proof. Let $\sqrt[3]{a}$ be a root of $x^3 - a$ in $\mathbf{F}_{2^n}$. Consider an elliptic curve $E : y^2 + y = a(x^3 + vx) + wx$ over $\mathbf{F}_{2^n}$. Observe that E is isomorphic to $E_1 : y^2 + y = x^3 + (\frac{va+w}{\sqrt[3]{a}})x$ by the linear transformation $(x, y) \mapsto (\sqrt[3]{a}x, y)$. If $Te[\frac{va+w}{\sqrt[3]{a}}] \neq 0$, E_1 has order $2^n + 1$ by Lemma 4, so $c(a \cdot F, l_w) = 0$. If $Te[\frac{va+w}{\sqrt[3]{a}}] = 0$, then $s^4 + s + \frac{va+w}{\sqrt[3]{a}}$ has a root, say s_0, in $\mathbf{F}_{2^n}$ by (3). By the linear transformation $(x, y) \mapsto (x + s^2, y + sx)$, we have that E_1 is isomorphic to $y^2 + y = x^3 + s_0^6$ which has order $2^n + 1 \pm 2^{n/2+1}$, so $c(a \cdot F, l_w) = \pm 2^{1-n/2}$.

On the other hand, if $a \notin \mathbf{F}_{2^n}^3$, then E is isomorphic to $y^2 + \gamma y = x^3 + \alpha$ for some nonzero $\alpha, \gamma \in \mathbf{F}_{2^n}$ which has order $2^n + 1 \pm 2^{n/2}$, so $c(a \cdot F, l_w) = \pm 2^{-n/2}$.

If we use $c(a \cdot F, l_w) = \widehat{\chi_{a \cdot F}}(w)/2^n$, we obtain the theorem.

Observe that $a \cdot F$ is a bent function for $a \notin \mathbf{F}_{2^n}^3$, regardless of a basis B of $\mathbf{F}_{2^n}$. In the case of $a \in \mathbf{F}_{2^n}^3$, $a \cdot F$ is balanced if and only if $Te[v \sqrt[3]{a}^2] \neq 0$ since $Te[w/\sqrt[3]{a}] = 0$ for $w = 0$.

5 Resilient Functions

In this section, we propose a method to generate a resilient function and give several examples.

Theorem 4. *Let n be an even integer, $b, v \in \mathbf{F}_{2^n}$, and $F_v(x) = x^3 + vx \in \mathbf{F}_{2^n}[x]$. Assume that $x, w \in \mathbf{Z}_2^n$ are embedded into $\mathbf{F}_{2^n}$ by a basis and its dual basis, respectively. Then $Tr[b^{-3}F_v(x)]$ is a 1-st order resilient function if and only if $Te[vb^{-2}] \neq 0$ and $Te[wb] \neq Te[vb^{-2}]$ for all elements w in the dual basis.*

Proof. Using Theorem 3, we have that

$$\widehat{\chi_{b^{-3} \cdot F}}(w) = 0 \quad \text{if and only if} \quad Te[vb^{-2} + wb] \neq 0.$$

Hence we have by Lemma 1 that $Tr[b^{-3}F(x)]$ is 1-st order resilient if and only if $Te[wa] \neq Te[vb^{-2}]$. Note that $x, w \in \mathbf{Z}_2^n$ are embedded into $\mathbf{F}_{2^n}$ by a basis and its dual basis, respectively. Since $wt(w) \leq 1$ implies that $w = 0$ or w is a basis element of the dual basis, we have the theorem.

Using this, we can generate resilient functions easily. See the following corollary.

Corollary 2. *Let $f_v(x) = Tr[x^3 + vx]$. Assume that $x, w \in \mathbf{Z}_2^n$ are embedded into $\mathbf{F}_{2^n}$ by a normal basis and its dual basis, respectively. Then f_v is a 1-st order resilient function for every v with $Te[v] = 1$, the number of which is 2^{n-2}.*

Proof. Take a normal basis $B = \{\theta, \theta^2, \cdots, \theta^{2^{n-1}}\}$ of $\mathbb{F}_{2^n}$ over $\mathbb{F}_2$. If we represent $x = \sum_{i=0}^{n-1} x_i \theta^{2^i}$
$(x_i \in \mathbb{F}_2)$ we have

$$Te[x] = \sum_{i=0}^{n-1} x_i' \theta^{2^i}, \quad \text{where } x_i = \begin{cases} x_1 + x_3 + \cdots + x_{n-1} & \text{for even } i \\ x_2 + x_4 + \cdots + x_n & \text{for odd } i \end{cases} \tag{6}$$

since $Te[x] = x + x^4 + x^{4^2} + \cdots + x^{4^{n/2-1}}$ and $x^2 = \sum_{i=0}^{n-1} x_{i-1} \theta^{2^i}$ (let $x_{-1} = x_{n-1}$). Note that a
dual basis of a normal basis is also a normal basis. Since $1 = \sum_{i=0}^{n-1} \theta^{2^i}$, we have that $Te[w] = 1$ if
and only if each sum of the odd and even components of w is 1. Hence any vector with Hamming
weight 0 or 1 does not satisfy $Te[w] = 1$, so $Tr[1 * (x^3 + vx)]$ is 1-st order resilient for every v with
$Te[v] = 1$. The last statement follows from the fact that a fourth of all elements of $\mathbb{F}_{2^n}$ satisfies
$Te[v] = 1$.

Using Theorem 4, we can generate a resilient function as the following procedure.

Procedure 1: Generate a Resilient Function

1. **Fix a basis B of $\mathbb{F}_{2^n}$.**
2. **Compute a dual basis $\hat{B}$ of B.**
3. **Calculate a set S**

$$S = \{b \in \mathbb{F}_{2^n} \mid Te[wb] \neq 1 \quad \text{for all } w \text{ in } \hat{B}\}.$$

4. **Find v_b for each $b \in S$ such that $Te[v_b b^{-2}] = 1$, which covers a fourth of all elements
   of $\mathbb{F}_{2^n}$.**
5. **Compute the function $Tr[b^{-3}(x^3 + v_b x)]$, which is a 1-st order resilient function.**

Note that the value of trace can be replaced by another nonzero element in $\mathbb{F}_{2^2}$.

When we use Theorem 4, we need to compute a dual basis of given basis. Now we introduce a
lemma to compute a dual basis of given polynomial basis.

Lemma 5. *[7] Let $B = \{1, \alpha, \cdots, \alpha^{n-1}\}$ be a polynomial basis of $\mathbb{F}_{2^n}$ over $\mathbb{F}_2$ and let $f(x)$ be the
minimum polynomial of α over $\mathbb{F}_{2^n}$. Let $f(x) = (x - \alpha)(\beta_0 + \beta_1 x + \cdots + \beta_{n-1} x^{n-1})$, $\beta_i, \alpha \mathbb{F}_{2^n}$.
Then the dual basis of B is $\hat{B} = \{\gamma_0, \gamma_1, \cdots, \gamma_{n-1}\}$ where*

$$\gamma_i = \frac{\beta_i}{f'(x)}, \quad i = 0, 1, \cdots, n-1.$$

Throughout this paper, we express an element of the finite field as a Boolean string. That is,
we denotes $\sum_{i=0}^{n-1} a_i \alpha_i$, $a_i \in \mathbb{F}_2$ by

$$\sum_{i=0}^{n-1} a_i 2^i = a_{n-1} \cdots a_2 a_1 a_0$$

where $\{\alpha_0, \alpha_1, \cdots, \alpha_{n-1}\}$ is a basis of $\mathbb{F}_{2^n}$ over $\mathbb{F}_2$. Also we denote a Boolean function $f(x)$ by its
value on $\mathbb{F}_{2^n}$, i.e.

$$\sum_{i=0}^{2^n-1} f(i) 2^i = f(2^n - 1) f(2^n - 1) \dots f(2) f(1) f(0)$$

where we use a Boolean string notation for elements of the finite field. When we represent such
binary string as a hexa-decimal expression, we denote by small x's. For example, $6a_x$ means 0110
1100 as a binary string.

Using the above procedure and Lemma 5, we generated several resilient functions as the following examples.

Example 1. Let $n = 4$. Let t be a root of an irreducible polynomial $x^4 + x + 1$ over $\mathbb{F}_2$. Take a basis $B = \{1, t, t^2, t^3\}$. Then we can compute its dual basis $\hat{B} = \{t^3 + 1, t^2, t, 1\}$ by Lemma 5. Using Procedure 1, we can get 4 resilient functions.

b_x	v_x	$f(x) = Tr[b^{-3}(x^3 + vx)]$
7_x	a_x	$665a_x$
d_x	1_x	$1de2_x$
e_x	1_x	$369c_x$
f_x	4_x	$1e78_x$

Example 2. Let $n = 6$. Let t be a root of an irreducible polynomial $x^6 + x + 1$ over $\mathbb{F}_2$. Take a basis $B = \{1, t, t^2, t^3, t^4, t^5\}$. Then we can compute its dual basis $\hat{B} = \{t^5 + 1, t^4, t^3, t^2, t, 1\}$ by Lemma 5. Using Procedure 1, we can get 11 resilient functions.

b	v	$f(x) = Tr[b^{-3}(x^3 + vx)]$
17_x	1_x	$2eb8\ 741d\ b8d1\ 1d8b_x$
19_x	1_x	$1be4\ d827\ 827d\ be41_x$
$1f_x$	4_x	$0cfc\ 3f30\ 5659\ 9a6a_x$
27_x	1_x	$2b71\ e8b2\ d48e\ 174d_x$
$2d_x$	1_x	$0f96\ a5c3\ 96f0\ 3ca5_x$
$2e_x$	1_x	$2e74\ 12b7\ 1db8\ de84_x$
32_x	4_x	$128b\ ed74\ b7d1\ 482e_x$
37_x	2_x	$3639\ 6c63\ 36c6\ 6c9c_x$
$3b_x$	20_x	$5665\ 596a\ 65a9\ 6aa6_x$
$3e_x$	2_x	$1ebb\ 4b11\ 44e1\ eeb4_x$
$3f_x$	3_x	$2471\ bd17\ e8bd\ 8e24_x$

6 Vector Resilient Functions

A function $F : \mathbb{Z}_2^n \rightarrow \mathbb{Z}_2^m$ is called *a vector Boolean function*. Note that there are unique Boolean function f_i such that $F = (f_1, f_2, \cdots, f_m)$. Any linear combination of f_i's are called a component function of F.

Definition 5. *A vector Boolean function is said to be k-th order resilient if and only if every component function is k-th order resilient. A vector Boolean function which is k-th order resilient is called a k-th order vector resilient function.*

Theorem 5. *Let n be an even integer and $F_v(x) = x^3 + vx \in \mathbb{F}_{2^n}[x]$ for $b, v \in \mathbb{F}_{2^n}$. Assume that $x, w \in \mathbb{Z}_2^n$ are embedded into $\mathbb{F}_{2^n}$ by a basis and its dual basis, respectively. For $e \in \mathbb{F}_{2^2}$, let*

$$S_e = \{b \in \mathbb{F}_{2^n} \mid Te[wb] \neq e \quad \text{for all } w \text{ in } \hat{B}\}. \tag{7}$$

If any linear combination of $b_1^{-3}, b_2^{-3}, \cdots, b_r^{-3}$ is equal to b^{-3} for some $b \in S$ and $Te[vb^{-2}] = e$ for such b, then $(f_1, f_2, \cdots, f_r)$ is a 1-st order vector resilient function where $f_i = Te[b_i^{-3}F_v(x)]$.

Proof. Let $f = \sum_{i=0}^r a_i f_i$, $a_i \in \mathbb{F}_2$ be a linear combination of the component functions of $(f_1, f_2, \cdots, f_r)$. Since $Te[\cdot]$ is a homomorphism, we have $f(x) = Te[(\sum_{i=1}^r b_i^{-3})F_v(x)]$. By assumption, we know that any linear combination of $b_1^{-3}, b_2^{-3}, \cdots, b_r^{-3}$ is equal to b^{-3} for some $b \in S_e$. Hence we have $f(x) = Te[b^{-3}F_v(x)]$. Since $Te[vb^{-2}] = e$ by assumption, we have $Te[vb^{-2}] \neq 0$ and $Te[wb] \neq Te[vb^{-2}]$ for all element w in the dual basis of the basis, which completes the proof.

Using Theorem 4, we can generate a vector resilient function as the following procedure.

Procedure 2: Generate a Resilient Function with Two Outputs

1. Fix a basis B of $\mathbb{F}_{2^n}$.

2. Compute a dual basis $\hat{B}$ of B.
3. Calculate a set S

$$S = \{b \in \mathbb{F}_{2^n} \,|\, Te[wb] \neq 1 \quad \text{for all } w \text{ in } \hat{B}\}.$$

4. Calculate a set $S_2 = \{(b_1, b_2, b_3)|b_1^{-3} + b_2^{-3} = b_3^{-3}, b_i \in S\}$.
5. Find v_b for each $(b_1, b_2, b_3) \in S_2$ such that $Te[v_b b_i^{-2}] = 1$.
6. Compute the function $f_i = Tr[b_i^{-3}(x^3 + v_b x)]$ for $i = 1, 2, 3$. Then (f_1, f_2) is a 1-st resilient function with two outputs.

Example 3. Let $n = 6$. Let t be a root of an irreducible polynomial $x^6 + x + 1$ over $\mathbb{F}_2$. Take a basis $B = \{1, t, t^2, t^3, t^4, t^5\}$. Then its dual basis is $\hat{B} = \{t^5 + 1, t^4, t^3, t^2, t, 1\}$ by Lemma 5. Using Procedure 2, we can get 2 vector resilient functions with two outputs.

- $b_1 = 17_x$, $b_2 = 3b_x$, $b_3 = 3e_x$, $v = 3c_x$
  - $f_1(x) = 74e2\ 2e47\ e28b\ 47d1_x$
  - $f_2(x) = 6aa6\ 65a9\ 596a\ 5665_x$
  - $f_3(x) = 1e44\ 4bee\ bbe1\ 11b4_x$
- $b_1 = 2e_x$, $b_2 = 37_x$, $b_3 = 3f_x$, $v = 2e_x$
  - $f_1(x) = 2e74,\ ed48,\ 1db8,\ 217b_x$
  - $f_2(x) = 6393,\ 39c9,\ 9c93,\ c6c9_x$
  - $f_3(x) = 4de7,\ d481,\ 812b,\ e7b2_x$

Example 4. Let $n = 8$. Let t be a root of an irreducible polynomial $x^8 + x^6 + x^5 + x + 1$ over $\mathbb{F}_2$. Take a basis $B = \{1, t, t^2, \cdots, t^7\}$. Then its dual basis is $\hat{B} = \{5d_x, ba_x, 4a_x, c9_x, f1_x, 81_x, 61_x, 9f_x\}$ by Lemma 5, we represent the elements of $\hat{B}$ by Boolean expression. Using Procedure 2, we can get 14 vector resilient functions with two outputs. We give three of them.

- $b_1 = 3d_x$, $b_2 = 7d_x$, $b_3 = f1_x$, $v = 22_x$
  - $f_1(x) = 69cc\ 3396\ 9633\ cc69\ 5a00\ ffa5\ a5ff\ 005a\ aaf0\ 0f55\ aaf0\ 0f55\ 66c3\ 3c99\ 66c3\ 3c99_x$
  - $f_2(x) = 3f56\ 9503\ c056\ 95fc\ cfa6\ 9a0c\ 30a6\ 9af3\ cfa6\ 65f3\ 30a6\ 650c\ c0a9\ 9503\ 3fa9\ 95fc_x$
  - $f_3(x) = 569a\ a695\ 5665\ 5995\ 95a6\ 65a9\ 9559\ 9aa9\ 6556\ 6aa6\ 9a56\ 6a59\ a66a\ a99a\ 596a\ a965_x$
- $b_1 = 5f_x$, $b_2 = f7_x$, $b_3 = ff_x$, $v = 83_x$
  - $f_1(x) = 0c3f\ cf03\ fc30\ 3f0c\ c0f3\ fc30\ 30fc\ 0c3f\ 9559\ 5665\ 9aa9\ 5995\ a66a\ 9aa9\ a99a\ 9559_x$
  - $f_2(x) = 36a0\ c95f\ f59c\ 0a63\ 5f36\ a0c9\ 63f5\ 9c0a\ 3950\ c6af\ 0593\ fa6c\ 50c6\ af39\ 93fa\ 6c05_x$
  - $f_3(x) = 3a9f\ 065c\ 09ac\ 356f\ 9fc5\ 5cf9\ 5309\ 9035\ ac09\ 90ca\ 9f3a\ a3f9\ f6ac\ 3590\ 3a60\ f95c_x$
- $b_1 = 65_x$, $b_2 = 6f_x$, $b_3 = be_x$, $v = e_x$
  - $f_1(x) = 4bd2\ 2d4b\ d2b4\ b42d\ 1e87\ 87e1\ 781e\ e178\ b42d\ d2b4\ d2b4\ b42d\ 1e87\ 87e1\ 87e1\ 1e87_x$
  - $f_2(x) = 5acc\ 99f0\ 33a5\ 0f66\ a5cc\ 66f0\ 335a\ 0f99\ 55c3\ 96ff\ 3caa\ 0069\ 553c\ 9600\ c3aa\ ff69_x$
  - $f_3(x) = 111e\ b4bb\ e111\ bb4b\ bb4b\ e111\ 4b44\ eee1\ e1ee\ 444b\ eele\ b444\ 4bbb\ 11e1\ 444b\ elee_x$

Example 5. Let $n = 10$. Let t be a root of an irreducible polynomial $x^{10} + x^3 + 1$ over $\mathbb{F}_2$. Take a basis $B = \{1, t, t^2, \cdots, t^7\}$. By similar procedure, we can get 33 vector Boolean functions with two outputs.

7 Conclusion

In this paper, we proposed a method to generate a vector resilient function with multi-output. Our approach is to derive a resilient function from a polynomial over a finite field, whose properties is closely related to the associated algebraic curves. In this paper, we analyze the properties of a cubic polynomial and associated elliptic curves to derive a 1-st order resilient function with multi-output. We expect this method may be generalized to higher degree polynomial and algebraic curves to get higher degree resiliency.

References

1. P. Camion, C. Carlet, P. Charpin, and N. Sendrier, *On Correlation Immune Functions*, in Proc. of CRYPTO'91, LNCS 576, Springer-Verlag, 1992, pp. 86-100.
2. S. Chee, S. Lee, and K. Kim, *Semi-bent functions*, in Proc. of Asiacrypt'94, LNCS 917, Springer-Verlag, 1995, pp 107 - 118.
3. S. Chee, S. Lee, K. Kim, and D. Kim, *Correlation Immune Functions with Controllable Nonlinearity*, in ETRI J., Vol. 19, No. 4, 1997, pp. 389-402.
4. J. Cheon, S. Chee and C. Park, *S-boxes with Controllable Nonlinearity*, in Proc. of Eurocrypt'99, LNCS 1592, Springer-Verlag, 1999, pp.286-294.
5. T. Satoh, T. Iwata and K. Kurosawa, *On Cryptographically Secure Vectorial Boolean Functions*, in Proc. of Asiacrypt'99, LNCS 1716, Springer-Verlag, 1999, pp. 20-28.
6. A. Menezes, **Elliptic Curve Public Key Cryptosystems**, Kluwer Academic Publishers, 1997.
7. A. Menezes, **Applications of Finite Fields**,
8. R. Rueppel, *Stream Ciphers*, in Contemporary Cryptology: The Science of Information Integraty, IEEE press, 1992, pp. 65-134.
9. J. Sebbery, X. Zhang, Y. Zheng, *Nonlinearity Balanced Boolean Functions and their Propagation Characteristics*, in Proc. of Crypto'93, LNCS 773, Springer-Verlag, 1994, pp. 49-60.
10. T. Siegenthaler, *Correlation-Immunity of Nonlinear Combining Functions for Cryptographic Applications*, in IEEE Transactions on Information Theory, IT-30(5), 1984, pp. 776-779.
11. J. H. Silverman, **The Arithmetic of Elliptic Curves**, Springer-Verlag, 1985.
12. M. Zhang and A. Chan, *Maximum Correlation Analysis of Nonlinear S-boxes in Stream Ciphers*' in Proc. of Cryto'00, LNCS 1880, Springer-Verlag, 2000, pp. 501 - 514.

Fast Universal Hashing with Small Keys and No Preprocessing: The **PolyR** Construction

Ted Krovetz[1] and Phillip Rogaway[1,2]

[1] Department of Computer Science
University of California, Davis CA 95616 USA

[2] Department of Computer Science, Faculty of Science
Chiang Mai University, Chiang Mai 50200 Thailand

Abstract. We describe a universal hash-function family, PolyR, which hashes messages of effectively arbitrary lengths in 3.9–6.9 cycles/byte (cpb) on a Pentium II (achieving a collision probability in the range 2^{-16}–2^{-50}). Unlike most proposals, PolyR actually hashes short messages faster (per byte) than long ones. At the same time, its key is only a few bytes, the output is only a few bytes, and no "preprocessing" is needed to achieve maximal efficiency. Our designs have been strongly influenced by low-level considerations relevant to software speed, and experimental results are given throughout.

Keywords: Universal hashing, software-optimized hashing, message authentication, UMAC.

1 Introduction

Ever since its introduction by Carter and Wegman in 1979 [6], universal hashing has been an important tool in computer science. Recent attention has been paid to universal hashing as a method to authenticate messages, an idea also proposed by these authors [12]. Its use in authentication has resulted in several very fast universal hash functions with low collision probabilities. But the implementations of these fastest universal hash functions tend to require either significant precomputed data, long keys or special-purpose hardware to achieve their impressive speeds.

Our contribution is a polynomial-based hash function we call PolyR. This hash function is not as fast as the fastest hash functions which have been designed for message authentication—speed is about 3.9–6.9 cpb. But that is still very fast, and, compared to the fastest of hash functions, PolyR has some different and desirable characteristics. First, it hashes messages of essentially *any* length (and varying lengths are fine). The key is short (say 28 bytes), independent of the message length. The key requires no preprocessing: the natural representation of the key is the desirable one for achieving good efficiency. Quite pleasantly, the hash function is fastest, per byte, on *short* messages—it actually gets slower, per byte, as the message gets longer (the rates are constant until particular

Hash Function	Collision Bound	Code + Data Size	Speed (cpb)	Output (bits)
This paper	$n2^{-28}$	$(124 + 8)$ bytes	3.9	32
This paper	$n2^{-49}$	$(409 + 16)$ bytes	6.9	64
Division Hash [10]	$n2^{-59}$	$\sim (? + 8)$ KB	7.5	64
UHASH-16 [3]	2^{-60}	$\sim (7 + 2)$ KB	1.0	64
UHASH-32 [3]	2^{-60}	$\sim (8 + 2)$ KB	2.0	64
hash127 [2]	$n2^{-127}$	$\sim (4 + 1.5)$ KB	4.3	127
MD5	unknown	1.7 KB	5.3	128
SHA-1	unknown	4.3 KB	13.1	160

Table 1. *Comparing the new constructions with some other hash families. Sizes marked with "$\sim$" are conservative estimates. All timings are for the fastest Pentium/Pentium II timings reported. To obtain smaller collision bounds one can hash twice or use the methods of this paper with $p(96)$ or $p(128)$.*

threshold lengths are crossed, like 2^{11} and 2^{33} bytes). This is the exact reverse of most optimized hash functions having short output lengths: they do better as the message gets longer. If used for authentication, working best for short messages is desirable insofar as *most* network traffic is short. Finally, implementation of our hash function family is simple and requires no special hardware (like floating-point units or multimedia execution units) to do well.

The hash function family PolyR was designed for use in a multi-layer hashing construction, to be used for fast message authentication. In such constructions a very fast first layer of hashing is applied to an incoming message to compress it to a small fraction of its original length. This compressed message is then passed to PolyR. When used as a second hash-layer in this manner, it can be expected that the *vast* majority of messages fed to PolyR will be short, since messages must be quite huge indeed before the second-layer compressed message gets long.

The hash function PolyR is a refinement to the classical suggestion of Carter and Wegman where one treats the message as specifying the coefficients of a polynomial, and one evaluates that polynomial at a point which is the key. Our refinements involve: (1) choosing the base field to be a prime just smaller than a power of 2^{32}, 2^{64}, or 2^{128} (this is a common trick); (2) using a simple "translation" trick to take care of the problem that some messages will now give rise to coefficients not in the field (because our field is just smaller than a power of two); (3) limiting the key space to a particular "convenient" subset of all the field points; and (4) using a "ramping-up" trick so that we don't have to pay in efficiency for short messages in order that the method can handle long ones. The result is a simple, flexible, fast-to-compute hash function. These various tricks, individually rather modest, work together to rise to a quite a nice hash-function family.

1.1 Related Work

Carter and Wegman introduced the ideas of universal hashing and using polynomials in universal hashing in 1979 [6]. Since that time polynomials have been used for fast hashing in many other works. In his "Cryptographic CRC", Krawczyk views messages to be hashed as polynomials over GF(2) which are divided modulo a random irreducible polynomial [8]. The division can be done quickly in hardware using linear feedback shift registers. Shoup describes several variants of polynomial hashing and provides implementation results [10]. His "generalized division hash", which bounds collisions between $64n$-bit messages as no more than $n2^{-59}$, views messages as polynomials over $GF(2^{64})$, uses an 8 KB precomputed table, and has a throughput of 7.5 cpb on a Pentium [10]. Afanassiev, Gehrmann and Smeets discuss fast polynomial hashing modulo random 3-, 4- and 5-nomials [1]. Their methods use small keys, but no implementation results are provided. Bernstein defines *hash127*, a polynomial hash defined over a large prime field, $\mathbb{Z}_{p(127)}$ [2]. Over $32n$-bit messages it has a collision probability of no more than $n2^{-127}$. Bernstein's implementation uses floating-point operations and a 1.5 KB precomputed table to achieve a throughput of 4.3 cpb on a Pentium II.

Other software efficient universal hash functions include Rogaway's bucket hash [9]; the MMH function of Halevi and Krawczyk [7]; and the NH function of Black, Halevi, Krawczyk, Krovetz and Rogaway [4]. The last is the current speed champion, providing collision probabilities of 2^{-60} with 4 KB of precomputed data and achieving throughput of 1.0 cpb on a Pentium II. using Pentium MMX instructions, and 1.9 cpb without MMX.

If one does not require the combinatoric certainties of universal hashing, one could employ cryptographic hashing to construct hash functions with short output lengths, short keys and little preprocessing. Bosselaers, Govaerts and Vandewalle report on optimized Pentium timing for several cryptographic hash functions: MD4 (3.8 cpb), MD5 (5.3 cpb) and SHA-1 (13.1 cpb) [5]. Simple methods can be used to convert these function into universal hash functions by, for example, keying their initial values [11]. We do not know what the collision probability would be for such constructions; for such a transformation to result in a good universal hash function, certain unproved assumptions must be made about the cryptographic hash function.

1.2 Notation

The algorithms described in this paper manipulate both bit-strings and integers. The i-th bit of string M is denoted $M[i]$ (bit-indices begin with 1). The substring consisting of the i-th through j-th bits of M is denoted $M[i \ldots j]$. The concatenation of string M_1 followed by string M_2 is denoted $M_1 \parallel M_2$. The length in bits of string M is $|M|$. The string of n zero-bits is denoted 0^n.

Given $b > 1$, the constant $p(b)$ is the largest prime smaller than 2^b. Given string M and $b > 0$, padonezero(M, b) returns the string $M \parallel 1 \parallel 0^n$, where n is the smallest number that makes the length of $M \parallel 1 \parallel 0^n$ divisible by b. Given

```
algorithm PolyCW [F] (k, m)
// Parameter: F is a finite field.
// Input: k ∈ F and m = (m₀, . . . , mₙ) where mᵢ ∈ F for 0 ≤ i ≤ n.
// Output: y ∈ F.
Let n be the number of elements in m
y = 0
for i ← 0 to n do
        y ← ky + mᵢ                        // Arithmetic in F
return y
```

Fig. 1. *The basic polynomial-hashing method of Carter and Wegman on which we build. The message* $\mathbf{m} = (m_0, \ldots, m_n)$ *is hashed to* $\sum_{i=0}^{n} m_i k^{n-i}$.

a string M, the function str2num(M) returns the integer that results when M is interpreted as an unsigned binary number. Similarly, num2str(n, b) produces the unique b-bit string which is the binary representation for the non-negative number n.

The number of elements in a set S is denoted $|S|$.

1.3 Organization

In the next few sections we develop a fast polynomial hash function. We build up to it in a couple of stages. In the appendix we generalize the hash function using arbitrary parameters. Theorems are given in both cases, but proven only for the concrete case. Proofs for the parameterized cases are straightforward adaptations of the ones for the concrete version, so they are omitted. Understanding the algorithms, theorems and proofs is easier in the concrete examples.

2 Carter-Wegman Polynomial Hashing: PolyCW

We begin by reviewing the "standard" approach for polynomial hashing. Let $\mathbb{F}$ be a finite field, let $k \in \mathbb{F}$ be a point in that field (the "key") and let $\mathbf{m} = (m_0, \ldots, m_n)$ be a vector of points in $\mathbb{F}$ that we want to hash (the "message"). We can hash message $\mathbf{m}$ to a point y in $\mathbb{F}$ (the "hash value") by computing $y = m_0 k^n + \cdots + m_{n-1} k^1 + m_n k^0$, where all arithmetic is done in $\mathbb{F}$. We denote this family of hash functions as PolyCW [$\mathbb{F}$]. The computation of this hash function (with $n+1$ multiplications in the field and $n+1$ additions in the field) is described in Figure 1.[1]

---

[1] All algorithms depicted in this paper which evaluate polynomials do so by using Horner's Rule which says that polynomial $m_0 k^n + \cdots + m_{n-1} k^1 + m_n k^0$ can be rewritten as $m_n + k(m_{n-1} + k(m_{n-2} + k(m_{n-3} + \cdots)))$. This allows for simple iteration with one multiplication and one addition for each element of the message.

```
algorithm PolyP32(k, m)
// Input: k ∈ K₃₂ and m = (m₁, ..., mₙ) where mᵢ ∈ Z_p(32) for 1 ≤ i ≤ n.
// Output: y ∈ Z_p(32).
Let n be the number of elements in m
p ← 2³² − 5                          // The largest prime smaller than 2³²
y = 0
for i ← 1 to n do
    y ← ky + mᵢ mod p
return y
```

Fig. 2. *The PolyP32 algorithm. A variant of the PolyCW hash, accelerated by choosing a field $Z_{p(32)}$ in which calculations can be performed quickly and choosing a key-set K_{32} which reduces arithmetic overflow on 32-bit processors. The **for** loop could be rewritten as the polynomial: $y = \sum_{i=1}^{n}(m_i k^{n-i}) \bmod p$.*

PolyCW [$\mathbb{F}$] is one of the most well-known universal hash-function families. It was described by Carter and Wegman in the paper that introduced that notion [6]. The main property it has is as follows. If $\mathbf{m} = (m_n, \ldots, m_0)$ and $\mathbf{m'} = (m'_n, \ldots, m'_0)$ are distinct vectors with the same number of components then $\Pr[H \leftarrow \text{PolyCW}[\mathbb{F}]; k \xleftarrow{\text{R}} \mathbb{F} : H_k(\mathbf{m}) = H_k(\mathbf{m'})] \leq \frac{n}{|\mathbb{F}|}$. This result is due to the Fundamental Theorem of Algebra which states that a nonzero polynomial of degree at most n can have at most n roots. Rewriting the above probability as $\Pr[k \xleftarrow{\text{R}} \mathbb{F} : \sum_{i=0}^{n} m_i k^{n-i} = \sum_{i=0}^{n} m'_i k^{n-i}] = \Pr[k \xleftarrow{\text{R}} \mathbb{F} : \sum_{i=0}^{n}(m_i - m'_i)k^{n-i} = 0]$, and applying the Fundamental Theorem, we see that there can be at most n values for k which cause $\sum_{i=0}^{n}(m_i - m'_i)k^{n-i}$ to evaluate to zero.

3   Making PolyCW [$\mathbb{F}$] Fast

Care must be taken in the implementation of PolyCW [$\mathbb{F}$]. A naive implementation is unlikely to perform well. Many choices of $\mathbb{F}$ and the set from which the hash-key is chosen can result in sub-optimal performance. We investigate the effect that shrewd choices for $\mathbb{F}$ and the key-set have on performance.

FIELD SELECTION. To make an efficient and practical hash function out of PolyCW [$\mathbb{F}$] we should carefully choose the finite field $\mathbb{F}$. Fields like GF[2^{64}] make natural candidates, because we are ultimately interested in hashing bit strings which are easily partitioned into 64-bit substrings. But arithmetic in GF[2^w] turns out to be less convenient for contemporary CPUs than a well-chosen alternative. In this paper we will do better by using prime fields in which the prime is just smaller than a power of two.

Consider first the use of the prime $p(32) = 2^{32} - 5$, which is the largest prime less than 2^{32}. To implement PolyCW [$\mathbb{Z}_{p(32)}$] efficiently, we need a good way to calculate $y \leftarrow ky + m \bmod p(32)$, where $y, k, m \in \mathbb{Z}_{p(32)}$. There are several options. One's first instinct is to use the native "mod" operand of a high-level programming language (like "%" in C), or to use a corresponding operator in the hardware architecture. But these choices are usually slow. For example, PolyCW $[\mathbb{Z}_{p(32)}]$, implemented in assembly using the native mod operator runs in **12.4** cpb (cycles/byte) on a Pentium II.

A faster method exploits the fact that since $p(32) = 2^{32} - 5$, the numbers 2^{32} and 5 are equivalent in the field $\mathbb{Z}_{p(32)}$, so $2^{33} = 10$, $2^{34} = 20$ and, more generally, $a2^{32} = 5a$ in $\mathbb{Z}_{p(32)}$. So, to calculate $ky \bmod p(32)$, first compute the 64-bit product $z = ky$ and separate z into a 32-bit high-word a and a 32-bit low-word b so that $z = a2^{32} + b$. We can then use the observation just made and rewrite $z \bmod p(32)$ as $5a + b$. This means that the calculation $y = ky + m \bmod p(32)$ can be done by computing $y = 5a + b + m \bmod p(32)$, which can be implemented more cheaply than the original approach because it does not require division to perform the modular reduction.

KEY-SET SELECTION. When implemented on a 32-bit architecture, the values a, b and m just discussed fit conveniently into 32-bit registers, making these quantities easy to manipulate. On most such architectures, the calculation of y is going to be fastest if it is done with minimal register overflow. To calculate $y = 5a + b + m \bmod p(32)$ using only 32-bit registers, we need one multiplication, two additions and then some additional instructions to handle register overflow. Each operation that can result in register overflow requires several instructions, including a conditional move or branch, to check and deal with the potential overflow event. To accelerate the calculation of y we reduce the number of potential overflows. Little can be done about overflow from the additions because both b and m can be nearly 2^{32}, but overflow from the multiplication can be eliminated. Only if a is larger than $\lfloor 2^{32}/5 \rfloor \approx 2^{29.7}$ can the term $5a$ overflow a 32-bit register. We can restrict a to safe values by restricting k to values less than 2^{29}. This allows for a faster implementation. The expense for this optimization is a higher collision probability because the key is chosen from a set of 2^{29} elements instead of a set of 2^{32} elements.

DIVISIONLESS MODULAR REDUCTION. Another optimization over a naive implementation is the elimination of division to calculate modular reductions. This technique is not new. In calculating $y = 5a + b + m \bmod p(32)$, each of the $5a$, b and m terms are less than 2^{32}. As we sum them using computer arithmetic with 32-bit registers, we can easily detect 32-bit overflows. Each such overflow indicates a 2^{32} term which is not accounted for in the resulting register. But, because $2^{32} \equiv 5$, these overflows are easily accounted for by adding 5 for each overflow to the resulting register. Done carefully, this observation results in a number y, derived without any division, which is representable in 32-bits (ie. $0 \le y < 2^{32}$). See Figure 3 for implementation details. Do we then need to reduce y to a number in $\mathbb{Z}_{p(32)}$? No. All of the discussion so far requires only that y be representable in a 32-bit register. Instead of reducing y to be in $\mathbb{Z}_{p(32)}$ after every intermediate calculation, we defer all such reductions until the end, when a final single reduction is performed.

```
; Calculate y = y * k + m mod p(32)
; Assume y is in register eax before and after code segment.
mul    k                    ; edx:eax = k * y
lea    edx, [edx*4+edx]     ; edx = 5 * edx
add    eax, edx             ; eax = edx + eax
lea    edx, [eax+5]         ; edx = eax + 5
cmovc  eax, edx             ; if (carried) then eax = edx
add    eax, m               ; eax = eax + m
lea    edx, [eax+5]         ; edx = eax + 5
cmovc  eax, edx             ; if (carried) then eax = edx
```

Fig. 3. *The $y = ky + m$ calculation of the PolyP32 algorithm written in Pentium II assembly. The flag "carried" is true only if the previous add instruction causes a register overflow. The conditional-move instruction (cmovc) is used to avoid any branches during execution of the routine, and the load-effective-address instruction (lea) is used for addition and multiplication of small constants. The result of the routine could possibly be in the range $p \leq y < 2^{32}$, which is outside of the field $\mathbb{Z}_{p(32)}$, but this is easily fixed with a single subtraction after hashing the final word of the entire message.*

SPEED. Taken together, the selection of a convenient prime field and the restriction of the key-set to keys which eliminate some register overflows allows a nice speed-up over a naive implementation of PolyCW. Figure 2 shows a version of polynomial hash based upon PolyCW which hashes over the field $\mathbb{Z}_{p(32)}$ and restricts key selection to the set $K_{32} = \{a : 0 \leq a < 2^{29}\}$. Our implementation of the core $y = ky + m \mod p(32)$ calculation uses just 8 lines of Pentium II assembly (Figure 3) and achieves a peak throughput of **3.69** cpb.

We state here the (simple) proposition establishing the collision bound of the PolyP32 hash function.

Proposition 1. *For any positive n and distinct messages $\mathbf{m} = (m_0, \ldots, m_n)$ and $\mathbf{m}' = (m'_0, \ldots, m'_n)$, consisting of elements from $\mathbb{Z}_{p(32)}$, the probability $\Pr[k \leftarrow K_{32} : \mathsf{PolyP32}(k, \mathbf{m}) = \mathsf{PolyP32}(k, \mathbf{m}')]$ is no more than $n/|K_{32}| = n2^{-29}$.*

64-BIT HASHING AND KEY RESTRICTION. We also implemented an analogous PolyP64 hash function whose core calculation is $y = ky + m \mod p(64)$ where $p(64) = 2^{64} - 59$ and k, y and m are all elements of $\mathbb{Z}_{p(64)}$. As in the 32-bit case, it is cheapest to calculate the result without using division. If we let $2^{32}k_h + k_\ell$ represent k and $2^{32}y_h + y_\ell$ represent y, then ky can be calculated as $ky = 2^{64}k_h y_h + 2^{32}(k_h y_\ell + k_\ell y_h) + k_\ell y_\ell$. Again, restricting the set of values that k can take on allows for faster implementations by eliminating some 32-bit register overflows. We define key-set $K_{64} = \{a2^{32} + b : 0 \leq a, b < 2^{25}\}$. This restriction allows an implementation of PolyP64 which has a collision probability of $(n/2^{50})$, uses 40 lines of assembly and has a peak throughput of **6.86** cpb.

```
algorithm PolyQ32(k, M)
// Input: k ∈ K₃₂ and M ∈ ({0,1}³²)⁺.
// Output: y ∈ Z_p(32).
p ← 2³² − 5                              // Largest prime smaller than 2³²
offset ← 5                              // For translating out-of-range words
marker ← 2³² − 6                        // For marking out-of-range words
n ← |M|/32
M₁ ‖ ... ‖ Mₙ ← M,                      // Break M into 32-bit chunks
    where |M₁| = ··· = |Mₙ| = 32
y ← 1                                   // Set highest coefficient to 1
for i ← 1 to n do
    m ← str2num(Mᵢ)
    if (m ≥ p − 1) then                 // If word is not in range, then
        y ← ky + marker mod p           // Marker indicates out-of-range
        y ← ky + (m − offset) mod p     // Offset m back into range
    else
        y ← ky + m mod p                // Otherwise hash in-range word
return y
```

Fig. 4. The PolyQ32 algorithm. The PolyP32 hash extended to hash strings instead of vectors of field elements and to allow good collision probabilities over two strings which differ in length.

4   Expanding the Domain to Arbitrary Strings

The hash function PolyP32 is not generally useful. It only works on same-length messages, and those messages must be made of elements from the field $\mathbb{Z}_{p(32)}$. We now remove these limitations and develop PolyQ32. The result, depicted in Figure 4, hashes most messages at a rate of **3.86** cpb.

ALLOWING VARIATIONS IN LENGTH. It is a trivial exercise to produce two different-length messages which collide when hashed with PolyP32 using *any* key: under PolyP32, the hash of a message $\mathbf{m} = (m_0, \ldots, m_n)$ using key k is simply $h(k, \mathbf{m}) = m_0 k^n + \cdots + m_n k^0 \bmod p(32)$, so prepending 0 to the vector $\mathbf{m}$ results in a message $\mathbf{m}' = (0, m_0, \ldots, m_n)$ which is hashed as $h(k, \mathbf{m}') = 0k^{n+1} + m_0 k^n + \cdots + m_n k^0 \bmod p(32)$ and is equal to $h(k, \mathbf{m})$ because the additional zero-term has no effect on the hash value. For the Fundamental Theorem of Algebra to guarantee a low number of roots (and hence a low collision probability), it is essential that the difference between $\mathbf{m}$ and $\mathbf{m}'$ be non-zero. This means that if the two vectors differ only in length, then at least one of the initial elements of the longer vector must be non-zero. To guarantee this we employ a standard trick and implicitly prepend a "1" to the vectors being hashed. Thus, the hash of $\mathbf{m} = (m_0, \ldots, m_n)$ implicitly becomes the hash of $\mathbf{m} = (1, m_0, \ldots, m_n)$, and the hash of $\mathbf{m}' = (0, m_0, \ldots, m_n)$ implicitly becomes the hash of $\mathbf{m}' = (1, 0, m_0, \ldots, m_n)$. The difference between these two vectors is non-zero. The following theorem assures that augmenting PolyP32 in this way results in a hash with nearly the same collision probability as PolyP32, but works over messages of different lengths.

Proposition 2. *Let $\ell < n$ be positive integers. Let $\mathbf{m} = (m_0, \ldots, m_\ell)$ and $\mathbf{m}' = (m_0', \ldots, m_n')$ be any two vectors of elements from the field $\mathbb{F}$. Then there are at most $n+1$ values for $k \in \mathbb{F}$ such that $k^{\ell+1} + \sum_{i=0}^{\ell} m_i k^{\ell-i} = k^{n+1} + \sum_{i=0}^{n} m_i' k^{n-i}$.*

Proof. Beginning with $k^{\ell+1} + \sum_{i=0}^{\ell} m_i k^{\ell-i} = k^{n+1} + \sum_{i=0}^{n} m_i' k^{n-i}$, and moving all of its terms to the right side of the equation we get $0 = k^{n+1} - k^{\ell+1} + \sum_{i=0}^{n} m_i' k^{n-i} - \sum_{i=0}^{\ell} m_i k^{\ell-i}$. But, the right side of this equations is now a non-zero polynomial, is of degree $n + 1$, and therefore has at most $n + 1$ roots. ◇

ALTERNATIVE METHOD. Another method of augmenting PolyP32 to allow variable length messages is to use a second key $k' \in \mathbb{Z}_{p(32)}$ and add it to each element of the message being hashed. Thus, $h(k, k', \mathbf{m})$ would be computed as $\sum_{i=0}^{\ell} (m_i + k') k^{\ell-i}$. This method requires an extra addition per message word being hashed and so the first method seems favorable.

ALLOWING BIT-STRINGS. To make the function PolyP32 of Figure 2 more useful, it must be adapted to allow bit-strings rather than only vectors from $\mathbb{Z}_{p(32)}$. The field $\mathbb{Z}_{p(32)}$ was chosen because it contains nearly all the numbers representable as 32-bit strings. Thus, when we desire to hash a bit-string, we may partition the string into 32-bit words and treat the partition as a vector of 32-bit numbers. PolyP32 can then hash the vast majority of the vector's elements without any modification. But, some of the 32-bit numbers may be in the range $p(32) \ldots 2^{32} - 1$, outside $\mathbb{Z}_{p(32)}$. What should be done with them?

One approach is to transform a vector of 32-bit numbers, which may have some elements outside of $\mathbb{Z}_{p(32)}$, into a vector which does not. The transformation must map distinct vectors into distinct vectors.

We solve this problem by examining a vector of 32-bit numbers and replacing each vector element m_i that is greater than $p(32) - 2$ with *two* numbers, $p(32) - 1$ and $m_i - 5$. Note that both of these numbers are in $\mathbb{Z}_{p(32)}$. Each such replacement lengthens the resulting vector by one element. Thus, the vector $\mathbf{m} = (4, 2^{32} - 3, 10)$, whose second element is greater than $p(32) - 2$, would be transformed into the vector $\mathbf{m}' = (4, 2^{32} - 6, 2^{32} - 8, 10)$. We call this transformation DoubleTransform : $(\mathbb{Z}_{2^{32}})^+ \rightarrow (\mathbb{Z}_{p(32)})^+$. The following proposition assures that DoubleTransform is correct.

Proposition 3. *For positive ℓ and n, and distinct messages $\mathbf{m} = (m_0, \ldots, m_\ell)$ and $\mathbf{m}' = (m_0', \ldots, m_n')$ made of elements from $\mathbb{Z}_{2^{32}}$, the transformed vectors DoubleTransform($\mathbf{m}$) and DoubleTransform($\mathbf{m}'$) consist of elements from $\mathbb{Z}_{p(32)}$ and are distinct.*

Proof. Let ℓ and n be positive integers and let $\mathbf{m} = (m_0, \ldots, m_\ell)$ and $\mathbf{m}' = (m_0', \ldots, m_n')$ be distinct vectors consisting of elements from $\mathbb{Z}_{2^{32}}$. Let $\mathbf{t} =$

DomainTransform($\mathbf{m}$) and $\mathbf{t}' = $ DomainTransform($\mathbf{m}'$). Let i be the smallest number such that $m_i \neq m_i'$. If such an i does not exist then one of $\mathbf{m}$ or $\mathbf{m}'$ must be a proper prefix of the other. In this case, any lengthening of the shorter vector by DoubleTransform must be mirrored by the transformation of the longer vector ensuring that the two remain different lengths after transformation.

If m_i and m_i' are both less than $p(32) - 1$, then after transformation $t_i = m_i$ and $t_i' = m_i'$, ensuring that $\mathbf{t} \neq \mathbf{t}'$. If only one of m_i and m_i' is less than $p(32) - 1$, say m_i, then after transformation $t_i = m_i$ and $t_i' = p(32) - 1$, again ensuring that $\mathbf{t} \neq \mathbf{t}'$. Finally, if both m_i and m_i' are greater than $p(32) - 2$, then after transformation $t_{i+1} = m_i - 5$ and $t_{i+1}' = m_i - 5'$ again ensuring that $\mathbf{t} \neq \mathbf{t}'$. $\Diamond$

ALTERNATIVE METHOD. There are many ways to patch PolyP32 to allow out-of-range elements. One probabilistic alternative is to offset every out-of-range number by a randomly chosen $k' \in \{5, \ldots, 2^{32} - 5\}$. All out-of-range numbers are in $\{2^{32} - 5, \ldots, 2^{32} - 1\}$, so k', when subtracted from an out-of-range number, will always yield a number in $\mathbb{Z}_{p(32)}$. This method has the advantage of not increasing message length upon transformation, but requires an extra key element, and in practice does not speed hashing with respect to the method of Proposition 3.

Together, Propositions 2 and 3 prove the following corollary.

Corollary 1. *For any positive integers $\ell \leq n$ and distinct messages $M \in \{0,1\}^{32l}$ and $M' \in \{0,1\}^{32n}$, $\Pr_{k \in K_{32}}[\mathsf{PolyQ32}(k, M) = \mathsf{PolyQ32}(k, M')] \leq 2n/|K_{32}| = n2^{-28}$.*

The discussion so far has focussed on PolyQ32, a hash function defined on 32-bit words. An analogous 64-bit variant, PolyQ64, yields the following bound.

Corollary 2. *For any positive integers $\ell \leq n$ and distinct messages $M \in \{0,1\}^{64l}$ and $M' \in \{0,1\}^{64n}$, $\Pr_{k \in K_{64}}[\mathsf{PolyQ64}(k, M) = \mathsf{PolyQ64}(k, M')] \leq 2n/|K_{64}| = n2^{-49}$.*

Two things are worth noting. First, the factor of two introduced in the $2n/|K_{32}|$ term is due to the potential doubling of message length by the DoubleTransform function. And, second, standard message padding techniques are not addressed in this paper. It is assumed that messages being hashed have been properly padded to a 32-bit boundary.

It should also be noted that the probability that PolyQ32 or PolyQ64 hash any message to a particular result is also low. Consider a message made of n 32-bit words $\mathbf{x} = (x_1, \ldots, x_n)$ and a constant c. If $c \geq p(32)$ then PolyQ32$(k, \mathbf{x})$ cannot hash to c, and if $c < p(32)$, then PolyQ32$(k, \mathbf{x})$ will hash to c only if PolyQ32$(k, \mathbf{x}')$ hashes to zero where $\mathbf{x}' = (x_1, \ldots, x_n - c)$. After the DoubleTransform transformation of $\mathbf{x}'$, the Fundamental Theorem tells us that there are no more than $2n$ keys which allow this to happen.

Claim. Let n and c be numbers, and let message M be an element from $\{0,1\}^{32\ell}$, then $\Pr_{k \in K_{32}}[\mathsf{PolyQ32}(k, M) = c] \leq 2n/|K_{32}|$.

```
algorithm PolyR32_64(k, M)
// Input: k = (k₁, k₂) with k₁ ∈ K₃₂ and k₂ ∈ K₆₄, and M ∈ {0,1}*.
// Output: Y ∈ {0,1}⁶⁴.
if (|M| ≤ 2¹⁴) then                                      // 2⁹ 32-bit words
    M ← padonezero(M, 32)
    y ← PolyQ32(k₁, M)                                   // Hash in Z_p(32)
else if (|M| ≤ 2³⁶) then                                 // 2³⁰ 64-bit words
    M₁ ← M[1 ... 2¹⁴]
    M₂ ← M[2¹⁴ + 1 ... |M|]
    M₂ ← padonezero(M₂, 64)
    y ← PolyQ32(k₁, M₁)                                  // Hash in Z_p(32)
    y ← PolyQ64(k₂, num2str(y, 64) ∥ M₂)                 // Hash in Z_p(64), prepending y
else
    return Error                                         // Message too long
Y ← num2str(y, 64)                                       // Convert to string
return Y
```

Fig. 5. *The PolyR32_64 algorithm. Combining the PolyP32 and PolyP64 hashes into a hash function which is fast on short messages but also performs well on long ones. PolyR32_64 also extends the domain to messages which are not a multiple of the constituent hashes word-lengths.*

5 PolyR: Overcoming Polynomial Hash Length Limitations

Taking a closer look at the bounds established for each of the polynomial hash functions, one can see that the collision bounds degrade linearly along with the length of the messages being hashed. This is a byproduct of the use of polynomials in hashing: As messages get longer, so do the degrees of the polynomials get higher, resulting in more potential collision-causing roots. This introduces a trade-off in application design. If one wants to guarantee some maximum collision probability ϵ and the hash-key is chosen from a set of k elements, then the length of messages to be hashed must be limited to around $k\epsilon$ words. The larger the key-set size k used in the hashing polynomial, the more words can be hashed before reaching the allowable collision probability q. But, to make the key-set size significantly larger requires the polynomial to be computed over a larger prime-field, and in general, as the prime p is increased, so is the time needed to evaluate the polynomials in $\mathbb{Z}_p$. As one can see by examining the timing results for PolyQ32 and PolyQ64, the move from a prime close to 2^{32} to one close to 2^{64} increases the number of cycles-per-byte required to hash a message by nearly 50%.[2]

---

[2] Some of this difference is an artifact of the fact that the Pentium II natively supports multiplication of 32-bit operands to a 64-bit result, but not the multiplication of 64-

Can we have the best of both worlds: a hash function which is as fast as PolyQ32 but can hash messages as long as PolyQ64, without having intollerably high collision probability? This is the goal which motivates this section. We approach the problem with the belief that most strings being hashed are short, but that a generalized hash function should be able to handle well long messages too.

Our idea is to hash short messages (up to some fixed number of bits ℓ) directly with PolyQ32, but hash messages longer than ℓ bits with a hybrid scheme. Let us say that message M is longer than ℓ bits. To hash M we first partition it into its ℓ bit prefix M_1 and the remainder M_2, so that $M_1 \parallel M_2 = M$. The hash of M under our hybrid scheme is then PolyQ64(k_2, PolyQ32(k_1, M_1) $\parallel M_2$). In this manner, the first ℓ bits of M is hashed with a fast hash function (which cannot safely hash long messages), and if there is any of the string left after hashing its prefix, the remainder is hashed with a slower hash function (which can safely hash longer messages). The parameter ℓ depends on the maximum desirable collision bound and how long a message can be before the fast hash function approaches this bound.

As an example, let us say that we want to hash messages and have a collision bound of no more than 2^{-16}. If we were to hash solely with PolyQ32, then we could hash no messages longer than around 2^{17} bits. Alternatively, we could hash with only PolyQ64 and would then be able to hash strings as long as 2^{39} bits before allowing 2^{-16} collision probability, but at a much slower rate than PolyQ32. Under our scheme, if a message M is shorter than 2^{17} bits, then the hash result is simply PolyQ32(M); whereas if M is longer than 2^{17} bits, then the hash is calculated as PolyQ64(PolyQ32(M_1) $\parallel M_2$) where M_1 is the 2^{17}-bit prefix of M. Such a construction is fast on short messages, but handles well long messages too. If messages were anticipated to be longer than 2^{39}, then a function PolyQ96, employing a 96-bit prime modulus, could be defined analogously and be employed as a third-stage polynomial. This ramping-up of the prime modulus used in the polynomial evaluations gives the construction its name: *Ramped polynomial hashing.*

One might expect the collision bound of such a hybrid approach to be approximately the *sum* of the collision bounds of each of its constituent functions, but as the following theorem shows, the overall collision bound is instead only the *maximum* of the functions.

The following theorem and proof address PolyR32_64, the ramped polynomial hash of Figure 5. This concrete hash function hashes up to 2^{14} bits (equivalent to 2^9 32-bit words) using the fast PolyQ32 function, and allows a total message length of up to 2^{36} bits (or 2^{30} 64-bit words). In the following theorem and proof, for increased generality, we use parameters ℓ and m instead of the numbers of words 2^9 and 2^{30}.

---

bit operands to a 128-bit result. Most processors will display this type of threshold behavior when operands exceed well-supported lengths.

Theorem 1. *Let $\ell = 2^9$ and $m = 2^{30}$. Let $M \neq M'$ be messages no longer than $64m$ bits. Then the probability $\Pr[k_1 \overset{\text{R}}{\leftarrow} K_{32}; k_2 \overset{\text{R}}{\leftarrow} K_{64} : \mathsf{PolyR32\_64}(k_1, k_2, M) = \mathsf{PolyR32\_64}(k_1, k_2, M')]$ is no more than $\max(\ell 2^{-28}, m2^{-49}) + 2^{-50}$.*

Proof. Let M and M' be messages, and imagine partitioning them into $M = M_1 \parallel M_2$ and $M' = M_1' \parallel M_2'$ so that M_1 and M_1' are the first 32ℓ bits of M and M'. If M is shorter than 32ℓ bits, then $M_1 = M$ and M_2 is empty. Likewise, if M' is shorter than 32ℓ bits, then $M_1' = M'$ and M_2' is empty. We ignore all padding issues in this discussion, assuming that standard padding techniques are used to bring M and M' to appropriate lengths. Let $k_1 \overset{\text{R}}{\leftarrow} K_{32}$ and $k_2 \overset{\text{R}}{\leftarrow} K_{64}$ be randomly chosen keys. We define the following values here for convenience, but all probabilities in this proof are assumed to be taken over these random choices of k_1 and k_2.

$$h_1 = \mathsf{num2str}(\mathsf{PolyQ32}(k_1, M_1), 64) \text{ and } h_1' = \mathsf{num2str}(\mathsf{PolyQ32}(k_1, M_1'), 64)$$
$$h_2 = \mathsf{num2str}(\mathsf{PolyQ64}(k_2, h_1 \parallel M_2), 64) \text{ and } h_2' = \mathsf{num2str}(\mathsf{PolyQ64}(k_1, M_2'), 64)$$

Depending on the lengths of M and M', the result of hashing M will be h_1 or h_2 and the result of hashing M' will be h_1' or h_2'. We examine several cases for the relative lengths of M and M'.

CASE 1: DIFFERENT LENGTHS, SAME RAMP. Here we examine the case where the messages M and M' are different lengths, but are both either longer than 32ℓ bits or both no longer. If both are no longer than 32ℓ bits then a collision occurs if $h_1 = h_1'$. But, $M_1 = M$ and $M_1' = M'$ differ in length which means (by Proposition 1) that $\Pr[h_1 = h_1'] \leq \ell 2^{-28}$. If both M and M' are longer than 32ℓ bits then a collision occurs if $h_2 = h_2'$. But, $h_1 \parallel M_2$ and $h_1' \parallel M_2'$ also differ in length which means (by Proposition 2) that $\Pr[h_2 = h_2'] \leq m2^{-49}$

CASE 2: DIFFERENT LENGTHS, DIFFERENT RAMP. If M is longer than 32ℓ bits and M' is not, then a collision occurs only if $h_1' = h_2$. Expanding the h_2 term, we see that a collision only occurs if $h_1' = \mathsf{num2str}(\mathsf{PolyQ64}(k_2, h_1 \parallel M_2), 64)$. If we fix k_1 to an arbitrary value, then h_1 and h_1' become fixed as well, and the probability of collision then depends only on the selection of k_2. The string $h_1 \parallel M_2$ is partitioned by the $\mathsf{PolyQ64}$ algorithm into 64-bit strings and then transformed by $\mathsf{DoubleTransform}$ into some sequence $x_0, x_1, \ldots, x_{n \leq 2m}$ of elements from $\mathbb{Z}_{p(64)}$. This sequence is then used in the summation $\sum_{i=0}^{n} x_i k_2^{n-i} \bmod p(64)$ to calculate the final hash result. A collision occurs if the result of this summation is h_1', or alternatively when $\sum_{i=0}^{n} x_i k_2^{n-i} - h_1' \bmod p(64) = 0$. The Fundamental Theorem of Algebra applies to this last polynomial, meaning there are no more than $n \leq 2m$ values for k_2 which satisfy it. Thus, $\Pr[h_1' = h_2] \leq 2m2^{-29} = m2^{-28}$.

CASE 3: EQUAL LENGTH MESSAGES, LAST RAMP DIFFERENT. If M and M' are equal length, longer than 32ℓ bits and $M_2 \neq M_2'$, then (by Proposition 2) $\Pr[h_2 = h_2'] \leq m2^{-49}$ because $h_1 \parallel M_2$ and $h_1' \parallel M_2'$ are distinct. Similarly, if M and M' are the same length, no longer than 32ℓ bits and $M_1 \neq M_1'$, then (by Proposition 1) $\Pr[h_1 = h_1'] \leq m2^{-28}$ because M_1 and M_1' are distinct.

CASE 4: EQUAL LENGTH MESSAGES, LAST RAMP SAME. If M and M' are equal length and longer than 32ℓ bits, and $M_1 \neq M_1'$ but $M_2 = M_2'$, then there are two opportunities for a collision to take place. First, if $\mathsf{PolyQ32}(k_1, M_1) = \mathsf{PolyQ32}(k_1, M_1')$, then the strings $h_1 \parallel M_2$ and $h_1' \parallel M_2'$ are equal, guaranteeing that h_2 and h_2' collide. The probability of this event is no more than $\ell 2^{-28}$. Second, if $h_1 \neq h_1'$, then a collision can still occur if $\mathsf{PolyQ32}(k_2, h_1 \parallel M_1) = \mathsf{PolyQ32}(k_2, h_2' \parallel M_2')$. One might think that this is an event with up to $m 2^{-49}$ probability, but it is not. Because $M_2 = M_2'$, the strings $h_1 \parallel M_1$ and $h_1' \parallel M_1'$ only differ in their first 64-bit word. The collision event when hashing such strings takes the form $(h_1 - h_1')k_2^n = 0$, which can only be satisfied if $k_2 = 0$, a 2^{-50} probability event. Thus, the total probability of collision in this case is bounded by $\ell 2^{-28} + 2^{-50}$. $\Diamond$

5.1 Security Notes

If a lower collision probability is desired, one can hash messages multiple times, using a different key for each message hash. A hash function which has an ϵ collision bound when hashing once with a random key, has an ϵ^2 collision bound when hashing twice with two random keys, and an ϵ^3 collision bound when hashed with three keys, etc.

Also, all of the theorems in this work have been stated in terms of collisions (ie. the difference between the result of evaluating the hash of two distinct messages is zero). It is a simple matter to tweak the algorithms and proofs to show that the probability that the difference between the hash of two distinct messages being a particular constant is bounded by the same ϵ. This version of universal hashing ("delta"-universal) is required in some message authentication schemes.

References

1. AFANASSIEV, V., GEHRMANN, C., AND SMEETS, B. Fast message authentication using efficient polynomial evaluation. In *Proceedings of the 4th Workshop on Fast Software Encryption* (1997), vol. 1267, Springer-Verlag, pp. 190–204.
2. BERNSTEIN, D. Floating-point arithmetic and message authentication. Unpublished manuscript, http://cr.yp.to/papers.html, 2000.
3. BLACK, J., HALEVI, S., KRAWCZYK, H., KROVETZ, T., AND ROGAWAY, P. UMAC: Fast and secure message authentication. In *Advances in Cryptology – CRYPTO '99* (1999), vol. 1666 of *Lecture Notes in Computer Science*, Springer-Verlag, pp. 216–233.
4. BLACK, J., HALEVI, S., KRAWCZYK, H., KROVETZ, T., AND ROGAWAY, P. UMAC: Fast and secure message authentication. In *Advances in Cryptology – CRYPTO '99* (1999), vol. 1666 of *Lecture Notes in Computer Science*, Springer-Verlag, pp. 216–233.
5. BOSSELAERS, A., GOVAERTS, R., AND VANDEWALLE, J. Fast hashing on the Pentium. In *Advances in Cryptology – CRYPTO '96* (1996), vol. 1109 of *Lecture Notes in Computer Science*, Springer-Verlag, pp. 298–312. Updated timing at http://www.esat.kuleuven.ac.be/ bosselae/fast.html.

6. CARTER, L., AND WEGMAN, M. Universal classes of hash functions. *J. of Computer and System Sciences 18* (1979), 143–154.

7. HALEVI, S., AND KRAWCZYK, H. MMH: Software message authentication in the Gbit/second rates. In *Proceedings of the 4th Workshop on Fast Software Encryption* (1997), vol. 1267, Springer-Verlag, pp. 172–189.

8. KRAWCZYK, H. LFSR-based hashing and authentication. In *Advances in Cryptology – CRYPTO '94* (1994), vol. 839 of *Lecture Notes in Computer Science*, Springer-Verlag, pp. 129–139.

9. ROGAWAY, P. Bucket hashing and its application to fast message authentication. In *Advances in Cryptology – CRYPTO '95* (1995), vol. 963 of *Lecture Notes in Computer Science*, Springer-Verlag, pp. 313–328.

10. SHOUP, V. On fast and provably secure message authentication based on universal hashing. In *Advances in Cryptology – CRYPTO '96* (1996), vol. 1109 of *Lecture Notes in Computer Science*, Springer-Verlag, pp. 313–328.

11. TSUDIK, G. Message authentication with one-way hash functions. *Computer Communications Review 22* (1992), 29–38.

12. WEGMAN, M., AND CARTER, L. New hash functions and their use in authentication and set equality. *J. of Computer and System Sciences 22* (1981), 265–279.

A Fully Parameterized: PolyQ, PolyR

The body of this paper developed a two-stage ramped polynomial hash function PolyR32_64 using polynomials over 32- and 64-bit prime fields. The concrete choices made for PolyR32_64 were designed especially for a message authentication code. In the MAC, we needed a universal hash function which would guarantee a collision bound of at most 2^{-16} and would typically be applied to messages no longer than a few dozen bytes. But, the hash function must also be able to process huge inputs, too, and still guarantee a bound of at most 2^{-16}. These requirements led to the development of ramped polynomial hashing in general, and in the choice of the 32- and 64-bit prime fields, and associated crossover points, used in the body of this paper.

Other collision bounds and message lengths not addressed by PolyR32_64 are likely, and so we present in this appendix fully parameterized versions of the hashes called PolyQ and PolyR. For each of the algorithms we state their collision bounds as a theorem, but give no proofs. The proofs are straightforward extensions of those given in the body of the paper.

Proposition 4. *Let v be any positive integer, let $K \subseteq \mathbb{Z}_{p(v)}$ be any subset of points in the field $\mathbb{Z}_{p(v)}$, and let $2^{v-1} \leq d \leq p(v)$. For any positive integers $\ell \leq n$ and distinct messages $M \in \{0,1\}^{lv}$ and $M' \in \{0,1\}^{nv}$, $\Pr[k \xleftarrow{\text{R}} K : \mathsf{PolyQ}[K,v,d](k,M) = \mathsf{PolyQ}[K,v,d](k,M')] \leq 2n/|K|$.*

Proposition 5. *Let all of the parameters from Figure 6 be fixed. For any distinct messages M and M', each shorter than $\sum_{1 \leq i \leq r} \ell_i v_i$ bits, $\Pr[\mathbf{k} \xleftarrow{\text{R}} \mathbf{K} : \mathsf{PolyR}(\mathbf{k}, M) = \mathsf{PolyR}(\mathbf{k}, M')] \leq$*

$$\max_{1 \le i \le r} \left\{ \frac{2\ell_i}{|K_i|} \right\} + \sum_{i=2}^{r} \frac{1}{|K_i|} .$$

$r \ge 1$	: Length of $\mathbf{v}$, l, $\mathbf{K}$ vectors used in PolyR.
$\mathbf{v} = (v_1, \ldots, v_r)$	: Word-lengths used in PolyR, with $1 < v_1 < \cdots < v_r$.
$\mathbf{l} = (\ell_1, \ldots, \ell_r)$	: Message lengths used in PolyR, with $\ell_i \ge 1$ for $1 \le i \le r$.
$\mathbf{d} = (d_1, \ldots, d_r)$	: Domain bounds used in PolyR, with $2^{v_i-1} \le d_i \le p(v_i)$.
$\mathbf{K} = (K_1, \ldots, K_r)$	: Key-sets used in PolyR, with $K_i \subseteq \mathbf{Z}_{p(v_i)}$ for $1 \le i \le r$.

Fig. 6. *Parameters used in the fully parameterized PolyR algorithm. Fixing these parameters fixes the algorithm definition specified in Figure 8.*

---

algorithm PolyQ$[K, v, d](k, M)$
// *Parameters: "Key set" $K \subseteq \mathbf{Z}_{p(v)}$, "word length" $v \ge 1$, "domain-bound" d.*
// *Input: $k \in K$ and $M \in (\{0,1\}^v)^+$.*
// *Output: $y \in \mathbf{Z}_{p(v)}$.*
$offset \leftarrow 2^v - p(v)$ // *For translating out-of-range words*
$marker \leftarrow p(v) - 1$ // *For marking out-of-range words*
$n \leftarrow |M|/v$
$M_1 \| \ldots \| M_n \leftarrow M,$ // *Break M into word size chunks*
     where $|M_1| = \cdots = |M_n| = v$
$y \leftarrow 1$ // *Set highest coefficient to 1*
for $i \leftarrow 1$ to n do
     $m \leftarrow \mathsf{str2num}(M_i)$
     if $(m \ge d)$ then   // *If word is not in range, then*
          $y \leftarrow ky + marker \bmod p(v)$ // *Marker indicates out-of-range*
          $y \leftarrow ky + (m - offset) \bmod p(v)$ // *Offset m back into range*
     else
          $y \leftarrow ky + m \bmod p(v)$ // *Otherwise hash in-range word*
return y

Fig. 7. *The PolyQ algorithm, parameterized on key-set K, word-length v and domain-bound d.*

---

algorithm PolyR($\mathbf{k}, M$)
// *Parameters: Uses "vector length" r, "word-length vector" $\mathbf{v}$,*
// *"message-length vector" l, "domain-bounds vector" $\mathbf{d}$, "key-set vector" $\mathbf{K}$.*
// *Input:* $\mathbf{k} = (k_1, \ldots, k_r)$ *with* $k_i \in K_i$ *for* $1 \le i \le r$ *and* $M \in \{0,1\}^*$.
// *Output:* $Y \in \{0,1\}^{v_r}$.
prepend $\leftarrow \varepsilon$ // *Initially no string to prepend*
$i \leftarrow 1$ // *Index for* $\mathbf{v}$, l, $\mathbf{K}$ *vectors*
while $(|M| > \ell_i v_i)$ **do**   // *While multiple ramp-levels remain*
    if $(i = r)$ **then return** Error   // *Message too long*
    $T \leftarrow M[1 \ldots \ell_i v_i]$ // *Extract string to hashed under* $p(v_i)$
    $M \leftarrow M[\ell_i v_i + 1 \ldots |M|]$
    $y \leftarrow$ PolyQ$[K_i, v_i, d_i](k_i, prepend \parallel T)$ // *Hash in* $\mathbf{Z}_{p(v_i)}$, *prepend previous*
    prepend $\leftarrow$ num2str(y, v_{i+1}) // *Update* prepend *for next ramp-level*
    $i \leftarrow i + 1$
$M \leftarrow$ padonezero(M, v_i) // *Final ramp needs bijective padding*
$y \leftarrow$ PolyQ$[K_i, v_i, d_i](k_i, prepend \parallel M)$ // *Hash in* $\mathbf{Z}_{p(v_i)}$, *prepend previous*
$Y \leftarrow$ num2str(y, v_r) // *Convert to string*
return Y

Fig. 8. *The* PolyR *algorithm. Parameters are described in Figure 6.*

Characterization of Elliptic Curve Traces under FR-Reduction

Atsuko Miyaji, Masaki Nakabayashi, and Shunzo Takano

Japan Advanced Institute of Science and Technology
{miyaji, manakaba}@jaist.ac.jp

Abstract. Elliptic curve cryptosystems([19, 25]) are based on the elliptic curve discrete logarithm problem(ECDLP). If elliptic curve cryptosystems avoid FR-reduction([11, 17]) and anomalous elliptic curve over $\mathbb{F}_q$ ([34, 3, 36]), then with current knowledge we can construct elliptic curve cryptosystems over a smaller definition field. ECDLP has an interesting property that the security deeply depends on elliptic curve traces rather than definition fields, which does not occur in the case of the discrete logarithm problem(DLP). Therefore it is important to characterize elliptic curve traces explicitly from the security point of view. As for FR-reduction, supersingular elliptic curves or elliptic curve $E/\mathbb{F}_q$ with trace 2 have been reported to be vulnerable. However unfortunately these have been only results that characterize elliptic curve traces explicitly for FR- or MOV-reductions. More importantly, the secure trace against FR-reduction has not been reported at all. Elliptic curves with the secure trace means that the reduced extension degree is always higher than a certain level.

In this paper, we aim at characterizing elliptic curve traces by FR-reduction and investigate explicit conditions of traces vulnerable or secure against FR-reduction. We show new explicit conditions of elliptic curve traces for FR-reduction. We also present algorithms to construct such elliptic curves, which have relation to famous number theory problems.

key words: elliptic curve cryptosystems, trace, FR-reduction, number theory

1 Introduction

Koblitz and Miller proposed independently a public key cryptosystem based on an elliptic curve E defined over a finite field $\mathbb{F}_q$ $(q = p^r)$([19, 25]). If elliptic curve cryptosystems satisfy so called FR-conditions ([24, 11, 17]) and avoid anomalous elliptic curve over $\mathbb{F}_q$ ([34, 3, 36]), then the only known attacks are the Pollard ρ-method ([27]) and the Pohlig-Hellman method ([26]). Hence with current knowledge, we can construct elliptic curve cryptosystems over a smaller definition field than the discrete logarithm problem (DLP)-based cryptosystems like the ElGamal cryptosystems ([13]) or the DSA ([12]) and RSA cryptosystems ([28]). Elliptic curve cryptosystems with a 160-bit key are thus believed to have the same security as both the ElGamal cryptosystems and RSA cryptosystems with a 1,024-bit key.

Recently some researches on comparing MOV and FR-reductions have been reported in [15, 18]. These attacks imbed a subgroup $< G > \subset E(\mathbb{F}_q)$ to $\mathbb{F}_{q^k}^*$ for an extension field $\mathbb{F}_{q^k}$ and reduce ECDLP based on $< G > \subset E(\mathbb{F}_q)$ to DLP based on a subgroup of $\mathbb{F}_{q^k}^*$, where $G \in E(\mathbb{F}_q)$ is called a basepoint for ECDLP. MOV-reduction reduces ECDLP to DLP by using the Weil pairing ([35]). Supersingular elliptic curves ([35]) have been reported to be vulnerable against MOV-reduction, which can be easily recognized by the trace t of the qth-power Frobenius endomorphism, $t = q + 1 - \#E(\mathbb{F}_q)$: an elliptic curve is supersingular if and only if $t \equiv 0 \pmod{p}$. On the other hand, FR-reduction reduces ECDLP to DLP by using the Tate pairing. FR-reduction can attack elliptic curves with trace 2 in addition to supersingular elliptic curves. In fact, these have been only results that characterize elliptic curve traces explicitly from a point of view of FR- and MOV-reductions. It is interesting that in the case of $E/\mathbb{F}_p$ over a prime field, dangerous elliptic curve traces happen to be equal to 0 (supersingular), 1 (anomalous) and 2, which can be easily recognized from other elliptic curves. Thus ECDLP has an interesting property that the security deeply depends on elliptic curve traces rather than definition fields, which does not occur in the case of DLP. Therefore it is important to characterize elliptic curve trace from the security point of view.

Balasubramanian and Koblitz investigate that extension degrees required to apply both reductions for ECDLP on $G \in E(\mathbb{F}_q)$ with order n are the same if $n \nmid q - 1$([4]). Therefore without loss of generality we deal with only FR-reduction. By FR-reduction, ECDLP on $G \in E(\mathbb{F}_q)$ with order n is reduced to DLP on $\mathbb{F}_{q^k}^*$ if and only if $n | q^k - 1$. The probability that elliptic curves are vulnerable against FR-reduction, i.e. the extension degree k is small, is shown to be highly unlikely ([4]): FR-reduction is considered not to be threat in a realistic sense. Nevertheless all but supersingular and trace 2 elliptic curves have not been proved to be secure in a sense that they are strong against FR-reduction. There might exist another trace of elliptic curves which is reduced to at most 6, seriously low, degree extension field, whose trace might not be simple like 0 or 2. In fact, supersingular elliptic curves have rather special properties compared with ordinary elliptic curves([35]), which is thought to cause such a weak factor. However also in the case of ordinary elliptic curves, non-special elliptic curves, there might exist elliptic curve traces with a weak factor.

More importantly, the secure trace against FR-reduction has not been reported yet. Elliptic curves with the secure trace means that the reduced extension degree is always higher than a certain level. This means that the security of ECDLP over $E/\mathbb{F}_q$ is guaranteed by the security of widely known DLP on $\mathbb{F}_{q^k}^*$ with higher k than a certain level since FR-reduction gives an isomorphism between ECDLP over $E/\mathbb{F}_q$ and DLP based on a subgroup of $\mathbb{F}_{q^k}^*$ ([20]). In another light, the secure trace against FR-reduction is useful for construction of elliptic curve cryptosystems. Let's consider the following requirements: it is desirable that a domain parameter such as an elliptic curve or a basepoint should be chosen independently by each entity or by each application in order to keep security high([1]), and that such an initialization could be done more easily over lower CPU power or smaller memory like a smart card. In such requirements, it would be certainly desirable that an elliptic curve is constructable at least as easy as generating a prime number, which is a dominant step of RSA-key generation([28]). This is why explicit conditions of secure elliptic-curve traces is useful since we can construct easily an elliptic curve with a given specific trace. Apparently SEA algorithm([30, 32, 7, 10]) is not suitable since it requires rather large memory.

In this paper, we aim at characterizing elliptic curve traces by FR-reduction and investigate explicit conditions of traces vulnerable or secure against FR-reduction. Here we summarize our results on new explicit conditions of elliptic curve traces against FR-reduction.

- Let $E/\mathbb{F}_q$ be an elliptic curve with prime order and the trace t.
  ○ ECDLP on $E/\mathbb{F}_q$ is reduced to DLP on $\mathbb{F}_{q^3}^*$ by FR-reduction
  $\Leftrightarrow$ (i)(q, t) can be represented by $q = 12l^2 - 1$ and $t = -1 \pm 6l$ $(l \in \mathbb{Z})$, or
  (ii) (q, t) can be represented by $q = p^r$ $(r$ is even$)$ and $t = \pm\sqrt{q}$ (i.e. supersingular elliptic curves).
  ○ ECDLP on $E/\mathbb{F}_q$ is reduced to DLP on $\mathbb{F}_{q^4}^*$ by FR-reduction
  $\Leftrightarrow$ (i) (q, t) can be represented by $q = l^2 + l + 1$ and $t = -l, l + 1$ $(l \in \mathbb{Z})$, or
  (ii) (q, t) can be represented by $q = 2^r$ $(r$ is odd$)$ and $t = \pm\sqrt{2q}$ (i.e. supersingular elliptic curves).
  ○ ECDLP on $E/\mathbb{F}_q$ is reduced to DLP on $\mathbb{F}_{q^6}^*$ by FR-reduction
  $\Leftrightarrow$ (i) (q, t) can be represented by $q = 4l^2 + 1$ and $t = 1 \pm 2l$ $(l \in \mathbb{Z})$, or
  (ii) (q, t) can be represented by $q = 3^r$ and $t = \pm\sqrt{3q}$ $(r$ is odd$)$ (i.e. supersingular elliptic curve).

Up to the present, it has not been reported whether there exist another elliptic curve trace, except supersingular and trace 2, reduced to at most 6-degree extension field or not. However, our explicit conditions mean that prime-order elliptic curves are reduced to at most 6-degree extension field if and only if they satisfy at least one of the above conditions.

- Let ECDLP on $E(\mathbb{F}_q)$ with the trace t be reduced to DLP on $\mathbb{F}_{q^k}^*$.
  ○ If $t \geq 3$, then the extension degree k satisfies

$$k \geq \frac{\log q}{\log (t - 1)} - \varepsilon,$$

where ε is a real number such that $\frac{1}{10} > \varepsilon > 0$.
  ○ Let $t = 3$. Then the extension degree k satisfies

$$k > \log q - \varepsilon.$$

Theses are the first explicit elliptic-curve-trace conditions on which reduced extension degrees are always higher than a certain level. In the case of $E/\mathbb{F}_p$, dangerous elliptic curve traces happen to be equal to 0, 1 and 2. To the contrary, our result shows that $E/\mathbb{F}_p$ with trace 3 is secure against FR-reduction.

Furthermore, we present an algorithm to construct elliptic curves with the above conditions and present some examples.

This paper is organized as follows. Section 2 summarizes MOV- and FR-reductions. Section 3 investigates the above new explicit conditions vulnerable or secure against FR-reduction. Section 4 shows algorithms to construct elliptic curves with new explicit conditions. Section 5 presents some examples.

2 MOV-reduction and FR-reduction

In this section, we summarize MOV- and FR-reductions against ECDLP on $G \in E(\mathbb{F}_q)$ with order n. Here the n-torsion subgroup is denoted by $E[n] = \{P \in E \mid nP = \mathcal{O}\}$.

We compare MOV-reduction with FR-reduction. In MOV-reduction, ECDLP on G is reduced to DLP for the smallest integer k such that $E[n] \subset E(\mathbb{F}_{q^k})$. Thus supersingular elliptic curves can be efficiently reduced to $\mathbb{F}_{q^k}^*$ for $k \le 6$. On the other hand, in FR-reduction ECDLP on G is reduced to DLP for the smallest integer k such that $n|q^k - 1$. If $E[n] \subset E(\mathbb{F}_{q^k})$, then $n|q^k - 1$ ([31]). Therefore such an elliptic curve vulnerable against MOV-reduction is also vulnerable against FR-reduction. In fact FR-reduction works also for elliptic curves with trace 2 efficiently in addition to supersingular elliptic curves.

Table 1. Known explicit conditions for FR-reduction

$\mathbb{F}_q$ $(q = p^r)$	trace(E)	extension degree
$p \not\equiv 1 \pmod 4$ if r is even	0	2
$p \not\equiv 1 \pmod 3$ if r is even	$\pm\sqrt{q}$	3
$p = 2$ and r is odd	$\pm\sqrt{2q}$	4
$p = 3$ and r is odd	$\pm\sqrt{3q}$	6
r is even	$\pm 2\sqrt{q}$	1
$\forall q$	2	1

Balasubramanian and Koblitz ([4]) show that if n is a prime and $n \nmid q - 1$, then

$$E[n] \subset E(\mathbb{F}_{q^k}) \iff n \mid q^k - 1.$$

As a result there is no difference between MOV-reduction and FR-reduction except elliptic curves with trace 2. Without loss of generality, we deal with the only FR-reduction in this paper.

Table 1 summarizes known explicit conditions of elliptic curve traces for FR-reduction, where the extension degree k means that ECDLP on $E(\mathbb{F}_q)$ is reduced to DLP on a subgroup of $\mathbb{F}_{p^k}^*$.

As for the probability such that ECDLP is reduced to the lower degree extension field by FR-reduction, Balasubramanian and Koblitz show the next theorem.

Theorem 1 ([4]). *Let (p, E) be a randomly chosen pair of a prime p in the interval $M/2 \leq p \leq M$ and an elliptic curve $E/\mathbb{F}_p$ with prime order n. The probability Pr of $n|p^k - 1$ for some $k \leq (\log p)^2$ satisfies*

$$Pr < C \frac{(\log M)^9 (\log \log M)^2}{M}$$

for $C > 0$. ∎

Theorem 1 says that FR-reduction is highly unlikely to be efficient attack against ECDLP. However we note that Theorem 1 does not describe whether there might exist another explicit criterion of an elliptic curve trace vulnerable or secure against FR-reduction or not. From Table 1, we see that such an explicit condition that gives the extension degree higher than a certain level has not been reported.

3 New explicit conditions for elliptic curve traces

In this section, we investigate new explicit conditions of elliptic curve traces for FR-reduction. Table 2 shows our results, which will be discussed in the following sections.

Table 2. New explicit conditions for FR-reduction

$\mathbb{F}_q (q = p^r)$	$t = \text{trace}(E)$	extension degree k
$12l^2 - 1$	$-1 \pm 6l$	3
$l^2 + l + 1$	$-l, l + 1$	4
$4l^2 + 1$	$1 \pm 2l$	6
$\forall q$	$t \geq 3$	$k \geq \frac{\log q}{\log (t-1)} - \varepsilon$

3.1 New explicit conditions vulnerable against FR-reduction

In this section, we investigate new conditions of which ECDLP on $E/\mathbb{F}_q$ is reduced to DLP on seriously low extension field like $\mathbb{F}_{q^3}$, $\mathbb{F}_{q^4}$, and $\mathbb{F}_{q^6}$, which just occurs in the case of supersingular elliptic curves. Supersingular elliptic curves have rather special properties compared with ordinary elliptic curves([35]), which would no doubt cause such vulnerable factor. Here we show that there exist also vulnerable conditions of traces in the case of ordinary elliptic curves.

Let $E/\mathbb{F}_q$ be an elliptic curve with order $n = \#E(\mathbb{F}_q) = q + 1 - t$, where t is the trace of E. Then we show the conditions of which ECDLP on $E/\mathbb{F}_q$ is reduced to DLP on $\mathbb{F}_{q^3}^*$ by FR-reduction.

Theorem 2. *Let $E/\mathbb{F}_q$ be an elliptic curve with prime order n ($q > 64$). ECDLP on $E/\mathbb{F}_q$ is reduced to DLP on $\mathbb{F}_{q^3}^*$ by FR-reduction if and only if one of the following conditions holds,*

(i) (q,t) can be represented by $q = 12l^2 - 1$ and $t = -1 \pm 6l$ ($l \in \mathbb{Z}$).
(ii) (q,t) can be represented by $q = p^r$ (r is even) and $t = \pm\sqrt{q}$ (i.e. super-singular elliptic curves).

proof: We assume that ECDLP on $E/\mathbb{F}_q$ with prime order n is reduced to DLP on $\mathbb{F}_{q^3}^*$ by FR-reduction. From the condition of FR-reduction, n satisfies that $n | q^3 - 1$ and $n \nmid q - 1$ since n is a prime. Therefore there is an integer λ such that $q^2 + q + 1 = \lambda n$. By setting $n = q + 1 - t$ and $q^2 + q + 1 = (q+1)^2 - t^2 + t^2 - q$, we get the following equation,

$$(q + 1 - t)(q + 1 + t - \lambda) = q - t^2. \tag{1}$$

By Hasse's Theorem, the trace t satisfies $|t| \le 2\sqrt{q}$. Hence, (1) satisfies

$$-3 \le (1 + \frac{1}{q} - \frac{t}{q})(q + 1 + t - \lambda) \le 1. \tag{2}$$

For the assumption of $q, t \in \mathbb{Z}$ and $q > 64$, we conclude that (q,t) satisfies one of the following equations,

$$q + 1 + t - \lambda = -3, -2, -1, 0, 1 \tag{3}$$

By substituting (3) to (1), we get that (q,t) satisfies the following equations,

$$t^2 + 3t - 4q - 3 = 0, \tag{4}$$
$$t^2 + 2t - 3q - 2 = 0, \tag{5}$$
$$t^2 + t - 2q - 1 = 0, \tag{6}$$
$$t^2 - q = 0, \tag{7}$$
$$t^2 - t + 1 = 0. \tag{8}$$

By simple discussion on the existence of integer solutions for congruence equations, we get that $(t, q) \in \mathbb{Z} \times \mathbb{Z}$ exists if and only if (t, q) satisfies (5) or (7).

In the case of (5), (t, q) is expressed by $t = -1 \pm 6l$ and $q = 12l^2 - 1$ for $l \in \mathbb{Z}$ since $q = p^r$ for a prime p, and $t \in \mathbb{Z}$ satisfies

$$t = -1 \pm \sqrt{3(q + 1)}.$$

In the case of (7), (t, q) is expressed by $t = \pm\sqrt{q} = \pm\sqrt{p^r}$ for even integers r. This is just a supersingular elliptic curve.

Conversely, if a prime-order elliptic curve $E/\mathbb{F}_q$ satisfies (i) or (ii) in Theorem 2, then $\#E(\mathbb{F}_q) = n$ satisfies $n | q^3 - 1$. Therefore ECDLP on $E/\mathbb{F}_q$ is reduced to DLP on $\mathbb{F}_{q^3}^*$. ∎

Note that possible order of elliptic curves is given by Deuring([9]) and Waterhouse([17]). In the case of $E/\mathbb{F}_p$, there exactly exists an elliptic curve of type (i) in Theorem 2. In the case of $\mathbb{F}_{2^r}$, there does not exist any elliptic curve of type (i) in Theorem 2, but in the case of $\mathbb{F}_{p^r}$ ($p \ge 3$) there exists.

We get the next corollary easily from Theorem 2.

Corollary 1 *Let $E/\mathbb{F}_q$ be an elliptic curve with trace t. If (q,t) can be represented by $q = 12l^2 - 1$ and $t = -1 \pm 6l(l \in \mathbb{Z})$, then ECDLP on $E(\mathbb{F}_q)$ is reduced to DLP on $\mathbb{F}_{q^3}^*$ by FR-reduction.*

proof: Here we set $\#E(\mathbb{F}_q) = n$ and let order of $G \in E(\mathbb{F}_q)$ be m. Then m divides n. From the assumption, $n = 12l^2 \pm 6l + 1$. This yields $12l^2 \equiv \pm 6l - 1 \pmod{n}$. Then by using the relation of both $12l^2 \equiv \pm 6l - 1 \pmod{n}$ and $q = 12l^2 - 1$, we get

$$
\begin{aligned}
q^3 - 1 &= (12l^2 - 2)((12l^2 - 1)^2 + 12l^2) \\
&\equiv (12l^2 - 2)((\pm 6l - 2)^2 + (\pm 6l - 1)) \pmod{n} \\
&\equiv (12l^2 - 2)(36l^2 \mp 18l + 3) \pmod{n} \\
&\equiv 0 \pmod{n} \\
&\equiv 0 \pmod{m}.
\end{aligned}
$$

Therefore ECDLP on $\forall < G > \subset E(\mathbb{F}_q)$ is reduced to DLP on $\mathbb{F}_{q^3}^*$ by FR-reduction. $\blacksquare$

Next we show the conditions of which ECDLP on $E/\mathbb{F}_q$ is reduced to DLP on $\mathbb{F}_{q^4}^*$ by FR-reduction.

Theorem 3. *Let $E/\mathbb{F}_q$ be an elliptic curve with prime order n ($q > 36$). ECDLP on $E/\mathbb{F}_q$ is reduced to DLP on $\mathbb{F}_{q^4}^*$ by FR-reduction if and only if one of the following conditions holds,*
(i) (q,t) can be represented by $q = l^2 + l + 1$ and $t = -l, l+1$ for $l \in \mathbb{Z}$.
(ii) (q,t) can be represented by $q = 2^r$ (r is odd) and $t = \pm\sqrt{2q}$ (i.e. supersingular elliptic curves).

proof: We assume that ECDLP on $E/\mathbb{F}_q$ with prime order n is reduced to DLP on $\mathbb{F}_{q^4}^*$ by FR-reduction. From the condition of FR-reduction, n satisfies that $n|q^4 - 1$ and $n \nmid q^2 - 1$ since n is a prime. Therefore there is an integer λ such that $q^2 + 1 = \lambda n$. In the same way as Theorem 2, we get the following equation,

$$(q + 1 - t)(q + 1 + t - \lambda) = 2q - t^2. \tag{9}$$

From Hasse's Theorem, (9) satisfies that

$$-2 \leq (1 + \frac{1}{q} - \frac{t}{q})(q + 1 + t - \lambda) \leq 2. \tag{10}$$

In the same discussion as Theorem 2, we get that $(t,q) \in \mathbb{Z} \times \mathbb{Z}$ exists if and only if (t,q) satisfies

$$t^2 - 2q = 0, \tag{11}$$

$$t^2 - t - q + 1 = 0. \tag{12}$$

In the case of (11), t satisfies $t = \pm\sqrt{2q} = \pm\sqrt{2p^r}$ for $p = 2$ and an odd positive integer r. This is just a supersingular elliptic curve. In the case of (12), (t,q) is expressed by $t = -l, l + 1$ and $q = l^2 + l + 1$ for $l \in \mathbb{Z}$ since $t \in \mathbb{Z}$ satisfies

$$t = \frac{1 \pm \sqrt{4q - 3}}{2}.$$

Apparently if a prime-order elliptic curve $E/\mathbb{F}_q$ satisfies (i) or (ii) in Theorem 3, then ECDLP on $E/\mathbb{F}_q$ is reduced to DLP on $\mathbb{F}_{q^4}$. ∎

The next corollary follows from Theorem 3.

Corollary 2 *Let $E/\mathbb{F}_q$ be an elliptic curve with trace t. If (q, t) can be represented by $q = l^2 + l + 1$ and $t = -l, l + 1$ for $l \in \mathbb{Z}$, then ECDLP on $E(\mathbb{F}_q)$ is reduced to DLP on $\mathbb{F}_{q^4}^*$ by FR-reduction.*

In the same way as Theorems 2 and 3, the explicit conditions of which ECDLP on $E/\mathbb{F}_q$ is reduced to DLP on $\mathbb{F}_{q^6}^*$ by FR-reduction are shown as follows.

Theorem 4. *Let $E/\mathbb{F}_q$ be an elliptic curve with prime order n. ECDLP on $E/\mathbb{F}_q$ is reduced to DLP on $\mathbb{F}_{q^6}^*$ by FR-reduction if and only if one of the following conditions holds,*
(i) (q, t) can be represented by $q = 4l^2 + 1$ and $t = 1 \pm 2l$ for $l \in \mathbb{Z}$.
(ii) (q, t) can be represented by $q = 3^r$ and $t = \pm\sqrt{3q}$ for an odd integer r (i.e. supersingular elliptic curve).

Corollary 3 *Let $E/\mathbb{F}_q$ be an elliptic curve with trace t. If (q, t) can be represented by $q = 4l^2 + 1$ and $t = 1 \pm 2l$ for $l \in \mathbb{Z}$, then ECDLP on $E/\mathbb{F}_q$ is reduced to DLP on $\mathbb{F}_{q^6}^*$ by FR-reduction.*

Remark 1 *Theorems 2, 3, and 4 use the fact that the k-th cyclotomic polynomial is decomposed into at most 2-degree irreducible polynomials over $\mathbb{Z}$ in the case of $k = 3, 4$, and 6, respectively. For other cases of k, the same discussion might be used if the k-th cyclotomic polynomial is decomposed into irreducible polynomials with rather small degrees over $\mathbb{Z}$.*

3.2 New explicit conditions secure against FR-reduction

In this section, from a secure point of view we investigate a new explicit condition of elliptic curve traces on which the reduced extension degree is always higher than a certain level. As for the known results on $E/\mathbb{F}_p$, dangerous elliptic curves happen to be small traces like 0, 1 and 2. However, on the contrary, our results of Theorems 2, 3 and 4 suggest that the elliptic curve trace whose order is near upper bound in Hasse's Theorem([35]) should be vulnerable. As a result, we show that the extension degree is higher than a certain level when the positive trace except for $t = 0, 1$ and 2 is small enough.

Theorem 5. Let $E/\mathbb{F}_q$ be an elliptic curve with prime order n ($q > 861$), ECDLP on $E(\mathbb{F}_q)$ be reduced to DLP on $\mathbb{F}_{q^k}^*$, and t be the elliptic curve trace. If $t \geq 3$, then the extension degree k satisfies

$$k \geq \frac{\log q}{\log(t-1)} - \varepsilon,$$

where ε is a real number such that $\frac{1}{10} > \varepsilon > 0$.

proof: ECDLP on $E(\mathbb{F}_q)$ is reduced to DLP on $\mathbb{F}_{q^k}$ if and only if

$$q^k \equiv 1 \pmod{n}. \tag{13}$$

By substituting $n = q + 1 - t$ to (13), we get that k is the smallest integer satisfying

$$(t-1)^k \equiv 1 \pmod{n}. \tag{14}$$

From the assumption and Hasse's theorem, t satisfies $3 \leq t \leq 2\sqrt{q} \ll q \approx n$. Therefore

$$1 < (t-1)^k < n < n+1$$

if $1 \leq k < \frac{\log n}{\log(t-1)}$. Then it follows that the smallest integer k such that $(t-1)^k \equiv 1 \pmod{n}$ is greater than or equal to $\frac{\log n}{\log(t-1)}$. Furthermore by substituting $n = q + 1 - t$, we get that

$$k \geq \frac{\log q}{\log(t-1)} - \varepsilon,$$

where $\varepsilon = -\log_{t-1}\left(1 - \frac{t-1}{q}\right)$. By using the relation of $3 \leq t \leq 2\sqrt{q}$, we get easily that

$$0 < \varepsilon < -\log_{t-1}\left(1 - \frac{2}{\sqrt{q}} + \frac{1}{q}\right) < \frac{1}{10},$$

if $q > 861$. Apparently the larger q is, the smaller ε is. Thus the lower bound of extension degree is given by

$$k \geq \frac{\log q}{\log(t-1)} - \varepsilon.$$

∎

The above theorem gives a lower bound of extension degree k in the case of small $t \geq 3$, which ensures the security of ECDLP over $E/\mathbb{F}_q$ by that of widely known DLP on $\mathbb{F}_{q^k}^*$.

The next corollary easily follows from Theorem 5.

Corollary 4 *Let $E/\mathbb{F}_q$ be an prime order elliptic curve with $t = 3$ ($q > 861$) and ECDLP on $E(\mathbb{F}_q)$ be reduced to DLP on $\mathbb{F}_{q^k}^*$. Then the extension degree k satisfies*

$$k > \log q - \varepsilon,$$

where ε is a real number such that $\frac{1}{10} > \varepsilon > 0$.

Remark 2 *The extension degree $k < \log q$ means that FR-reduction gives a subexponential attack against ECDLP under the index calculus method([8]), which runs over any field $\mathbb{F}_q$ in time $L_q[1/2, c] = \exp((c + O(1))(\log q)^{1/2}(\log\log q)^{1/2})$. On the other hand, the extension degree $k < (\log q)^2$ means that FR-reduction gives a subexponential attack against ECDLP under the number field sieve([14]) which runs over some fields $\mathbb{F}_q$ in time*

$$L_q[1/3, c] = \exp((c + O(1))(\log q)^{1/3}(\log\log q)^{2/3}).$$

Therefore in order to construct enough secure elliptic curve cryptosystems it would be desirable that $k \geq (\log q)^2$. However the condition of $k \geq \log q$ in Corollary 4 is not highly optimistic if we estimate under a rather realistic assumption of the discrete logarithm algorithm for definition fields of elliptic curves([29, 8]).

In the case of prime-order elliptic curves $E/\mathbb{F}_p$ with $t = 3$, we will easily see that the following strict condition also holds: the extension degree is just exponential.

Corollary 5 *Let $E/\mathbb{F}_p$ be a prime-order elliptic curve with $t = 3$ (i.e. $\#E(\mathbb{F}_p) = p - 2$ is prime). If 2 is a primitive root in $\mathbb{F}_{p-2}$, then the extension degree k such that ECDLP on $E(\mathbb{F}_p)$ is reduced to DLP on $\mathbb{F}_{p^k}^*$ satisfies $k = p - 3$.*

4 Algorithm

In this section, we describe algorithms to construct elliptic curves vulnerable or secure against FR-reduction in Section 3 and confirm that such elliptic curves exist in a realistic sense (i.e. constructable). From the point of view of theoretical interest, each construction is deeply related to each famous number theory problem: the former is a problem of finding integer solutions of Pell's equation([16]), and the latter is a problem of finding twin prime numbers.

4.1 Construction of elliptic curves reducible to lower extension degree

Here we present an algorithm to construct elliptic curves over $\mathbb{F}_p$ in Corollary 1 since Theorem 2 is a special case of Corollary 1. By using the CM-method([2])[1],

---

[1] The procedure of the CM-method includes a step of computing the Hilbert class polynomials([23]), $P_d(x)$. The computation of the Hilbert class polynomials are not so easy if the degree of the Hilbert class polynomial, $\deg(P_d(x))$, namely the class number is large. Therefore we usually fix d and so $P_d(x)$ beforehand in order to avoid the computation of $P_d(x)$ as we will see in Algorithm 2. In another way, we may make use of the recent researches([5, 6]) on the construction of the CM elliptic curves by both the CM tests and liftings instead of the CM-method.

the dominant step of construction of elliptic curves with both $p = 12l^2 - 1$ and $t = -1 \pm 6l(l \in \mathbb{Z})$ is finding integer solutions (l, y) of $12l^2 \pm 12l - 5 = dy^2$ for a given positive integer $d \equiv 3 \pmod 4$, which is easily transformed into finding integer solutions of an indeterminate equation

$$x^2 - 3dy^2 = 24. \tag{15}$$

From the elementary number theory([37]), all integer solutions (x, y) of (15) is given by

$$x + y\sqrt{3d} = (x_1 + y_1\sqrt{3d})(t_0 + u_0\sqrt{3d})^n,$$

where (t_0, u_0) is the *minimum positive integer solution* on $\epsilon = t_0 + u_0\sqrt{3d} > 0$ of Pell's equation,

$$T^2 - 3dU^2 = 1, \tag{16}$$

and (x_1, y_1) is an integer solution of (15) in the following domain *Dom*,

$Dom = \{(x, y) | \sqrt{24} \le x < t_0\sqrt{24}, \, 0 \le x < u_0\sqrt{24}\}.$

Here we call two integer solutions (x, y) and (x', y') of (15) are associated if

$$x + y\sqrt{3d} = \pm(x' + y'\sqrt{3d})(t_0 + u_0\sqrt{3d})^n$$

for $\exists n \in \{0, \pm 1, \pm 2, \cdots\}$.

After finding an integer solution (x, y) of (15) in the above procedure, the construction of elliptic curves $E/\mathbb{F}_p$ with the trace t easily follows the CM-method. In order to find integer solutions efficiently, we need some techniques specific to (15). Here we show only specific techniques, all of which are proved by simple discussion on the existence of integer solutions for congruence equations.

Lemma 1. *If there exists an integer solution* (l, y) *of* $12l^2 \pm 12l - 5 = dy^2$, *then* $d \equiv 19 \pmod{24}$.

proof: From $dy^2 = 12l^2 \pm 12l - 5 = 12l(l \pm 1) - 5 \equiv 19 \pmod{24}$, we get $dy^2 \equiv 19 \pmod{24}$. By using the fact of $y^2 \equiv 0, 1, 4, 9, 12, 16 \pmod{24}$, we get that $d \equiv 19 \pmod{24}$ if there exists an integer solution of $dy^2 \equiv 19 \pmod{24}$. ∎

Lemma 2. *Let* $d \in \mathbb{Z}$ *be* $d \equiv 19 \pmod{24}$. *If there exists an integer solution* (x_0, y_0) *of (15), then* $\gcd(x_0, y_0) = 1$.

proof: Let (x, y) be an integer solution of (15) and $\gcd(x, y) = g > 1$. Then $g = 2$ since $g^2 | 24$. So we can set $x = 2x'$ and $y = 2y'$ $(x', y' \in \mathbb{Z})$ with $\gcd(x', y') = 1$. From the assumption of $d \equiv 19 \pmod{24}$, (x', y') satisfies $x'^2 + 3y'^2 \equiv 6 \pmod{12}$. This is contradictory because there does not exist any integer solution (x, y) of $x^2 + 3y^2 \equiv 6 \pmod{12}$. ∎

Corollary 6 *Let $d \in \mathbb{Z}$ be $d \equiv 19$ (mod 24). If there exists an integer solution (x_0, y_0) of (15), then both x_0 and y_0 are odd.*

proof: This follows from Lemma 2. ∎

Lemma 3. *Let $d \in \mathbb{Z}$ be $d \equiv 19$ (mod 24) and (x_0, y_0) be a set of integer solutions of (15). Then both (x_0, y_0) and $(x_0, -y_0)$ are not associated.*

proof: Two solutions (x, y) and (x', y') of (15) are associated if and only if $xy' - x'y \equiv 0$ (mod 24) (see Section 34 in [37]). Therefore if both (x_0, y_0) and $(x_0, -y_0)$ are associated, then $2x_0 y_0 \equiv 0$ (mod 24). This is contradictory to Corollary 6. ∎

Lemma 4. *Let $d \in \mathbb{Z}$ be $d \equiv 19$ (mod 24). Then there are at most two integer solutions in Dom for (15).*

proof: From Lemma 2, there exist an integer solution s satisfying the following conditions:
$12d = s^2 - 96m$, $\gcd(24, s, m) = 1$, $s^2 \equiv 12d$ (mod 96), and $-24 \leq s < 24$,
if there exist an integer solution (x, y) in Dom for (15)(see Section 35 in [37]). From the simple discussion on the existence of integer solutions for congruence equations, there are at most two integer solutions s satisfying the above conditions. Therefore there are at most two integer solutions in Dom for (15). ∎
The next proposition follows from Lemmas 3 and 4.

Proposition 1 *Let $d \in \mathbb{Z}$ be $d \equiv 19$ (mod 24). Then there exist just two sets of integer solutions in Dom for (15) if there exist.*

Here we give the algorithm as follows:

Algorithm 1 Given the upper bound $UP > 0$ on a prime p, this algorithm outputs (p, d, l), or *fail* if such a (p, d, l) does not exist.

1. Choose a positive integer d such that $d \equiv 19$ (mod 24).
2. Find the minimum positive integer solution (t_0, u_0) of (16).
3. Find an integer solution $(x, y) \in Dom$ of (15), if exists. Otherwise, output *fail* and terminate the algorithm.
4. For $n \geq 1$, set x_n, y_n in such a way that
   $x_n + y_n \sqrt{3d} := (x + y\sqrt{3d})(t_0 + u_0\sqrt{3d})^n$.
5. Set $l_{1,n} := (x_n - 3)/6$, $l_{2,n} := (x_n + 3)/6$, $p_{1,n} := 12l_{1,n}^2 - 1$, and $p_{2,n} = 12l_{2,n}^2 - 1$.
6. If $p_{1,n} > UP$ and $p_{2,n} > UP$, then output *fail* and terminate the algorithm.
7. If $p_{1,n}$ or $p_{2,n}$ is prime, then output $(p_{1,n}, d, l_{1,n})$ or $(p_{1,n}, d, l_{2,n})$ respectively, and terminate the algorithm. Otherwise goto 4.

4.2 Construction of elliptic curves reducible to higher extension degree

Here we present an algorithm to construct elliptic curves $E/\mathbb{F}_p$ with $t = 3$ in Corollary 4, in which the CM-method is also used in the same way as Section 4.1. By using the CM-method, the dominant steps of construction of prime-order elliptic curves $E/\mathbb{F}_p$ with $t = 3$, namely $\#E(\mathbb{F}_p) = p - 2$, are finding a prime number $p = dl^2 + dl + \frac{d+9}{4}$ with $l \in \mathbb{Z}$ for an given positive integer $d \equiv 3 \pmod 4$, and checking $p - 2$ is also prime.

In this case we can easily show the following condition of d.

Lemma 5. *Let* $p \in \mathbb{Z}$ *be* $p = dl^2 + dl + \frac{d+9}{4}$ *with a positive integer* $d \equiv 3$ (mod 4). *If both* p *and* $p - 2$ *are prime, then* $d \equiv 19 \pmod{24}$.

proof: For the assumption of $d \equiv 3 \pmod 4$, we set $d = 3 + 4m$ ($m \in \mathbb{Z}$). Then

$$p = dl^2 + dl + \frac{d+9}{4}$$
$$= dl(l+1) + (m+3) \tag{17}$$
$$\equiv m+1 \pmod 2. \tag{18}$$

Since p is prime, $m \equiv 0 \pmod 2$ from (18). So we can set $d = 3 + 8m'$ ($\exists m' \in \mathbb{Z}$). On the other hand, we get $p \equiv 1 \pmod 6$ since both p and $p - 2$ are prime and also get easily $l(l+1) \equiv 0, 2 \pmod 6$ for $\forall l \in \mathbb{Z}$. If $l(l+1) \equiv 0 \pmod 6$, then $m' \equiv 2 \pmod 3$ from (17). This yields $d \equiv 19 \pmod{24}$. If $l(l+1) \equiv 2 \pmod 6$, then this yields contradictory. In this way we get $d \equiv 19 \pmod{24}$. ∎

Here we give the algorithm as follows:

Algorithm 2 Given the upper bound $UP > 0$ on a prime p, this algorithm outputs the prime-order elliptic curve $E/\mathbb{F}_p$ with $t = 3$, or *fail* if such an $E/\mathbb{F}_p$ does not exist.

1. Choose a positive integer d such that $d \equiv 19 \pmod{24}$.
2. Set $p = dl(l+1) + \frac{d+9}{4}$ for $\mathbb{Z} \ni l > 0$ such that $l \equiv 0, 2 \pmod 3$.
3. If $p > UP$, then output *fail* and terminate the algorithm. Otherwise goto step 4.
4. If both p and $p - 2$ are prime, then goto step 5. Otherwise goto step 2 and try the next l.
5. Compute the Hilbert class polynomial $P_d(x)$.
6. Solve a root j_0 of $P_d(x) \equiv 0 \pmod p$.
7. Construct two elliptic curves E_{j_0} and E'_{j_0},
   $E_{j_0} : y^2 = x^3 + a_{j_0} x + b_{j_0}$, $E'_{j_0} : y^2 = x^3 + a_{j_0} c^2 x + b_{j_0} c^3$,
   where $a_{j_0} = \frac{3j_0}{1728-j_0} \pmod p$, $b_{j_0} = \frac{2j_0}{1728-j_0} \pmod p$,
   and c is any quadratic non-residue in $\mathbb{F}_p$.
8. Output $E \in \{E_{j_0}, E'_{j_0}\}$ with $\#E(\mathbb{F}_p) = p - 2$ and terminate the algorithm.

Note that the step 8 can be performed easily: output E such that $(p-2)G = \mathcal{O}$ for $E(\mathbb{F}_p) \ni \exists G \neq \mathcal{O}$.

5 Experimental results

In this section, we present some examples in both vulnerable and secure cases.

5.1 Elliptic curves reducible to lower extension degree

We present one example which satisfies the condition of Corollary 1. We searched elliptic curves $E/\mathbb{F}_p$ in the range of $0 < p < 2^{1000}$ by using Algorithm 1. Our modulo arithmetic uses the GNU MP Library GMP([38]). The platform is an Alpha 21264(500 MHz/C Compiler for Digital UNIX). It took on the average 0.101 sec to find an elliptic curve $E/\mathbb{F}_p$ in the case of $d = 19$. We have also confirmed experimentally that vulnerable elliptic curves with new explicit conditions are constructable systematically in the same way as supersingular or trace 2 elliptic curves. This means that even in the case of ordinary elliptic curves, we must check FR-conditions.

Recently some researches([21, 22]) on a protocol using an elliptic curve $E/\mathbb{F}_p$ with the *computable* FR-reduction have been proposed, in which an elliptic curve $E/\mathbb{F}_p$ reduced to $\mathbb{F}_{p^k}$ with the computable lower extension degree is desired. Our approach is also deeply related to their researches.

Example 1
$E/\mathbb{F}_p : x^3 + ax + b$
$\quad p = 9\ 08761\ 00379\ 04279\ 08077\ 54895\ 57583\ 80356\ 67582\ 90265\ 31247\ \textit{(170-bit)}$,
$\quad a = 8\ 18416\ 34259\ 48882\ 91485\ 04408\ 88116\ 40789\ 05308\ 57899\ 75506$,
$\quad b = 6\ 66070\ 44332\ 39783\ 49780\ 03588\ 18034\ 13282\ 86571\ 48420\ 57992$,
$\quad t = -5\ 22138\ 20118\ 54029\ 93899\ 01413$,
$\#E(\mathbb{F}_p) = 7^2 * 313 * n$,
$\quad n = 59\ 25285\ 28258\ 73893\ 72612\ 30363\ 15589\ 78126\ 20544\ 05453\ \textit{(156-bit)}$.

5.2 Elliptic curves reducible to higher extension degree

We present experimental results and some examples of elliptic curves in Corollaries 4 and 5. We have confirmed that secure elliptic curves with new explicit conditions are constructible systematically. Table 3 shows numerical results of twin primes $(p, p - 2)$ with $p = dl^2 + dl + \frac{d+9}{4}$, which was searched in the range of $2^{76} - 2^{20} \le l \le 2^{76} + 2^{20}$. Our modulo arithmetic uses the GNU MP Library GMP([38]). The platform is an Alpha 21264(500 MHz/C Compiler for Digital UNIX). It took on the average 0.053 sec to find a pair of $(p, p - 2)$ in the case of $d = 163$. For other cases of d, we could find such a pair of primes on the average $0.064 \sim 1.402$ sec. Fig.1 shows the plot of Table 3 from the point of view of $\deg(P_d(x))$ and the size of d on $P_d(x)$. From our experimental result, we have found a heuristic property that the number of twin primes are closely related to two factors, $\deg(P_d(x))$ and the size of d on $P_d(x)$. If we fix the size of d, then the larger $\deg(P_d(x))$ is, the less twin primes are found. If we fix $\deg(P_d(x))$, then the larger the size of d is, the more twin primes are found. s

To make a comparison to RSA key generation, we have compared twin-prime-generation times and RSA-prime-generation times. From the point of view of the

Table 3. The number of twin primes $(p, p-2)$

d	$\deg(P_d(x))$	# twin primes	times (sec)
19	1	190	0.550
43	1	1,157	0.094
67	1	1,902	0.064
91	2	450	0.365
115	2	1,036	0.209
139	3	139	0.323
163	1	5,158	0.053
187	2	1,402	0.107
211	3	292	1.401
235	2	2,523	0.089
259	4	247	0.348
283	3	645	0.234
307	3	696	0.134
331	3	1,458	0.103
355	4	635	0.261
379	3	1,583	0.074
403	2	3,392	0.069

$$p = dl^2 + dl + \tfrac{d+9}{4} \ (2^{76} - 2^{20} \le l \le 2^{76} + 220)$$

same security level, we consider the following three conditions of bit size on (elliptic curve cryptosystem, RSA): (160, 1024), (224, 2,048) and (256, 3,072)([33]). Table 4 shows both of twin-prime-generation times and RSA-prime-generation times, where the size of RSA-prime is just half size of the above security level. As for the twin-prime generation, we dealt with four cases of $d = 163, 427, 907, 1555$ that correspond to $\deg(P_d(x))$=1, 2, 3, 4 respectively. These characters are also used in Table 4 and Fig 2. We searched for 1,000 twin primes by Algorithm 2 and computed the average times. As for the RSA-prime generation, we searched for 1,000 RSA primes by simply performing a primality test among odd numbers, and computed the average times. The platform is also an Alpha 21264 (500 MHz / C compiler for Digital UNIX). For the primality test, we made use of Miller-Rabin's probablistic test in GNU MP Library GMP. Fig 2 shows the plot of Table 4. Note that the vertical axis is represented in logarithm. We can easily see that the generation of twin primes is faster than that of RSA primes in any case.

We present $E/\mathbb{F}_p : y^2 = x^3 + ax + b$ with $t = 3$ in the following. In Examples 2 ~ 4, 2 is a primitive root in $\mathbb{F}_{p-2}$.

Example 2

$E_1/\mathbb{F}_p : y^2 = x^3 + a_1 x + b_1$, $(|p| = 159 - bit)$

$p = 519\ 51816\ 01449\ 69382\ 38659\ 23754\ 49686\ 02163\ 04833\ 66071$,

$n = 519\ 51816\ 01449\ 69382\ 38659\ 23754\ 49686\ 02163\ 04833\ 66069$,

$a_1 = 35\ 29380\ 82819\ 03345\ 16798\ 59515\ 21747\ 57876\ 817006\ 32697$,

$b_1 = 408\ 46477\ 52610\ 12095\ 24877\ 04686\ 28212\ 53233\ 12948\ 77155$,

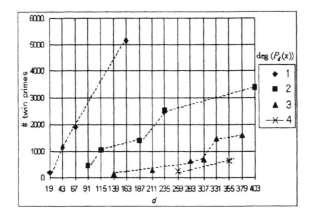

Fig. 1. Relations between # twin primes and $P_d(x)$

Example 3
$E_1/\mathbb{F}_p : y^2 = x^3 + a_1 x + b_1$, $(|p| = 159 - bit)$

$p = 793\ 54971\ 71445\ 13671\ 92705\ 06772\ 26939\ 83458\ 80422\ 30471$,

$n = 793\ 54971\ 71445\ 13671\ 92705\ 06772\ 26939\ 83458\ 80422\ 30469$,

$a_1 = 622\ 32433\ 75781\ 36504\ 38145\ 80347\ 56708\ 57012\ 73203\ 93428$,

$b_1 = 679\ 39946\ 41002\ 62226\ 89665\ 55822\ 46785\ 65828\ 08943\ 39109$,

Example 4
$E/\mathbb{F}_p : y^2 = x^3 + ax + b$, $(|p| = 240 - bit)$

$p = 112\ 49846\ 54526\ 86189\ 73518\ 65205\ 55113\ 42541\ 99281\ 27068\ 83806\ 23265$
$87119\ 55023\ 07023$,

$n = 112\ 49846\ 54526\ 86189\ 73518\ 65205\ 55113\ 42541\ 99281\ 27068\ 83806\ 23265$
$87119\ 55023\ 07021$,

$a = 52\ 37381\ 80880\ 77183\ 56601\ 62811\ 25609\ 08710\ 91667\ 71974\ 15904\ 90057$
$09224\ 69377\ 60775$,

$b = 34\ 91587\ 87253\ 84789\ 04401\ 08540\ 83739\ 39140\ 61111\ 81316\ 10603\ 26704$
$72816\ 46251\ 73850$.

Example 5
$E_1/\mathbb{F}_p : y^2 = x^3 + a_1 x + b_1$, $(|p| = 240 - bit)$

Table 4. Times of twin-prime generation and RSA-prime generation (sec)

bit size (twin primes, RSA)		(160, 1024)	(224, 2048)	(256, 3072)
RSA		0.098	0.826	16.274
	1	0.047	0.130	0.242
Twin primes	2	0.058	0.164	0.265
	3	0.057	0.274	0.401
	4	0.057	0.175	0.272

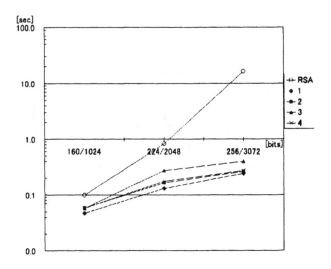

Fig. 2. Times of twin-prime generation and RSA-prime generation

$p = $ 145 62684 79172 80895 91487 33486 94032 72646 08218 46342 12380 03553 12226 43548 52871,

$n = $ 145 62684 79172 80895 91487 33486 94032 72646 08218 46342 12380 03553 12226 43548 52869,

$a_1 = $ 144 44371 02824 33267 37769 11780 11326 91187 09134 83450 79361 18648 91066 43377 85210,

$b_1 = $ 50 11979 94855 57136 68786 73438 08285 32827 34850 99302 48151 81056 65622 14743 74505,

6 Conclusion

In this paper, we have shown some new explicit conditions of elliptic curve traces vulnerable or secure against FR-reduction. We have also presented algorithms to construct elliptic curves with our new explicit conditions. Especially our new secure elliptic curve realizes rather light initialization, which sets up a pair of elliptic curve and basepoint.

Acknowledgments

The authors are grateful to anonymous referees for invaluable comments.

References

1. R. Anderson and R. Needham, "Robustness principles for public key protocols", *Advances in Cryptology-Proceedings of CRYPTO'95*, Lecture Notes in Computer Science, **963**(1995), Springer-Verlag, 236-247.

2. A. O. L. Atkin and F. Morain, "Elliptic curves and primality proving", *Math. of Computation*, **61**(1993), 29-68.

3. K. Araki and T. Satoh "Fermat quotients and the polynomial time discrete log algorithm for anomalous elliptic curves", *Commentarii Math. Univ. St. Pauli.*, vol. **47** (1998), 81-92.

4. R. Balasubramanian and N. Koblitz, "The Improbability That an Elliptic Curve Has Subexponential Discrete Log Problem under the Menezes-Okamoto-Vanstone Algorithm", J. Cryptology, **11** (1998), 141-145.

5. J. Chao, O. Nakamura, K. Sobataka, and S. Tsujii, "Construction of secure elliptic curves with CM tests and lifting", *Advances in Cryptology-Proceedings of ASIACRYPT'98*, Lecture Notes in Computer Science, **1514**(1998), Springer-Verlag, 95-109.

6. J. Chao, M. Hosoya, K. Sobataka, and S. Tsujii, "Construction of Elliptic Cryptosystems Using Ordinary Lifting", *Proceeding of the 1999 Symposium on Cryptography and Information Security*, 163-166.

7. J. M. Couveignes and F. Morain, "Schoof's algorithm and isogeny cycles", *Proceedings of the ANTS-I*, Lecture Notes in Compute Science, **877** (1994), Springer-Verlag, 43-58.

8. T. Denny, O. Schirokauer and D. Weber, "Discrete logarithms: the effectiveness of the index calculus method", *Proceedings of ANTSII*, Lecture Notes in Computer Science, **1122**(1996), Springer-Verlag, 337-361.

9. M. Deuring, "Die typen der multiplikatorenringe elliptischer funktionenkörper", *Abh. Math. Sem. Hamburg*, **14**(1941), 197-272.

10. N. D. Elkies, "Explicit isogenies", Preprint, 1991

11. G. Frey and H. G. Rück, "A remark concerning m-divisibility and the discrete logarithm in the divisor class group of curves", *Mathematics of computation*, **62**(1994), 865-874.

12. "Proposed federal information processing standard for digital signature standard (DSS)", *Federal Register*, **56** No. 169, 30 Aug 1991, 42980–42982.

13. T. ElGamal, "A public key cryptosystem and a signature scheme based on discrete logarithms", *IEEE Trans. Inform. Theory*, **IT-31** (1985), 469–472.

14. D. M. Gordon, "Discrete logarithms in $GF(p)$ using the number field sieve", *SIAM J. on Discrete Math.*, **6**(1993), 124-138.

15. R. Harasawa, H. Imai, J. Shikata, J. Suzuki, "Comparing the MOV and FR Reductions in Elliptic Curve Cryptography", *Advances in Cryptology-Proceedings of EUROCRYPT '99*, Lecture notes in Computer Science, **1592** (1999), 190-205.

16. K. Ireland and M. Rosen, *A classical introduction to modern number theory*, GTM 84, Springer-Verlag, New-York, 1982.

17. *IEEE P1363 Working Draft*, June 16, 1998.

18. N. Kanayama, T. Kobayashi, T. Saito, and S. Uchiyama "Remarks on elliptic curve discrete logarithm problems", *IEICE Trans.*, Fundamentals. vol. E83-A, No.1(2000), 17-23.

19. N. Koblitz, "Elliptic curve cryptosystems", *Mathematics of Computation*, **48** (1987), 203–209.

20. N. Koblitz, "An elliptic curve implementation of the finite field digital signature algotirhm", *Advances in Cryptology-Proceedings of CRYPTO'98*, Lecture Notes in Computer Science, **1462**(1998), Springer-Verlag, 327-337.

21. M. Kasahara, K. Ohgishi, and R. Sakai "Notes on ID-based key sharing systems on elliptic curve", *IEICE Japan Tech. Rep.*, **ISEC99-57**(1999-11), 37-42.

22. M. Kasahara, K. Ohgishi, and R. Sakai "Cryptosystems based on pairing", *The 2000 Symposium on Cryptography and Information Security*, **SCIS2000-C20**, Jan. 2000.

23. S. Lang, *Elliptic Functions*, GTM112, Springer-Verlag, New York, 1987.

24. A. Menezes, T. Okamoto and S. Vanstone, "Reducing elliptic curve logarithms to logarithms in a finite field", *Proceedings of the 22nd Annual ACM Symposium on the Theory of Computing* (1991), 80–89.

25. V. S. Miller, "Use of elliptic curves in cryptography", *Advances in Cryptology-Proceedings of Crypto'85*, Lecture Notes in Computer Science, **218** (1986), Springer-Verlag, 417-426.

26. S. C. Pohlig and M. E. Hellman, "An improved algorithm for computing logarithms over $GF(p)$ and its cryptographic significance", *IEEE Trans. Inf. Theory*, **IT-24** (1978), 106–110.

27. J. Pollard, "Monte Carlo methods for index computation (mod p)", *Mathematics of Computation*, **32** (1978), 918–924.

28. R. Rivest, A. Shamir and L. Adleman, "A method for obtaining digital signatures and public-key cryptosystems", *Communications of the ACM*, **21** No. 2 (1978), 120–126.

29. T. Saitoh and S. Uchiyama, "A Note on the Discrete Logarithm Problem on Elliptic Curves of Trace Two", *Technical Report of IEICE*, ISEC98-27(1998), 51-57.

30. R. Schoof, "Elliptic Curves Over Finite Fields and the Computation of Square Roots mod p", *Mathematics of computation*, **44** (1985), 483–494.

31. R. Schoof, "Nonsingular plane cubic curves over finite fields", *Jornal of Combination Theory*, vol. A. **46** (1987), 183-211.

32. R. Schoof, "Counting points on elliptic curve over finite fields", *Journal de Théorie des Nombres de Bordeux*, **7** (1995), 219–254.

33. Standards for Efficient Cryptography Group. http://www.secg.org/

34. I. A. Semaev "Evaluation of discrete logarithms in a group of p-torsion points of an elliptic curve in characteristic p", *Mathematics of computation*, **67** (1998), 353-356.

35. J. H. Silverman, *The Arithmetic of Elliptic Curves*, GTM **106**, Springer-Verlag, New York, 1986.

36. N. P. Smart "The discrete logarithm problem on elliptic curves of trace one", J. Cryptology, **12** (1999), 193–196.

37. T. Takagi, *Syotou seisuuronn kougi*, Kyouritu Syuppan, 1971, (in Japanese).

38. Torbjorn Granlund, THE GNU MP LIBRARY, version 3.1, August 2000. ftp://ftp.gnu.org/gnu/gmp/gmp-3.1.tar.gz

A Multi-party Optimistic Non-repudiation Protocol

Olivier Markowitch and Steve Kremer
{omarkow, skremer}@ulb.ac.be

Université Libre de Bruxelles
Computer Science Department
Bd du Triomphe C.P. 212
1050 Brussels Belgium

Abstract. In this paper we consider the optimistic approach of the non-repudiation protocols. We study a non-repudiation protocol with off-line trusted third party and we keep on with the definition of the multi-party non-repudiation, compare it to multi-party fair exchange and show some fundamental differences between these two problems. Finally, we generalize our protocol and propose a multi-party non-repudiation protocol with off-line trusted third party.

1 Introduction

The impressive growth of open networks during the last decade has given more importance to several security related problems. The non-repudiation problem is one of them. Non-repudiation services must ensure that when Alice sends some information to Bob over a network, neither Alice nor Bob can deny having participated in a part or the whole of this communication. Therefore a non-repudiation protocol has to generate non-repudiation of origin evidences intended to Bob, and non-repudiation of receipt evidences destined to Alice. In case of a dispute (e.g. Alice denying having sent a given message or Bob denying having received it) an adjudicator can evaluate these evidences and take a decision in favor of one of the parties without any ambiguity.

In comparison to other security issues, such as privacy or authenticity of communications, non-repudiation has not been studied intensively. However many applications such as electronic commerce, fair exchange, certified electronic mail, etc. are related to non-repudiation. Non-repudiation of origin can easily be provided by signing the sent information. A digital signature provides an irrefutable non-repudiation of origin evidence: a digital signature creates a link between a message and a public verification key. A certification authority assures the correspondance between this public signature verification key and an identity. Non-repudiation of receipt is more difficult to achieve: therefore Alice and Bob have to follow a protocol that assures both services. Such a protocol is said to be fair if either Alice receives a non-repudiation of receipt evidence and Bob receives a non-repudiation of origin evidence or none of them obtains any valid evidence.

First solutions to those problems involve a trusted third party (TTP) that acts as an intermediary between the participating entities. The major disadvantage of this approach is the communication bottleneck created at the TTP. Therefore more efficient solutions have been proposed. Two different approaches have been considered: one consists in designing protocols without a TTP, the other tries to minimize its involvement.

The approach without TTP involvement is often based on a gradual release of the knowledge. However it generally requires that all involved parties have the same computational power. Another disadvantage is the important number of transmitted messages. In [6] a protocol for digital contract signing without TTP is proposed: the probability that the contract has been signed, is increased each round until reaching one. The assumption of same computational power is not needed. Another recently presented probabilistic non-repudiation protocol [13], also succeeded in relaxing the condition on the computational power. Here the idea is that the recipient does not know a priori which transmission will contain the message. The probability of guessing the transmission including the message is arbitrarily small. However to decrease the probability the number of messages has to be increased.

The other approach, trying to minimize the TTP involvement has got more attention in literature during the last years. Asokan et al. presented, in the context of fair exchange, the optimistic approach[4, 1]: usually all participants are honest and only in the case of a misbehaving party, the TTP has to be involved. The protocols inspired by this approach are said to be protocols using an off-line TTP.

The most complete non-repudiation protocols with off-line TTP have been presented in [16] and independently in [12]. The optimistic approach assumes that in general Alice and Bob are honest, i.e. they correctly follow the protocol, and the TTP only intervenes, by the mean of a recovery protocol, when a problem arises.

Most of the proposed non-repudiation protocols are two-party protocols. In fair exchange first works have been done to generalize them to the case of n participants [3, 2, 9, 5]. Considering non-repudiation protocols, the only to us known work generalizing these protocols has been presented in [11] and concerns a non-repudiation protocol with on-line TTP (which intervenes in each session of the protocol, even if the parties behave correctly).

The protocol presented here considers an off-line TTP. To the best of our knowledge, the only comparable work is the multi-party certified mail protocol proposed by Asokan et al. in [2]. However the here proposed protocol is more general: as it will be outlined later in more details, the certified mail protocol only continues if the whole set of receivers is willing to do so. Our protocol leaves the choice to the sender to finish with only the subset of the responding receivers, or to stop in the case of a non-responding receiver, as the certified mail protocol does.

We first remind the properties required by non-repudiation protocols. Then, we go on presenting a two-party non-repudiation protocol with off-line TTP.

Afterwards, we define the multi-party non-repudiation problem, showing some differences with multi-party fair exchange. The requirements of a multi-party non-repudiation protocol are defined as well. Finally, we present a generalization of the previously presented two-party non-repudiation protocol to the case of n parties, using a group encryption scheme.

2 Properties

2.1 The Communication Channels

In the framework of such exchange protocols, we can distinguish three classes of communication channels: unreliable channels, resilient channels and operational channels. No assumptions have to be made about unreliable channels: data may be lost. A resilient channel delivers data after a finite, but unknown amount of time. Data may be delayed, but will eventually arrive. When using an operational channel data arrive after a known, constant amount of time. Operational channels are however rather unrealistic in heterogenous networks.

2.2 Requirements on Non-repudiation Protocols

A first property non-repudiation protocols must provide is fairness. One can distinguish between the two notions of *strong fairness* and *weak fairness*.
A non-repudiation protocol is said to provide *strong fairness* if at the end of the protocol Alice has got a complete non-repudiation of receipt evidence if and only if Bob has got the message with a complete corresponding non-repudiation of origin evidence.
A non-repudiation protocol provides *weak fairness* if either the protocol provides strong fairness, or the protocol provides evidences that can prove to an adjudicator until which state the protocol has been executed.

A second property we require is timeliness: a protocol must be finished after a finite amount of time for each participating entity that is behaving correctly with respect to the protocol.

Each protocol must ensure both properties to be acceptable.

3 A Two-Party Non-repudiation Protocol with Off-line TTP

3.1 Introduction

We will now present a two-party non-repudiation protocol with an off-line TTP [12]. In this protocol, Alice wants to exchange a message m and its corresponding non-repudiation of origin evidence against a non-repudiation of receipt evidence, issued by Bob.

This protocol results from modifications made on the Zhou-Gollman optimistic protocol [18] in which an operational channel is needed between the TTP and Alice in order to assure fairness. The here presented protocol only needs a resilient channel between the TTP and respectively Alice and Bob. The channel between Alice and Bob may even be unreliable. The protocol is similar to the independently developed autonomous two-party non-repudiation protocol proposed earlier by Zhou et al. in [16].

The protocol is composed of three sub-protocols: the main protocol, the recovery protocol and the abort protocol. The main protocol consists of messages exchanged directly between Alice and Bob. In case of problems during this main protocol, two (mutually exclusive) possibilities are offered to the entities. Either Alice contacts the TTP to abort the protocol in order to cancel the exchange, or Alice or Bob contacts the TTP to launch the recovery protocol in order to complete the exchange.

3.2 Notations

We use the following notation to describe the protocol:

- $X \rightarrow Y$: transmission from entity X to entity Y
- $X \leftrightarrow Y$: ftp get operation performed by X at Y
- $h()$: a collision resistant one-way hash function
- $E_k()$: a symmetric-key encryption function under key k
- $D_k()$: a symmetric-key decryption function under key k
- $S_X()$: the signature function of entity X
- m: the message sent from A to B
- k: the message key A uses to cipher m
- $c = E_k(m)$: the cipher of m under the key k
- $l = h(m, k)$: a label that in conjunction with (A, B) uniquely identifies a protocol run
- f: a flag indicating the purpose of a message
- $\mathsf{EOO} = S_A(f_{\mathsf{EOO}}, B, l, h(c))$
- $\mathsf{EOR} = S_B(f_{\mathsf{EOR}}, A, l, h(c))$
- $\mathsf{Sub} = S_A(f_{\mathsf{Sub}}, B, l, E_{TTP}(k))$
- $\mathsf{EOO}_k = S_A(f_{\mathsf{EOO}_k}, B, l, k)$
- $\mathsf{EOR}_k = S_B(f_{\mathsf{EOR}_k}, A, l, k)$
- $\mathsf{Rec}_X = S_X(f_{\mathsf{Rec}_X}, Y, l)$
- $\mathsf{Con}_k = S_{TTP}(f_{\mathsf{Con}_k}, A, B, l, k)$
- $\mathsf{Abort} = S_A(f_{\mathsf{Abort}}, B, l)$
- $\mathsf{Con}_a = S_{TTP}(f_{\mathsf{Con}_a}, A, B, l)$

3.3 Main Protocol

The basic idea of the main protocol is to first exchange the cipher of the message m against a receipt for this cipher. Secondly, we exchange the decryption key against a receipt for this key. Each transmission is associated to some maximum time-out. Once this time-out value is exceeded the recipient supposes the transmission will not arrive and initiates either a recovery or an abort protocol.

1. $A \rightarrow B$: $f_{\mathsf{EOO}}, f_{\mathsf{Sub}}, B, l, c, E_{TTP}(k), \mathsf{EOO}, \mathsf{Sub}$
2. $B \rightarrow A$: $f_{\mathsf{EOR}}, A, l, \mathsf{EOR}$ (time-out: abort)
3. $A \rightarrow B$: $f_{\mathsf{EOO}_k}, B, l, k, \mathsf{EOO}_k$ (time-out: recovery$[X := B, Y := A]$)
4. $B \rightarrow A$: $f_{\mathsf{EOR}_k}, A, l, \mathsf{EOR}_k$ (time-out: recovery$[X := A, Y := B]$)

Alice starts the protocol by sending the cipher of the message, as well as the decryption key, ciphered under the public key of the TTP, to Bob. The message does also contain Alice's signature on the encrypted key and the hash of the cipher. These signatures serve as evidences of origin for the ciphers. In this first message we use two purpose flags as it conveis two proofs (EOO and Sub).

If Bob receives the first message he replies with a receipt to confirm the arrival of the first message. This receipt contains Bob's signature on the hash of the cipher c and serves to Alice as an evidence of receipt for the cipher.

In the case that Alice does not receive message 2 from Bob before a given time-out, she initiates the abort protocol. Note that Alice cannot perform a recovery at this moment, as the recovery protocol requires EOR, the evidence of receipt for the cipher.

If message 2 arrives to Alice before the time-out, she sends to Bob the decryption key k, as well as her signature on this key. This signature is used as an evidence of origin for the key. The evidence of origin for the cipher c, together with the evidence of origin for the key k, form together the non-repudiation of origin evidence of the message m.

Message 3 has to arrive to Bob before a given time-out. Otherwise Bob initiates the recovery protocol with the TTP.

If message 3 arrives in time, Bob sends a receipt for the key to Alice: his signature on the key k. The signature serves as the evidence of receipt for the key. Together with the evidence of receipt for the cipher c, they form the non-repudiation of receipt evidence of the message m.

Alice may also initiate the recovery protocol with the TTP if this last message does not arrive in time.

3.4 Recovery Protocol

To launch the recovery protocol Alice or Bob has to send to the TTP the hash of c, the key k ciphered for the TTP, the evidence of origin for the cipher c, EOO, the evidence of origin for the encrypted key, Sub, the evidence of receipt for the cipher c, EOR, as well as the evidence of origin for the recovery request, Rec_X (where X may take the values A or B). Note that the recovery protocol can only be executed once per protocol run and is mutually exclusive with the abort protocol.

By the mean of these evidences the TTP can be sure that Alice sent the cipher c to Bob and that Bob really received it.

1. $X \rightarrow TTP$: $f_{\mathsf{Rec}_X}, f_{\mathsf{Sub}}, Y, l, h(c), E_{TTP}(k), \mathsf{Rec}_X, \mathsf{Sub}, \mathsf{EOR}, \mathsf{EOO}$

if aborted or recovered then stop

else recovered=true

    2. $TTP \rightarrow A$: $f_{\mathsf{Con}_k}, A, B, l, k, \mathsf{Con}_k, \mathsf{EOR}$
    3. $TTP \rightarrow B$: $f_{\mathsf{Con}_k}, A, B, l, k, \mathsf{Con}_k$

When the first message arrives, the TTP checks wether an abort protocol or a recovery protocol has already been started for this protocol run: a protocol run is uniquely identified by the label $l = h(m, k)$ and the identities (A, B). If either an abort or a recovery protocol has already been initiated the TTP halts. The resilient channels assure that some message ending the protocol will eventually arrive at X. If the TTP accepts to perform a recovery protocol, the TTP sends to Alice the confirmation of submission of the key, as well as the evidence of receipt for the cipher EOR. It is important to include EOR, as Bob can initiate the recovery protocol after having received the cipher, without having sent a receipt for it. The TTP sends to Bob the key k, as well as the confirmation of the submission of the key, serving to Bob as an evidence of origin for k.

If the recovery protocol is executed, the key confirmation evidence Con_k will make part of the non-repudiation evidences for the message m. It is used to replace both the evidence of origin for the key as well as the evidence of receipt for the key.

3.5 Abort Protocol

Alice has the possibility to run an abort protocol. If she decides to do so she sends a signed abort request, including label l, to the TTP.

If the TTP accepts the request (neither a recovery nor an abort has yet been initiated), the TTP sends to both Alice and Bob a signed abort confirmation.

if recovered or aborted then stop

    aborted=true

    1. $A \rightarrow TTP$: $f_{\mathsf{Abort}}, l, B, \mathsf{Abort}$
    2. $TTP \rightarrow A$: $f_{\mathsf{Con}_a}, A, B, l, \mathsf{Con}_a$
    3. $TTP \rightarrow B$: $f_{\mathsf{Con}_a}, A, B, l, \mathsf{Con}_a$

Note that Alice could specify a wrong B in her abort request. However this would refer to a different protocol run—a protocol run is identified by l and (A, B)—and would enable Bob to launch a recovery protocol.

3.6 Dispute Resolution

The non-repudiation of origin and receipt evidences for message m are the following :

– $\mathsf{NRO} = (\mathsf{EOO}, \mathsf{EOO}_k)$ or $\mathsf{NRO} = (\mathsf{EOO}, \mathsf{Con}_k)$
– $\mathsf{NRR} = (\mathsf{EOR}, \mathsf{EOR}_k)$ or $\mathsf{NRR} = (\mathsf{EOR}, \mathsf{Con}_k)$

Repudiation of Origin. When Alice denies the origin of the message, Bob has to present to the judge EOO, EOO_k or Con_k, l, c, m and k. The judge verifies that

- $EOO = S_A(f_{EOO}, B, l, c)$,
- $EOO_k = S_A(f_{EOO_k}, B, l, k)$ or $Con_k = S_{TTP}(f_{Con_k}, A, B, l, k)$,
- $l = h(m, k)$,
- $c = E_k(m)$.

If Bob can provide all the required items and all the checks hold, the adjudicator claims that Alice is at the origin of the message.

Repudiation of Receipt. When Bob denies receipt of m, Alice can prove his receipt of the message by presenting EOR, EOR_k or Con_k, l, c, m and k to a judge. The judge verifies that

- $EOR = S_B(f_{EOR}, A, l, h(c))$,
- $EOR_k = S_B(f_{EOR_k}, A, l, k)$ or $Con_k = S_{TTP}(f_{Con_k}, A, B, l, k)$,
- $l = h(m, k)$,
- $c = E_k(m)$.

If Alice can present all of the items and all the checks hold, the adjudicator concludes that Bob received the message.

3.7 Fairness and Timeliness

If Bob stops the protocol after having received the first message, Alice may perform the abort protocol, in order to avoid Bob to initiate a recovery later. As neither Bob nor Alice received complete evidences the protocol remains fair. If Bob had already initiated the recovery protocol, the TTP sends all the missing evidences to Alice and Bob. Note that the TTP also sends the EOR to Alice, as she has not received it yet. Thus the protocol stays fair.

If Alice does perform step 3, Bob receives a complete non-repudiation of origin evidence. There are two ways to finish the protocol: either Bob sends message 4 of the main protocol and Alice receives a complete non-repudiation of receipt evidence or Alice performs the recovery protocol. As the channels between the TTP and both Alice and Bob are resilient, all data sent by the TTP to Alice and Bob eventually arrive. In both cases all entities receive valid evidences and the protocol finishes providing fairness.

If Alice does not send message 3 during the main protocol, Alice and Bob may initiate the recovery protocol. Fairness is still guaranteed, as during the recovery protocol, Alice and Bob receive all expected evidences.

In the case where Alice sends a wrong key k' either to Bob or to the TTP $(E_{TTP}(k'))$ the evidences will not be valid as $l \neq h(m, k')$. If Alice also transmits the label $h(m, k')$, the generated evidence will correspond to the message $m' = D_{k'}(c)$. As m' is the message Bob actually received, fairness is still provided.

Note that the protocol achieves strong fairness.

When looking at the timeliness, three situations may arrive: the main protocol ends up successfully (without any time-out); Alice aborts the protocol and the abort confirmation signed by the TTP arrives at Alice and Bob after a finite amount of time, as the channels between the TTP and both Alice and Bob are resilient; a recovery protocol is performed and Alice and Bob receive the evidences after a finite amount of time because of the resilience of the channels.

4 Multi-party Non-repudiation

In literature, different kinds of multi-party fair exchange have been considered. In [9] a classification has been proposed. One can differ between single-unit and multi-unit exchanges. Moreover different topologies are possible: [9] and [5] concentrated on a ring topology. Each entity e_i ($0 \leq i \leq n - 1$) desires an item (or a set of items) from entity $e_{i \boxminus 1}$ and offers an item (or a set of items) to entity $e_{i \boxplus 1}$, where $\boxplus$ and $\boxminus$ respectively denote addition and subtraction modulo n. Another topology is the more general matrix topology, where each entity may desire items from a set of entities and offer items to a set of entities. Such protocols have been proposed by Asokan et al. in [3] and [2].

A fundamental difference between non-repudiation and fair exchange is the following. In non-repudiation, the originator sends some data with a non-repudiation of origin evidence to a recipient, who has to respond with a non-repudiation of receipt evidence. The sent data is generally not known to the recipient a priori. In a fair exchange each entity offers an a priori known item and receives another item, also known a priori. In a multi-party fair exchange protocol one can imagine sending an item to one entity and receiving an item from a different one. In non-repudiation it does not make sense that one entity receives some data and a distinct entity sends the corresponding receipt. Thus a ring topology is not sound. The most natural and here considered generalization seems to be a one-to-many protocol, where one entity sends a message to $n-1$ receiving entities who respond to the sender. However other possibilities for generalization exist (many-to-one, many-to-many) although they seem to be less natural.

The expectations we have towards such a protocol are rather similar to the properties required in two-party non-repudiation. A multi-party non-repudiation protocol is said to provide *strong fairness* if at the end of the protocol the sender has got a complete non-repudiation of receipt evidence for a given recipient if and only if this recipient has got the message with a complete corresponding non-repudiation of origin evidence.

A multi-party non-repudiation protocol is said to provide *weak fairness* if either the protocol provides strong fairness, or the protocol provides evidences that can prove to an adjudicator until which state the protocol has been executed with each of the receivers.

Here we can clearly see the difference with the certified mail protocol proposed by Asokan et al. Their protocol requires that at the end of the protocol *all* receivers have got the message with corresponding origin evidence and that the sender has got a receipt for *every* receiver, or none of them gained any valuable information. A last required property is the timeliness property, defined as for two-party protocols.

5 A Multi-party Optimistic Non-repudiation Protocol

We propose a generalization of the presented two-party non-repudiation protocol, using an off-line TTP. We suppose that the channels between the TTP and both Alice and all possible receivers are resilient. The channels between Alice and any receiver may be unreliable.

5.1 Notations

The following notation will be used:

- $\mathcal{B}$: the set of receiving entities
- $A \Rightarrow \mathcal{B}$: multicast from Alice to the set of entities $\mathcal{B}$
- $E_{\mathcal{E}}()$: a group encryption scheme E, that can be deciphered by each party $P \in \mathcal{E}$
- $\mathcal{B}'$: the set of receiving entities having sent an evidence of receipt for the cipher to Alice
- $l = h(m, k)$: a label that in conjunction with the identity A uniquely identifies a protocol run
- $\mathsf{EOO} = S_A(f_{\mathsf{EOO}}, \mathcal{B}, l, t, h(c))$
- $\mathsf{EOR}_i = S_{B_i}(f_{\mathsf{EOR}}, A, l, h(c))$
- $\mathsf{Sub} = S_A(f_{\mathsf{Sub}}, \mathcal{B}, l, E_{TTP}(k))$
- $\mathsf{EOO}_k = S_A(f_{\mathsf{EOO}_k}, \mathcal{B}', l, h(k))$
- $\mathsf{EOR}_{i,k} = S_{B_i}(f_{\mathsf{EOR}_k}, A, l, h(k))$
- $\mathsf{Rec}_X = S_X(f_{\mathsf{Rec}_X}, A, \mathcal{B}, l)$
- $\mathsf{Con}_k = S_{TTP}(f_{\mathsf{Con}_k}, A, \mathcal{B}', l, h(k))$
- $\mathsf{Early} = S_{TTP}(f_{\mathsf{Early}}, l)$
- $\mathsf{Set} = S_A(f_{\mathsf{Set}}, \mathcal{B}', l)$

5.2 Main Protocol

In the main protocol Alice sends a cipher of the message to all potential receivers $\mathcal{B}$. Several of these receivers $(\mathcal{B}')$ are sending a receipt for this cipher. Alice continues the protocol with the receivers in $\mathcal{B}'$ by sending them the key corresponding to the cipher of message 1.

1. $A \Rightarrow \mathcal{B}:$ $f_{\mathsf{EOO}}, f_{\mathsf{Sub}}, \mathcal{B}, l, t, c, E_{TTP}(k), \mathsf{EOO}, \mathsf{Sub}$
2. $B_i \to A:$ $f_{\mathsf{EOR}}, A, l, \mathsf{EOR}_i$
             where $B_i \in \mathcal{B}$ and $i \in \{1, \ldots, |\mathcal{B}|\}$
3. $A \Rightarrow \mathcal{B}':$ $f_{\mathsf{EOO}_k}, \mathcal{B}', l, E_{\mathcal{B}'}(k), \mathsf{EOO}_k$
4. $B_i' \to A:$ $f_{\mathsf{EOR}_k}, A, l, \mathsf{EOR}_{i,k}$
             where $B_i' \in \mathcal{B}'$ and $i \in \{1, \ldots, |\mathcal{B}'|\}$

The first message destined to all receivers in $\mathcal{B}$ includes the label, the cipher, as well as the key k ciphered using the public key of the TTP (this information is used by the TTP in the case of a recovery). Alice also sends a time-out t: a recovery may only be performed after t. If one of the receivers does not accept the time-out he may stop the protocol.

After having sent the cipher in message 1, Alice decides of the moment to continue the protocol. All receipts arriving after this moment are not considered any more, without any risk for the receivers of losing fairness. To realize a service similar to the certified mail presented in [2], Alice may stop the protocol if not all of the receivers in $\mathcal{B}$ have answered.

Afterwards, when Alice sends the deciphering key it is crucial that only the recipients in $\mathcal{B}'$ receive it. Therefore we need to use a cipher (this is not needed in a two-party protocol). In order to cipher only once and to use multicasting, Alice uses a group encryption scheme. The idea is that the key can be ciphered in a way such that only recipients $B_i' \in \mathcal{B}'$ can decipher it. Examples of such ciphering methods can be found in [8] and [10].

5.3 Recovery Protocol

At each moment during the main protocol Alice and Bob have the possibility to launch the recovery protocol with the TTP. The recipients in $\mathcal{B}'$ launch a recovery if Alice does not multicast the ciphered key; Alice initiates the recovery protocol, if not all recipients in $\mathcal{B}'$ send a receipt for the ciphered key.

1. $X \rightarrow TTP$: $f_{\mathsf{Rec}_X}, f_{\mathsf{Sub}}, A, \mathcal{B}, l, t, h(c), E_{TTP}(k), \mathsf{Rec}_X, \mathsf{Sub}, \mathsf{EOO}$
if recovered then
$\quad TTP \rightarrow X$: $\quad f_{\mathsf{Con}_k}, A, \mathcal{B}', l, E_{\mathcal{B}'}(k, S_{TTP}(k)), \mathsf{Con}_k$
else if before t then
$\quad TTP \rightarrow X$: $\quad f_{\mathsf{Early}}, \mathsf{Early}$
else
$\quad$ recovered=true
$\quad$ 2. $TTP \leftrightarrow A$: $\quad\quad\quad\quad f_{\mathsf{Set}}, \mathcal{B}', l, \mathsf{Set}$
$\quad$ 3. $TTP \rightarrow A$: $\quad\quad\quad\quad f_{\mathsf{Con}_k}, A, \mathcal{B}', l, \mathsf{Con}_k$
$\quad$ 4. $TTP \Rightarrow \mathcal{B}' \cup \{X\} \backslash \{A\}$: $f_{\mathsf{Con}_k}, A, \mathcal{B}', l, E_{\mathcal{B}'}(k, S_{TTP}(k)), \mathsf{Con}_k$

Before t (t is specified by Alice during the first message in the main protocol and it is resent together with Alice's signature during the recovery request) the recovery cannot be initiated.

If someone tries to execute a recovery before t, which is resent together with Alice's signature during the recovery request, the TTP sends a message to notify that the recovery cannot yet be initiated.

When the recovery protocol is initiated the first time after t, the TTP uses ftp get to fetch the set $\mathcal{B}'$ at Alice. For this purpose Alice maintains a read-only accessible directory containing the set of users and her signature on this set. If the directory is not accessible the TTP supposes $B' = \emptyset$.

As soon as the TTP made the ftp get on Alices's public directory, Alice considers that a recovery is in execution and stops execution of the main protocol. Otherwise a malicious B_i could be inserted in B' after the ftp get, and B_i could benefit of a race condition to cheat Alice.

Now the TTP sends to Alice the confirmation of receipt of the key that may be used to substitute $EOR_{i,k}$ for all i such that $B_i \in B'$. The TTP sends to all receiver $B_i \in B'$ the confirmation of the key, that substitutes EOO_k, as well as the signed key ciphered for B' using a group encryption scheme.

If a receiver $B_i \notin B'$ wants to perform a recovery, after the recovery has already been performed for the first time (the TTP uniquely identifies each protocol run by l and A), the TTP sends to this entity the same message as to each $B_i \in B'$. This message is however useless as k has been ciphered for the set B' and only informs the recipient that he does not belong to this set.

5.4 Dispute Resolution

At the end of a successful protocol execution, each recipient $B_i \in B'$ and Alice receive the following non-repudiation of origin respectively receipt evidence for message m:

- NRO $= (EOO, EOO_k)$ or NRO $= (EOO, Con_k)$
- $NRR_i = (EOR_i, EOR_{i,k})$ or NRR $= (EOR_i, Con_k)$

Two kinds of disputes can arise: repudiation of origin and repudiation of receipt. Repudiation of origin arises when a recipient B_i claims having received a message m from Alice, who denies having sent it. Repudiation of receipt arises when Alice claims having sent a message m to a recipient B_i who denies having received it.

Repudiation of Origin. When Alice denies the origin of the message, B_i has to present to the judge EOO, EOO_k or Con_k, l, m, k, B and B'.

The message m and the key k have to be sent to the judge via a secure channel, for example using encryption. Otherwise a recipient $B_i \in B \backslash B'$ can recover the transmission. If any of these informations cannot be provided the recipient's claim is rejected.

The judge validates the recipient's claim if he can successfully verify that:

- $EOO = S_A(f_{EOO}, B, l, h(c))$ after having computed $c = E_k(m)$ and $h(c)$,
- $EOO_k = S_A(f_{EOO_k}, B', l, h(k))$ or $Con_k = S_{TTP}(f_{Con_k}, A, B', l, k)$,
- $l = h(m, k)$.

Repudiation of Receipt. When B_i denies receipt of m, A can prove his receipt of the message by presenting EOR_i, $\text{EOR}_{i,k}$ or Con_k, l, m, k and B' to a judge. As above a secure channel is needed to transmit m and k.

To accept Alice's claim, the judge verifies that

- $\text{EOR}_i = S_{B_i}(f_{\text{EOR}}, A, l, h(c))$ after having computed $c = E_k(m)$ and $h(c)$,
- $\text{EOR}_k = S_{B_i}(f_{\text{EOR}_k}, A, l, h(k))$ or $\text{Con}_k = S_{TTP}(f_{\text{Con}_k}, A, B', l, h(k))$,
- $l = h(m, k)$.

5.5 Fairness and Timeliness

Our generalized protocol provides strong fairness. In fact when the main protocol ends without problems, the non-repudiation evidences have been exchanged and all receivers in B' got the message m. When a problem arises both Alice and Bob can initiate the recovery protocol at each moment following the transmission of message 1. As outlined in the previous section, Alice maintains a read-only accessible directory that is consulted by the TTP during the recovery to get the description of the set B'. The TTP then sends the missing evidences to both Alice and the entities in B'.

If Alice tries to cheat by submitting a wrong key k', the generated evidences will not be valid as $l \neq h(m, k')$. If Alice also transmits $l = h(m, k')$, the evidences will be generated for the message $m' = D_{k'}(c)$ and Alice can only prove that B_i received m'. Alice could also try to cheat by publishing a wrong set $\widetilde{B}'$. Publishing a set smaller than B' does not provide an advantage as Alice would not receive the confirmation of receipt of the key for the entities in $B' \backslash \widetilde{B}'$. If $\widetilde{B}'$ is bigger than B', Alice would harm herself as all the entities in $\widetilde{B}' \backslash B'$ receive k with a confirmation of origin for k, while Alice does not have an evidence of receipt for the cipher of those entities. Hence Alice does not have any interest in publishing a different set.

Now consider a scenario where at the second step several receivers send the evidence of receipt. After a fixed amount of time, Alice continues the protocol. Now, if a group of late receipts arrive at Alice, she will possess an evidence of receipt for the cipher c from these receivers. However fairness is not threatened by this scenario as Alice will not receive a confirmation of the receipt for the key from those entities. If Alice would include these receivers in B', they would also receive k and the corresponding evidences of origin. So strong fairness is still provided.

We shall now show that timeliness is also respected. Alice has two possibilities: either she finishes the main protocol or she has to initiate a recovery protocol. The timeliness in the later case is assured by the resilient channels. A recipient B_i can either finish the main protocol or launch a recovery. If he launches a recovery several cases may arise. He may be too early (before t) and he has to relaunch the recovery after t (t has been known a priori when the recipient agreed to continue the protocol). Otherwise, if B_i successfully initiates the protocol, the TTP will send him the ciphered key with the corresponding confirmation of origin. As the channels between the TTP and $B_i \in \mathcal{B}$ are resilient the protocol finishes after a finite time preserving timeliness.

6 Conclusion

We have defined multi-party non-repudiation and presented a generalization of an optimistic two-party non-repudiation protocol to an n-party non-repudiation protocol. To the best of our knowledge this is the first *optimistic* multy-party non-repudiation protocol. We have shown that the generalized protocol provides strong fairness and respects timeliness.

Acknowledgments. The authors would like to thank the anonymous referees for their helpful comments on the draft version of this paper.

References

1. N. Asokan. *Fairness in Electronic Commerce.* PhD thesis, University of Waterloo, May 1998.
2. N. Asokan, B. Baum-Waidner, M. Schunter, and M. Waidner. Optimistic synchronous multi-party contract signing. Research Report RZ 3089, IBM Research Division, Dec. 1998.
3. N. Asokan, M. Schunter, and M. Waidner. Optimistic protocols for multi-party fair exchange. Research Report RZ 2892 (# 90840), IBM Research, Dec. 1996.
4. N. Asokan, M. Schunter, and M. Waidner. Optimistic protocols for fair exchange. In T. Matsumoto, editor, *4th ACM Conference on Computer and Communications Security*, pages 6, 8–17, Zurich, Switzerland, Apr. 1997. ACM Press.
5. F. Bao, R. Deng, K. Q. Nguyen, and V. Vardharajan. Multi-party fair exchange with an off-line trusted neutral party. In *DEXA'99 Workshop on Electronic Commerce and Security*, Florence, Italy, Sept. 1999.
6. M. Ben-Or, O. Goldreich, S. Micali, and R. L. Rivest. A fair protocol for signing contracts. *IEEE Transactions on Information Theory*, 36(1):40–46, 1990.
7. C. Boyd. Some applications of multiple key ciphers. In C. G. Günther, editor, *Advances in Cryptology—EUROCRYPT 88*, volume 330 of *Lecture Notes in Computer Science*, pages 455–467, Davos, Switzerland, May 1988. Springer-Verlag.
8. G. Chiou and W. Chen. Secure broadcasting using the secure lock. *IEEE Transactions on Software Engineering*, 15(8):929–934, Aug. 1989.
9. M. Franklin and G. Tsudik. Secure group barter: Multi-party fair exchange with semi-trusted neutral parties. *Lecture Notes in Computer Science*, 1465, 1998.
10. T. Hwang. Cryptosystem for group oriented cryptography. In I. B. Damgård, editor, *Advances in Cryptology—EUROCRYPT 90*, volume 473 of *Lecture Notes in Computer Science*, pages 352–360, Aarhus, Denmark, May 1990. Springer-Verlag, 1991.
11. S. Kremer and O. Markowitch. A multi-party non-repudiation protocol. In *SEC 2000: 15th International Conference on Information Security*, IFIP World Computer Congress, pages 271–280, Beijing, China, Aug. 2000. Kluwer Academic.
12. S. Kremer and O. Markowitch. Optimistic non-repudiable information exchange. In J. Biemond, editor, *21st Symp. on Information Theory in the Benelux*, pages 139–146, Wassenaar (NL), May 25-26 2000. Werkgemeenschap Informatie- en Communicatietheorie, Enschede (NL).

13. O. Markowitch and Y. Roggeman. Probabilistic non-repudiation without trusted third party. In *Second Conference on Security in Communication Networks'99*, Amalfi, Italy, Sept. 1999.

14. A. J. Menezes, P. C. van Oorschot, and S. A. Vanstone. *Handbook of applied cryptography*. CRC Press series on discrete mathematics and its applications. CRC Press, 1996. ISBN 0-8493-8523-7.

15. J. Zhou. *Non-repudiation*. PhD thesis, University of London, Dec. 1996.

16. J. Zhou, R. Deng, and F. Bao. Evolution of fair non-repudiation with TTP. In *ACISP: Information Security and Privacy: Australasian Conference*, volume 1587 of *Lecture Notes in Computer Science*, pages 258–269. Springer-Verlag, 1999.

17. J. Zhou and D. Gollmann. Observations on non-repudiation. In K. Kim and T. Matsumoto, editors, *Advances in Cryptology—ASIACRYPT '96*, volume 1163 of *Lecture Notes in Computer Science*, pages 133–144, Kyongju, Korea, 3–7 Nov. 1996. Springer-Verlag.

18. J. Zhou and D. Gollmann. An efficient non-repudiation protocol. In *PCSFW: Proceedings of The 10th Computer Security Foundations Workshop*. IEEE Computer Society Press, 1997.

Secure Matchmaking Protocol

Byoungcheon Lee and Kwangjo Kim

Information and Communications University,
58-4, Hwaam-dong, Yusong-gu, Taejon, 305-348, Korea
{sultan,kkj}@icu.ac.kr

Abstract. Matchmaking protocol is a procedure to find matched pairs in registered groups of participants depending on their choices, while preserving their privacy. In this study we define the concept of matchmaking and construct a simple and efficient matchmaking protocol under the simple rule that two members become a matched pair only when they have chosen each other. In matchmaking protocol, participant's privacy is of prime concern, specially losers' choices should not be opened. Our basic approach to achieve privacy is finding collisions among multiple secure commitments without decryption. For this purpose we build a protocol to find collisions in ElGamal ciphertexts without decryption using Michels and Stadler's protocol [MS97] of proving the equality or inequality of two discrete logarithms. Correctness is guaranteed because all procedures are universally verifiable.

Keywords: matchmaking, secure multiparty computation, proof of knowledge, proving the equality or inequality of two discrete logarithms, finding collisions without decryption, public commitment.

1 Introduction

Consider a set of parties who trust neither other entities nor the channels by which they communicate. The parties wish to correctly compute some common function of their local inputs, while keeping their local data as private as possible. This is the problem of secure multiparty computation [Yao82], [GMW87], [CCD88], [BGW88], [Can96], [Gol98], [Cra99], which has fundamental importance in cryptography and is relevant to many distributed cryptographic applications such as electronic cash, voting, auction, and so on.

In this study we consider a problem of finding matched pairs in registered groups of participants depending on their choices, while preserving their privacy. This is an interesting example of secure multiparty computation where the common function is finding matched pairs in the registered groups of participants and local data are participants' choices.

There is a popular TV program in Korea which performs matchmaking between two registered groups of men and women. In the program all participants commit their choices to a host secretly, but in the opening stage the host opens all the choices and decides couples. The established couple members are OK for opening their choices, but losers may wish that their choices might not be opened if possible. A loser can have another chance to participate in matchmaking and in that case his or her previous choice is important private information. So secrecy of commitment is a prime security issue in this case.

Another example can be found in setting up project teams in a class. The lecturer of the class tries to permit students to form project teams of two members if they want each other. The choices of established team members are published, but losers who could not form a team may wish that their choices might not be opened if possible.

In this study we consider a typical model of secure matchmaking protocol which is used to set up couples among m male members $M_i(i = 1, ..., m)$ and n female members $F_j(j = 1, ...n)$. The basic rule of matchmaking is that in commitment stage each participant commits a single choice to TTP secretly and then in opening stage two participants are decided as a couple only when they have chosen each other. Our basic approach to provide secrecy of commitment is finding collisions in ElGamal ciphertexts without decryption. For this purpose we use Michels and Stadler's protocol [MS97] of proving the equality or inequality of two discrete logarithms. Our design provides the secrecy of commitment and guarantees the correctness of results.

This paper is organized as follows. In section 2, we describe our model of matchmaking and its security requirements. Then in section 3, we describe some building blocks such as public commitment, proving the correctness of decryption, and finding collisions in ElGamal ciphertexts. Using these primitives we construct a simple and efficient matchmaking protocol in section 4 and provide its security analysis in section 5.

2   Model of matchmaking

2.1   Definition of terms

In this paper the following terms are used in a specific sense, so we need to define them more rigorously.

Matchmaking is a protocol to find matched pairs $< a_i, b_j >$ between two registered groups of participants $A = (a_1, ..., a_m)$ and $B = (b_1, ..., b_n)$ which satisfy $< a_i, b_j >\in R$ for a special relation R we try to find.

Established couple is a matched pair $< a_i, b_j >\in R$ which is found using the matchmaking protocol.

Loser is a participant who was not established as a couple. All participants except the established couple members are losers.

Registered group of participants is the members registered in the matchmaking system who participate in the matchmaking process and try to find a partner there.

Public commitment is a commitment scheme with which a participant commits his or her choice to the public. The correctness of result should be publicly verifiable.

Collision in ciphertexts is the case that two ciphertexts are probabilistic encryptions of the same message.

Proof of coupling is a proof for the correctness of the established couple.

2.2 Participants and tools

Our matchmaking protocol has the following participants and tools. All participants are assumed to have own public/private key pairs certified by a certificate authority(CA).

Male participants: m male members $M_i (i = 1, ..., m)$ want to find partners among the female members.

Female participants: n female members $F_j (j = 1, ..., n)$ want to find partners among the male members.

TTP(Host): A trusted third party T is a host of matchmaking and mediates the matchmaking procedure, and can be modeled as a probabilistic polynomial-time Turing machine. T receives participants' public commitments as input and outputs the results of matchmaking and proofs of them. W.l.o.g., it is assumed that T does not collude with any specific participant.

Bulletin board: It is a public communication channel which can be read by anybody. Only legitimate participant is allowed to post his or her message on the specified region of the bulletin board. All participants communicate via the bulletin board.

2.3 Rule of matchmaking

In this study we consider the following simplified model of matchmaking between two registered groups of men and women.

- Each participant in a group has a single choice among the participants of the other group.
- Two participants are decided as a couple only when they have chosen each other.

In the real world there can be a variety of situations and rules for matchmaking. Some possible examples are committing multiple choices, setting up a team of multiple partners, *etc*, and each can be a good model for the study of secure multiparty computation.

2.4 Security requirements

The main security concern of matchmaking is the secrecy of participants' choices. A participant wants to keep the secrecy of his or her choice as private as possible while trying to get any information on others' choices. Even the established couple members may want to know who else have chosen him or her except the current partner. A loser may want to know who have chosen him or her, because it can be important information for the next round of matchmaking while it is also private information for the loser. Therefore secrecy of commitment should be provided such that malicious participants cannot get any partial information on others' choices.

Another scenario is that a participant A can try to help other participant B, i.e., A helps B to setup a couple $< B, C >$. If this is possible, B gets to have two choices against the fairness of the matchmaking rule. So the authenticity of commitment is required and the correctness of result should be publicly verifiable.

Still another scenario is that a member of an established couple may want to change his or her mind later and tries to repudiate the result. It is clear that this should not be allowed.

The security requirements of secure matchmaking protocol can be listed as follows.

Secrecy(privacy) The choices of participants should not be exposed to others including TTP in the whole process of matchmaking. Only the choices of established couple members are published in the opening stage.

Fairness In the commitment stage, anyone cannot be in advantageous position than others, i.e., anyone cannot have any partial information on other's choice.

Correctness The correctness of the result of matchmaking should be publicly verifiable.

Authenticity The authenticity of commitment should be provided such that each registered participant has committed a single choice can be verified.

Non-repudiation The established couple members should not be able to repudiate their commitments later.

3 Building blocks

In this section we describe some building blocks used in the proposed matchmaking protocol.

3.1 Notation

The basic cryptographic primitives used in our protocol are ElGamal public key encryption, digital signature, and zero-knowledge proof, which are commonly based on the discrete logarithm setting. Let Z_p^* denote the multiplicative group of prime modulo p. We consider a cyclic subgroup of Z_p^* of prime order q with $q|(p-1)$. Let g be the published generator of the subgroup of order q. TTP has a certified public key $y = g^x$ and its corresponding private key x. In this paper we use the following notation.

$Sig_i(m)$ is a digital signature of a participant i for a message m where the signature scheme is secure against existential forgery.

$E_j(m)$ is a probabilistic ElGamal encryption of a message m with the participant j's public key.

$Env(i, j, m) = [c||s] = [E_j(m)||Sig_i(c)]$ is an enveloped message of a plaintext m sent by a sender i to a recipient j. The plaintext m is signed by the sender i and encrypted with the recipient j's public key.

$H()$ is a collision resistant hash function. For security proof, it is considered as a random oracle.

3.2 Public commitment scheme

Commitment scheme is a digital implementation of a sealed box which is opened later. In a two party computation, commitment scheme is a two-phase protocol of commitment and reveal stages, and should satisfy the requirement of secrecy and unambiguity. The situation in matchmaking protocol is a multiparty computation, so we define a new concept of *public commitment scheme* where a message is committed to the public, revealed by the help of TTP, and is publicly verified.

Definition 1 (Public commitment scheme).
Public commitment scheme is a 3-stage protocol between multiple participants $P_1, ..., P_n$ and a TTP which consists of:

1. *Commitment: Multiple participants commit their secure messages to the public in a secure way such that only TTP can open it.*
2. *Reveal: TTP opens the commitments and provides the proofs of results.*
3. *Verification: Anyone verifies the correctness of the results.*

And it satisfies the following requirements:

1. *Secrecy: At the end of stage 1, anyone except the sender cannot tell what value is being sent.*
2. *Unambiguity: Given a commitment of stage 1, there is at most one value everybody may accept as valid. It means that the commitment is bound to a value.*
3. *Non-malleability: Given a commitment of stage 1, anyone except the sender cannot generate another legal commitment which has a message related with the original message.*
4. *Non-repudiation: Once a participant committed a message to the public in stage 1, he cannot repudiate it later.*

In this study we implement the public commitment scheme using the enveloped message defined above. In commitment stage, a participant P_i commits a message m_i to the public as $Env(i, T, m_i)$ which is encrypted with TTP's public key. In reveal stage, TTP recovers the message m_i and publishes it with a proof of the correctness of decryption. In verification stage, anyone verifies the correctness of the message.

Lemma 1. *Let $P_1, ..., P_n$ be multiple participants and T be a TTP. A commitment scheme using an enveloped message $Env(i, T, m) = [c||s] = [E_T(m)||Sig_i(c)]$ is a public commitment scheme between P_i and T.*

Proof. (sketch) The probabilistic public key encryption scheme provides secrecy and unambiguity assuming that T does not collude with participants. Considering the result of [TY98] and [SJ00], the enveloped message(a combination of a public key encryption and a digital signature) can be considered as a non-malleable encryption scheme. So non-malleability is satisfied. The usage of digital signature provides non-repudiation. □

3.3 Proving the correctness of decryption

Firstly we describe the zero-knowledge proof of knowledge protocol for the equality of two discrete logarithms [CP93], [CGS97]. For universal verifiability, we consider its non-interactive version using the Fiat-Shamir heuristics [FS87] and a collision resistant hash function. The 3-move interactive protocol is an honest verifier zero-knowledge and the security of the non-interactive version is obtained under the random oracle model [CGS97].

Let α and β be two independent elements of the cyclic group Z_p^* of order q, i.e., nobody knows the relative discrete logarithm $\log_\alpha \beta$.

Protocol 1. Proving the equality of two discrete logarithms

The prover wants to prove that two elements $y = \alpha^x$ and $z = \beta^x$ have the same discrete logarithm for base α and β, respectively, without exposing x, i.e., he wants to prove that $\log_\alpha y = \log_\beta z$ holds. (α, y, β, z) are given as common input.

Prover
- Chooses $k \in_R Z_q^*$.
- Computes $r_\alpha = \alpha^k$, $r_\beta = \beta^k$.
- Computes $v = H(\alpha, y, \beta, z, r_\alpha, r_\beta)$.
- Computes $s = k - vx$.
- Publishes (r_α, r_β, s).

Verifier(anyone)
- Computes $v = H(\alpha, y, \beta, z, r_\alpha, r_\beta)$.
- Verifies $r_\alpha \overset{?}{=} \alpha^s y^v$, $r_\beta \overset{?}{=} \beta^s z^v$.

Now let the public key of a recipient R be $y = g^x$. When the recipient gets an ElGamal ciphertext $(a, b) = (g^k, y^k m)$, he can recover the message m by decrypting it with his private key x and prove its correctness by exposing x. If he wants to prove the correctness of decryption without exposing x, he can do it using the following protocol.

Protocol 2. Proving the correctness of decryption

Let the public key of a recipient R be $y = g^x$. He wants to prove that m is the plaintext of a ciphertext $(a, b) = (g^k, y^k m)$ without exposing the private key x. He can prove it by showing $\log_g y = \log_a(b/m)$ using *Protocol 1*.

Lemma 2. *Protocol 2 proves the correctness of decryption.*

Proof. (sketch) The public key $y = g^x$ of the recipient R is publicly known. So showing $\log_g y = \log_a(b/m)$ is equivalent to showing $b/m = a^x$, which proves the correctness of decryption. □

3.4 Proving the equality or inequality of two discrete logarithms

Assume that the prover knows the discrete logarithm x of $y = \alpha^x$ for the base α. He wants to allow the verifier to decide whether $\log_\alpha y = \log_\beta z$ or $\log_\alpha y \neq \log_\beta z$ for given group elements β and z, such that no partial information about the logarithm x is leaked.

Protocol 1 provides only the proof of equality of two discrete logarithms. If this proof fails, all the arguments of prover are invalid. [MS97] provides the proof of equality or inequality at the same time. Zero-knowledge interactive proof system convinces only the verifier and requires interaction, but in our situation of matchmaking the prover(host of matchmaking) wants to convince all the participants, so we use the non-interactive version of their scheme which provides universal verifiability.

Protocol 3. Proving the equality or inequality of two discrete logarithms

The prover wants to convince the public whether the two elements y and z have the same discrete logarithm or not for the base α and β, respectively, without exposing the discrete logarithm, i.e., he wants to prove that either $\log_\alpha y = \log_\beta z$ or $\log_\alpha y \neq \log_\beta z$ holds. (α, β, y, z) are given as common input.

Prover(TTP)
- Chooses $k, k' \in_R Z_q^*$.
- Computes $r_\alpha = \alpha^k$, $r_\beta = \beta^k$, $r'_\alpha = \alpha^{k'}$ and $r'_\beta = \beta^{k'}$.
- Computes $v = H(\alpha, y, \beta, z, r_\alpha, r_\beta, r'_\alpha, r'_\beta)$.
- Computes $s = k - vx$ and $s' = k' - vk$.
- Publishes $(r_\alpha, r_\beta, r'_\alpha, r'_\beta, s, s')$.

Verifier(anyone)
- Computes $v = H(\alpha, y, \beta, z, r_\alpha, r_\beta, r'_\alpha, r'_\beta)$.
- Verifies $r_\alpha \stackrel{?}{=} \alpha^s y^v$, $r'_\alpha \stackrel{?}{=} \alpha^{s'} r_\alpha^v$, $r'_\beta \stackrel{?}{=} \beta^{s'} r_\beta^v$. If the verification fails, stop the verification process.
- If $r_\beta = \beta^s z^v$, then $\log_\alpha y = \log_\beta z$. Else if $r_\beta \neq \beta^s z^v$, then $\log_\alpha y \neq \log_\beta z$.

3.5 Finding collisions in ElGamal ciphertexts

When TTP has a public key $y = g^x$ and the corresponding private key x, the ElGamal encryption of a message m with TTP's public key y is given by $(a, b) = (g^k, y^k m)$ where $k \in_R Z_q^*$. Now assume that TTP has received two ciphertexts $(a_1, b_1) = (g^{k_1}, y^{k_1} m_1)$ and $(a_2, b_2) = (g^{k_2}, y^{k_2} m_2)$. Of course TTP can recover two messages m_1 and m_2 using his private key x and publish them, but TTP wants to convince the public whether the two ciphertexts have the same message or not, without exposing any partial information on the messages or his private key x. This task can be obtained as follows using *Protocol 3*.

Protocol 4. Finding collisions in ElGamal ciphertexts

Let $y = g^x$ be TTP's public key and x be his private key. Given two ElGamal ciphertexts $(a_1, b_1) = (g^{k_1}, y^{k_1} m_1)$ and $(a_2, b_2) = (g^{k_2}, y^{k_2} m_2)$, TTP wants to prove whether the two ciphertexts have the same message or not, without exposing any partial information on the messages or his private key x.

Prover(TTP)
- Computes $(a_3, b_3) \equiv (a_1/a_2, b_1/b_2) = (g^{k_1 - k_2}, y^{k_1 - k_2} m_1/m_2)$.
- Proves that $\log_g y = \log_{a_3} b_3$ or $\log_g y \neq \log_{a_3} b_3$ using *Protocol 3*.

Verifier(anyone)
- If $\log_g y = \log_{a_3} b_3$, then $m_1 = m_2$. Else if $\log_g y \neq \log_{a_3} b_3$, then $m_1 \neq m_2$.

Lemma 3. *Protocol 4 proves whether two ciphertexts (a_1, b_1) and (a_2, b_2) have the same plaintext or not.*

Proof. (sketch) TTP's public key $y = g^x$ is publicly known. So proving $\log_g y \overset{?}{=} \log_{a_3} b_3$ is equivalent to proving $a_3^x \overset{?}{=} b_3$, which proves $m_1 \overset{?}{=} m_2$. $\qquad\square$

Protocol 4 can be used in wide range of applications where just the proof of equality or inequality of plaintext is required while the plaintext should not be recovered.

4 Proposed matchmaking protocol

In this section we describe our matchmaking protocol which is constructed using the primitives described above. The participants are m male members $M_i (i = 1, ..., m)$, n female members $F_j (j = 1, ..., n)$ and a host T. They use the bulletin board as a public communication channel.

The proposed matchmaking protocol consists of 5 stages: registration, commitment, opening, proof of coupling, and verification. In the sequel we will describe the case that M_i and F_j have chosen each other for the ease of description.

Stage 1. Registration

Any applicant who wants to participate in matchmaking should register to T. Each participant does the following steps:

- M_i chooses a random number $a_i \in_R Z_q^*$ and computes his temporal ID $m_i = g^{a_i}$. Likewise F_j chooses a random number $b_j \in_R Z_q^*$ and computes her temporal ID $f_j = g^{b_j}$.
- M_i presents to T his name, his public key with certificate, and his temporal ID. Likewise F_j presents to T her name, her public key with certificate, and her temporal ID.

– T publishes these information on the bulletin board.

Stage 2. Commitment (by participants)

In commitment stage each participant chooses a member from the other group as a possible partner, generates a couple ID for the choice, generates a public commitment by encrypting the couple ID with TTP's public key, and posts it on the bulletin board. Note that each participant is allowed to commit a single choice.

To generate the couple ID, we use the Diffie-Hellman key agreement technique. A couple ID between M_i and F_j is given by $CID(M_i, F_j) = H(m_i^{b_j}) = H(f_j^{a_i}) = H(g^{a_i b_j})$, and it can be computed only by M_i and F_j.

Each participant does the following steps:

– Chooses a possible partner from the other group.
– Computes a couple ID for the choice. If M_i has chosen F_j, he generates his couple ID with his random number a_i and the partner's published temporal ID f_j such that $CID(M_i, F_j) = H(f_j^{a_i})$.
– Generates a public commitment for the choice and posts it on the bulletin board. M_i's public commitment is given by $Env(i, T, CID(M_i, F_j))$.

Stage 3. Opening (by TTP)

When the deadline of commitment has passed, TTP tries to find couples using *Protocol 4* for every possible pairs of participants (M_i, F_j) where $i = 1, ..., m$ and $j = 1, ..., n$. TTP posts the results and their proofs on the bulletin board.

For the established couple members, TTP additionally decrypts their commitments and publishes the couple IDs, which demonstrate that the two commitments are equal. TTP proves the correctness of his decryption using *Protocol 2*.

TTP does the following steps:

– TTP tries to find couples for every possible pairs (M_i, F_j) using *Protocol 4* and posts all the results and proofs on the bulletin board.
– For the established couple members, TTP decrypts their public commitments to get the colliding couple ID and publishes them on the bulletin board. He provides proofs of correctness of the decryptions using *Protocol 2*.

Stage 4. Proof of coupling (by established couple members)

The participants who were published as a couple in the opening stage have to show that the colliding couple ID is authentic. A simple way to prove the authenticity of couple ID is showing the random number corresponding to participant's temporal ID. $CID(M_i, F_j)$ can be proven authentic by verifying $CID(M_i, F_j) = H(f_j^{a_i})$ if M_i opens a_i or by verifying $CID(M_i, F_j) = H(m_i^{b_j})$ if F_j opens b_j.

If the established couple members do not want to open their secret random numbers, the authenticity of $CID(M_i, F_j)$ can be proven using *Protocol 2*. M_i can prove $\log_g m_i = \log_{f_j}(CID(M_i, F_j))$ without exposing a_i.

One of the couple members cannot repudiate his or her commitment while the other member does not want to repudiate, because both of them can prove the authenticity of $CID(M_i, F_j)$. If both of them refuse to prove it(changed their mind together), they will not be decided as a couple. Based on their proofs of coupling, everyone can verify the correctness of coupling.

Stage 5. Verification (by anyone)

Anyone can verify the correctness of results by:

- Verify the correctness of TTP's opening in stage 3.
- Verify the authenticity of the colliding couple ID published in stage 4.
- Verify the couple member's signature for the public commitments posted in stage 2.

5 Security analysis

Our proposed matchmaking protocol satisfies all the listed security requirements.

Theorem 1. *The proposed matchmaking protocol is secure in the sense that it satisfies secrecy, fairness, correctness, authenticity, and non-repudiation.*

Proof. (sketch)

Secrecy(privacy) All the commitments are encrypted with TTP's public key. So if TTP does not help, any participant cannot get any partial information on the choices of others. Although TTP can decrypt a commitment, he cannot get any information on the choice without help of the specific couple members. Under the computational Diffie-Hellman assumption, identifying the couple (M_i, F_j) from $H(g^{a_i b_j})$ is computationally infeasible. Only the choices of established couple members are published by TTP in the opening stage.

Fairness If TTP does not help, anyone cannot be in advantageous position than others in the commitment stage. If TTP helps a participant by secretly opening other group member's commitment, he can check whether there is any choice for him by trying all the possible pairs.

Correctness Correctness of result is guaranteed because all the procedures are publicly verifiable. In opening stage, TTP tries to find couples for every possible pairs of participants and provides the proofs of result using *Protocol 4* whose correctness is given by Lemma 3. For the established couple members, TTP additionally decrypts their commitments and publishes the colliding couple ID. TTP proves the correctness of his decryption using *Protocol 2* whose correctness is given by Lemma 2. In proof of coupling stage, the established couple members prove the authenticity of the colliding couple ID by showing their random number.

Authenticity In commitment stage, public commitment is signed by the sender. It guarantees the authenticity of commitment and the fact that each registered participant has committed a single choice is verified.

Non-repudiation In commitment stage, the committed message is signed by the sender. So the established couple members cannot repudiate their commitments later. In proof of coupling stage, one of the couple member cannot repudiate the commitment while the other member does not want to repudiate because both of them can prove the authenticity of the colliding couple ID.

□

We can consider a situation when TTP is not honest. If TTP decrypts the public commitments and exposes them intentionally, each pair of participants can check whether the other participant has chosen him or her because both of them can generate the corresponding couple ID. Therefore TTP should not decrypt the commitments and expose them.

Our proposed matchmaking scheme requires $O(mn)$ computation in opening stage because TTP has to try to find couples for every possible pairs of participants. Enhancing the efficiency of the protocol is left as further study.

6 Conclusion

We introduce a secure matchmaking protocol as a new application of secure multiparty computation and construct a simple and efficient protocol under the simple rule that two participants become a couple only when they have chosen each other. To implement secure matchmaking, we use various primitives of zero-knowledge proofs: proving the correctness of decryption, proving the equality or inequality of two discrete logarithms, and finding collisions in ElGamal ciphertexts. We also define the concept of public commitment scheme and show that enveloped message(a combination of public key encryption and digital signature) can be used for public commitment. Our basic approach is that participants commit their choices secretly by encrypting them with TTP's public key and TTP tries to find collisions in ElGamal ciphertexts without decryption.

The main security issue is the honesty of TTP which was assumed in this study. If TTP colludes with a participant, he can get partial information on other participants' choices. The computational load of TTP in opening stage is $O(mn)$ because he has to try to find couples for every possible pairs of participants. In this sense more intensive study to enhance the efficiency of the protocol will be challenging very much.

To the best of our knowledge, this is the first trial which applies cryptographic primitives to the problem of matchmaking. We believe that our result can play a significant role for narrowing the gap between cryptographic theory and real multiparty applications.

Acknowledgements

We are grateful to Moti Yung for valuable comments on commitment scheme and to the anonymous referees for their helpful comments.

References

[BGW88]   M. Ben-Or, S. Goldwasser and A. Wigderson, "Completeness theorems for non-cryptographic fault-tolerent distributed computation", 20th STOC, pages 1–10, 1988.

[Can96]   R. Canetti, "Studies in Secure Multiparty Computation and Applications", PhD Thesis, The Weizmann Institute of Science, 1996.

[CCD88]   D. Chaum, C. Crepeau and I. Damgard, "Multiparty unconditionally secure protocols", 20th STOC, pages 11–19, 1988.

[CGS97]   R. Cramer, R. Gennaro, and B. Schoenmakers, "A secure an optimally efficient multi-authority election schemes", In *Advances in Cryptology – Eurocrypt '97, LNCS Vol. 1233*, pages 103–118, Springer-Verlag, 1997.

[CP93]   D. Chaum and T. Pedersen, "Wallet databases with observers", In *Advances in Cryptology – Crypto'92, LNCS Vol. 740*, pages 89–105, Springer-Verlag, 1993.

[Cra99]   R. Cramer, "Introduction to Secure Computation", In Lectures on Data Security - Modern Cryptology in Theory and Practice, Ivan Damgaard (Ed.), Springer LNCS Tutorial, vol.1561, pages 16–62, 1999.

[FS87]   A. Fiat and A. Shamir, "How to prove yourself: Practical solutions to identification and signature problems", In *Advances in Cryptology - Crypto'86*, pages 186–194, Springer-Verlag, 1987.

[Gol98]   O. Goldreich, "Secure Multi-Party Computation", Manuscript version 1.1, 1998.

[GMW87]   O. Goldreich, S. Micali and A. Wigderson, "How to play any mental game", 19th STOC, pages 218–229, 1987.

[MS97]   M. Michels and M. Stadler, "Efficient convertible undeniable signature", Proc. of 4th annual workshop on selected areas in cryptography, 1997.

[SJ00]   C.P. Schnorr and M. Jakobsson, "Security of signed ElGamal encryption", In *Advances in Cryptology - Asiacrypt'2000, LNCS V. 1976*, pages 73–89, Springer-Verlag, 2000.

[TY98]   Y. Tsiounis and M. Yung, "On the Security of ElGamal Based Encryption", *PKC'98, LNCS 1431*, pages 117-134, Springer-Verlag, 1998.

[Yao82]   A. Yao, "Protocols for Secure Computation", 23th FOCS, pages 160–164, 1982.

An Improved Scheme of the Gennaro-Krawczyk-Rabin Undeniable Signature System Based on RSA

Takeru Miyazaki

Kyushu Institute of Technology,
680-4 Kawazu Iizuka 820-8502, Japan.
E-mail:takeru@capricorn.cse.kyutech.ac.jp

Abstract. Digital signatures are important applications of public key cryptography in today's digital networks. However, they have a problem that anyone can verify the signature even if a signer wants to restrict the verifiers to confirm his signatures. D. Chaum et al.[1] proposed undeniable signatures to solve this problem. These signatures are based on the discrete logarithm problem and are extended to those with different properties [2–4]. After that, R.Gennaro et al.[5] proposed another undeniable signature scheme based on RSA. However, this scheme also has following problems. Firstly, the undeniable signature of them cannot be converted into a usual signature individually. So if a user wants to use both of the undeniable signature and the usual signature, he must prepare separate parameters for each type of signatures. Secondly, the denial protocol is not deterministic because it uses a zero knowledge interactive proof. So it is not efficient. Thirdly, their signature system cannot resist hidden verifier attack[7]. In this paper we will propose an improved scheme to solve these problems.

1 Introduction

Nowadays the use of public key cryptosystems is important not only for the encryption but also for the realization of a digital signature system. The digital signature can produce the authentication of digital contents, i.e., anybody can verify whether the signer is the owner of these digital contents by checking their signature.

Usual digital signature allows that anybody can verify the signature. This is not desirable in case a signer wants to restrict the verifier.

Undeniable signatures proposed by D. Chaum and H. V. Antwerpen[1] is useful for signer to restrict the verifier, since signature cannot be verified without the signer's cooperation.

The signature scheme has both confirmation protocol and denial protocol, so that a signer does not refuse the request for verifying his signature. The signer uses the confirmation protocol to verify the valid signature and the denial protocol to prove the falseness of the invalid signature.

Chaum's undeniable signature scheme is based on the discrete logarithm problem. Its variations[2–4] also exist with different additional properties.

After that, R. Gennaro, H. Krawczyk and T. Rabin[5] proposed another undeniable signature scheme that is based on RSA, i.e., the factoring problem. The scheme can be used only for the undeniable signature, since the public key for encryption in usual RSA system is used as a private key in the scheme. The undeniable signature can be converted into in the usual RSA signature. However, by this type of conversion, all the previous undeniable signatures are converted to the usual RSA signatures which can be verified by anyone. This is one problem of the undeniable signature by Gennaro-Krawczyk-Rabin. Another problem is that it suffers from hidden verifier attack[6, 7]. The denial protocol is not deterministic and needs large costs for calculations and communications, since it is based on ZKIP.

In this paper we'll propose an improved scheme of Gennaro-Krawczyk-Rabin's one to solve these problems. At first, we'll extend RSA key pair to a product of a public key and two private keys. When a signer wants to generate a usual RSA signature, he uses the product of two private keys as a private key of the RSA system. On the other hand, when he wants to generate our undeniable signature, he uses the product of the public key and one of the private keys as a key to sign a message. We'll improve the probabilistic denial protocol in the Gennaro-Krawczyk-Rabin's scheme to the deterministic one proposed on Chaum's scheme. In addition, we'll also improve the commitment protocol in order to resist hidden verifier attack.

The paper is organized as follows. Some of definitions and assumptions are given in Section 2. Protocols of undeniable signature scheme proposed by Gennaro et al. is given in Section 3. Some problems of their scheme is given in Section 4. Our new undeniable signature scheme is given in Section 5.

2 Some of Definitions and Assumptions

In this section, we summarize some definitions and assumptions about security of cryptosystem. They are necessary to guarantee the security of protocols in this paper.

2.1 Factoring Problem

In this paper, we define "large primes p, q" satisfying following assumptions.

1. No Efficient Factoring Scheme

   Anyone who has no information about neither p nor q cannot calculate both p and q from $n = pq$ efficiently.

2. Impossibility of Exhausting Factoring

   Anyone doesn't have calculating power to execute exhausting factoring of n.

2.2 Security Assumption of RSA

Let p, q are large primes. Let $n = pq, p \neq q$ and $L = \text{LCM}((p-1), (q-1))$. Here we assume the limit of the calculation power as follows.

Let e be a number satisfying the equation $(e, L) = 1$. Then anyone who knows only n and e cannot calculate $d = e^{-1} \bmod L$ without the information about L.

2.3 Proper Public Information

In this paper, anyone can publish any types of public information properly, i.e., he can provide his public information to others without altering and lacking. This assumption is useful to realize PKI(Public Key Infrastructure) systems[8].

2.4 Proof of the Validity of n

We define p, q, p', q' as distinct large primes satisfying

$$p = 2p' + 1, q = 2q' + 1. \tag{1}$$

In this paper, we assume that anyone who publish only $n(= pq)$ can prove that n is the product of large primes p, q and p, q satisfy the equation (1) without publishing any information about p and q. This assumption is useful to realize the Zero-Knowledge proof system[9].

3 Gennaro-Krawczyk-Rabin's Undeniable Signatures

In this section, we summerize the undeniable signature scheme proposed by R. Gennaro, H. Krawczyk and T. Rabin[5].

3.1 System and Key Generation

Gennaro et al. proposed an undeniable signature scheme based on RSA. In this scheme, the keys of each user are generated by the system that is trusted by the user.

The system first generates the set $\mathcal{N}$ as,

$$\mathcal{N} = \{n \mid n = pq, \ p < q, \ p = 2p' + 1, q = 2q' + 1 \ (p, q, p', q' : \text{large primes}) \}$$

and prepares following parameters for each of his users;

1. Choose $n \in \mathcal{N}$.
2. Choose e, d so that $ed \equiv 1 \pmod{\phi(n)}$
3. Choose (w, S_w) as $w \in Z_n{}^*$, $w \neq 1$, $S_w \equiv w^d \pmod{n}$.

Then he opens (n, w, S_w) to the public as public keys and gives only the user (e, d) as private keys.

P : Prover	V : Verifier
	1, **Choose** i, j where $1 \leq i, j \leq n$ at random **Calculate** Q : $$Q \stackrel{\triangle}{\equiv} \hat{S}_m^i S_w^{\,j} \pmod{n}$$
$\Longleftarrow Q$	
2, **Calculate** A from Q $$A \stackrel{\triangle}{\equiv} Q^e \pmod{n}$$	
$A \Longrightarrow$	
	3, **Check** $A \stackrel{?}{\equiv} m^i w^j \pmod{n}$ If equality holds then V accepts $\hat{S}_m$ as P's signature for m.

Fig. 1. Gennaro-Krawczyk-Rabin's Signature Confirmation Protocol

3.2 Generation of Signatures

The process for generating of signatures by this scheme is similar to that by the usual RSA signature. A signer prepares a message m which is the hash value of a plain text. Then he calculates the signature S_m as,

$$S_m \equiv m^d \pmod{n}. \tag{2}$$

3.3 Signature Confirmation

The confirmation of signature needs the signer's cooperation. This property allows the signer to restrict the verification of his signature. Then the receiver must execute a signature confirmation protocol to confirm validity of the received signature.

We define a signer of a signature as P(Prover) and a verifier of the signature as V(Verifier). Assume that V has a $(m, \hat{S}_m)$, where P asserts that $\hat{S}_m$ is his signature for the message m. Figure 1 shows Gennaro-Krawczyk-Rabin's signature confirmation protocol.

If $\hat{S}_m$ is a valid signature for m, i.e., $\hat{S}_m \equiv m^d \pmod{n}$, then the value A calculated by P becomes

$$A \equiv Q^e \equiv (\hat{S}_m^i S_w^{\,j})^e \equiv (m^{di} w^{dj})^e \equiv (m^i w^j)^{ed} \equiv m^i w^j \pmod{n},$$

so the equation in phase 3 in Figure 1 is right.

3.4 Denial of Signature

Here we present the denial protocol of Gennaro-Krawczyk-Rabin's scheme. This protocol is different from their signature confirmation protocol and is executed when a signer wants to prove that the signature shown by the verifier is false. Now we define the false signature $\hat{S}_m$ of a message m as $\hat{S}_m \not\equiv m^d \pmod{n}$. This protocol shows that $\hat{S}_m$ isn't the valid signature for m. So if a signer cannot execute this protocol, he cannot prove that $\hat{S}_m$ is the false signature for a message m.

Assume that V has $(m, \hat{S}_m)$, where $\hat{S}_m$ is considered as the P's signature of the m but P wants to show that $\hat{S}_m$ is not a valid signature for m. Figure 2 shows Gennaro-Krawczyk-Rabin's denial protocol.

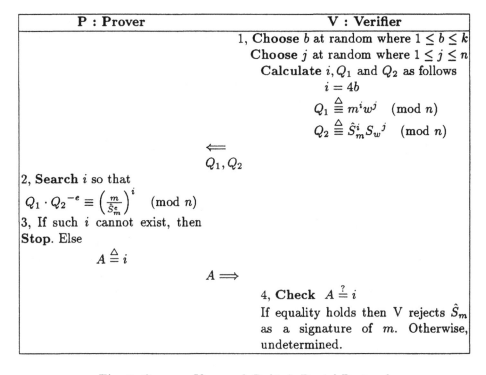

Fig. 2. Gennaro-Krawczyk-Rabin's Denial Protocol

The size of k is at most $1024 = 2^{10}$ because P specifies i by exhaustive search in phase 2.

If P is dishonest and the signature $\hat{S}_m$ is valid, then the equation in phase 2 becomes

$$Q_1 Q_2^{-e} \equiv 1 \pmod{n}$$

and is useless for finding i. The probability is $1/k$ that P will succeed in finding such as i as $A = i$ in phase 3. The probability of P's cheating V is at most $1/2^{100}$ when V executes this protocol at least ten times.

On the other hand, though V doesn't send the right value Q, he cannot get any information from P because P can stop this protocol at phase 3.

4   Some Problems of Gennaro-Krawczyk-Rabin's Scheme

In this section we point out some problems in Gennaro-Krawczyk-Rabin's scheme.

1. Convertibility

   The way of converting the undeniable signature into usual one is to open the private key e to the public. This approach makes all his previous and future undeniable signatures to usual signatures.

   However, the convertibility of the undeniable signature schemes based on the discrete logarithm problem includes to convert each of all undeniable signatures into usual ones individually.

2. Resistance to the Hidden Verifier Attack

   Y. Desmedt and M. Yung pointed out existence of the hidden verifier in the undeniable signature[6]. We define the entity that isn't allowed to verify the signer P's undeniable signature as the hidden verifier H, and the entity that is allowed to verify that signature as the verifier V. H can verify P's signature by threatening V to verify it for H, although P doesn't accepts verifying request directly from H.

   M. Jakobsson, K. Sako and R.Impagliazzo proposed the protocol that can resist hidden verifier attacks[7]. However, Gennaro-Krawczyk-Rabin's scheme doesn't include this protocol. So we adapt their protocol to this scheme.

3. Deterministic Denial Protocol

   The denial protocol of their scheme is based on ZKIP, so P and V must execute the denial protocol many times. This means that they must have large costs of calculations and communications. However, an undeniable signature schemes based on the discrete logarithm problem with deterministic denial protocol exists[10].

5   New Scheme

In this section we propose new undeniable signature scheme based on the factoring problem.

5.1   Key Generating System

In Gennaro-Krawczyk-Rabin's scheme, the calculation of key generation is executed by the system. This means that each user can use this scheme without execution of it. It also means that the system has information of private keys which he has assigned his users. So all of his user's private information are also leaked, if the secret information of the system has been leaked. Then the system is required high security against some attacks.

We consider another way to generate keys. In our proposal scheme, some programs executed by each user realize the system of key generation. This means that each user generates his keys by himself.

The system first generates the same set $\mathcal{N}$ as the Gennaro-Krawczyk-Rabin's scheme,

$$\mathcal{N} = \{n \mid n = pq, \ p < q, \ p = 2p' + 1, q = 2q' + 1 \ (p, q, p', q' : \text{large primes}) \}$$

and each user generates following parameters.

1. Choose n as $n \in \mathcal{N}$.
2. Calculate $L = \mathbf{LCM}(p - 1, q - 1) = 2p'q'$
3. Choose two odd numbers e, d_2 where $3 \leq e, d_2 \leq L - 1$
4. Calculate d_1 as $d_1 \equiv (ed_2)^{-1} \pmod{L}$
5. Calculate d as $d \equiv d_1 d_2 \pmod{L}$
6. Choose (w, S_w) as $w \in Z_n{}^*$, $w \neq 1$, $S_w \equiv w^d \pmod{n}$

Then the user opens (e, n, w, S_w) to the public as public keys and keeps (d_1, d_2, d) secret as private keys.

The relation of keys e, d_1, d_2 and d becomes

$$^\forall M \in Z_n \ ; \ M^{ed} \equiv M^{ed_1 d_2} \equiv M \pmod{n}, \tag{3}$$

i.e., the relation between the public key e and the private key d is the same as that between the key pair of the usual RSA and private key d is product of two keys d_1, d_2.

The cost of generating these keys is negligible.

5.2   Generation of a Usual RSA Signature

Our scheme can generate the same signature as that of a usual RSA because it has all of parameters used in the usual RSA.

The generation and the verification of the usual signature for a message m are as follows.

– Generation of the usual signature:

$$S \equiv M^d \pmod{n}$$

– Verification of the usual signature:

$$M \stackrel{?}{\equiv} S^e \pmod{n}$$

5.3   Generation of an Undeniable Signature

The undeniable signature S_m for a message m can be computed by using the public key e and one of the private key d_1 as follows.

$$S_m = m^{ed_1} \pmod{n} \tag{4}$$

5.4 Signature Confirmation

Here we will present a confirmation protocol of our undeniable signature scheme. We define a signer as P : the prover of his signature and a verifier as V : the verifier of P's signature. A prover P must have some parameters which are defined in section 5.1 but V doesn't have to generate them. V must calculate only n_v which is the product of two large primes p_v, q_v and open the value of n_v to the public. This means that V's parameters are allowed not only RSA's one but also one of other RSA-type encryption schemes (like Rabin, Wiliams, and so on). Of course, V's parameters belong to those defined in section 5.1. We define the parameters of P and V as follows.

	Public Information	Private Information
P	n_p, e_p, w_p, S_{wp}	$p_p, q_p, d_p, d_{1p}, d_{2p}$
V	$n_v, (e_v, w_v, S_{wv})$	$p_v, q_v, (d_v, d_{1v}, d_{2v})$

The parameters in parentheses are not necessary to execute our scheme.

V has already got $(m, \hat{S}_m)$, where P has asserted that $\hat{S}_m$ is a signature for the message m. From the equation (4), the P's valid signature S_m for the message m is

$$S_m \equiv m^{e_p d_{1p}} \pmod{n_p}.$$

Then Figure 3 exhibits our signature confirmation protocol.

This protocol uses the equation as follows,

$$(M^{e_p d_{1p}})^{d_{2p}} \equiv M^{e_p d_{1p} d_{2p}} \equiv M^{e_p d_p} \equiv M \pmod{n_p}.$$

Now we consider some situations about this protocol.

If $\hat{S}_m$ is valid, i.e., $\hat{S}_m \equiv m^{e_p d_{1p}} \pmod{n_p}$ in phase 2 of Figure 3, then the value a which P calculates becomes as follows.

$$a \equiv Q^{d_{2p}} \pmod{n_p}$$
$$\equiv S_m{}^{i d_{2p}} S_{wp}{}^{j d_{2p}} \pmod{n_p}$$
$$\equiv m^{i e_p d_{1p} d_{2p}} w_p{}^{j e_p d_{1p} d_{2p}} \pmod{n_p}$$
$$\equiv m^i w_p{}^j \pmod{n_p}$$

In phase 3, P defines A as the odd one of $\pm a \pmod{n}$ and sends V the value C which is a random number x to the A-th power. V receives i and j after getting C. When x is one of $\{1, -1$ or $0\}$, the equalities in phase 5 always holds. So x must satisfy $2 \le x \le (n_v - 2)$ and V must check that x sent by P satisfies the relation in phase 5. In phase 4, P checks the value Q from i and j. If Q is right, then he sends A and x. In phase 5, V verifies the signature to check C and A^2. The reason of checking A^2 instead of A is to satisfy the relation both the value $\pm a$ which is selected by P.

Next we consider the case that dishonest P wants to deceive verifier V, i.e., P wants to deceive V into recognizing invalid signature $\hat{S}_m \not\equiv m^{e_p d_{1p}} \pmod{n_p}$ as valid signature for a message m. At first, P signs another message m' and generates its signature $\hat{S}_m \equiv m'^{e_p d_{1p}} \pmod{n_p}$ where $m \ne m'$. Then he asserts

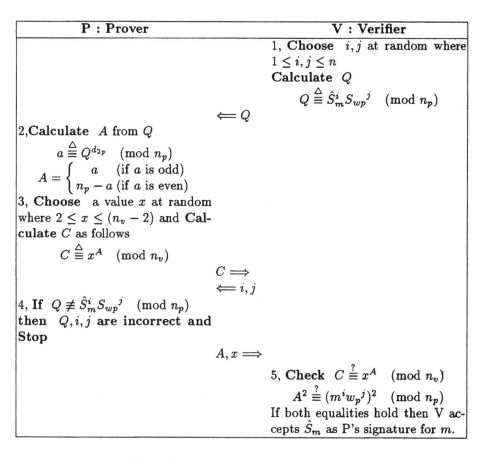

Fig. 3. Signature Confirmation Protocol

V that $\hat{S}_m$ is the signature for m. V wants to execute this confirmation protocol to check P's assertion. V chooses i and j and sends $Q \equiv \hat{S}_m^i S_{wp}^{\ j}$ (mod n_p). P can calculate $a \equiv Q^{d_{2p}}$ (mod n_p) as follows.

$$a \equiv Q^{d_{2p}}$$
$$\equiv (\hat{S}_m^i S_{wp}^{\ j})^{d_{2p}}$$
$$\equiv (m'^{e_p d_{1p} i} w^{e_p d_{1p} j})^{d_{2p}}$$
$$\equiv (m'^i w^j)^{e_p d_{1p} d_{2p}}$$
$$\equiv m'^i w^j \quad (\text{mod } n_p)$$

On the other hand, he must receive the value A as follows,

$$A \equiv \pm m^i w_p^{\ j} \quad (\text{mod } n_p)$$

to pass the check of this protocol. However, he cannot calculate such A because he cannot specify both i and j from Q, m and m'. So no one can deceive any verifier into recognizing invalid signature $\hat{S}_m$ as a valid signature with this protocol.

In addition, P doesn't also select appropriate values C, A and x to pass the check in phase 5 because the calculation of C demands both A and x and the calculation of A demands both i and j. However, P must send C before he gets i and j. So he cannot calculate these appropriate values.

Next we consider a situation that V is dishonest. In this case, the dishonest V cannot request incorrect value Q because if Q isn't true value, i.e., $Q \not\equiv \hat{S}_m^i S_{wp}^j$ (mod n_p), then P finds this fact from the check in phase 4.

Finally, we consider another situation that a dishonest V wants to generate prover P's signature for another message $\overline{m}$ by using this protocol. The relation between Q and A is

$$A^2 \equiv Q^{2d_{2p}} \pmod{n_p}$$
$$A^{2e_p d_{1p}} \equiv Q^{2e_p d_{1p} d_{2p}} \equiv Q^2 \pmod{n_p}$$

i.e., Q is the signature for either of $\pm A$. If one of $\pm A$ is a message $\overline{m}$, then V can assert Q as the P's signature $S_{\overline{m}}$. However because of the following reason, it is difficult for V to calculate such a Q. From $A \equiv \overline{m}$, we can show that

$$m^i w_p^j \equiv \overline{m} \pmod{n_p}$$

so that if V can specify both i and j satisfying this condition, he can calculate Q as $Q \equiv S_m^i S_w^j \pmod{n}$. However, specifying these values is as difficult as the discrete logarithm problem. So no one can fabricate any prover's signature by using this protocol.

Generally the commitment scheme used in this protocol has indicated an attack using a product lots of numbers. The probability that this attack has effects is decided by $\#(A)$, i.e., the number of different A's. We have

$$\#(A) = \frac{n_p + 1}{2}$$

because A is an odd number in $1 \leq A \leq n_p$. So in this protocol, it is easy to see that the probability is very small, i.e., this attack isn't serious menace to break our proposal scheme.

5.5 Denial of Signature

Here we present the denial protocol of our scheme. This protocol is different from the signature confirmation protocol shown above and is executed when a signer wants to prove that the signature shown by verifier is false. Now we define the false signature $\hat{S}_m$ of a message m as $\hat{S}_m \not\equiv m^{ed_1} \pmod{n}$. This protocol shows that $\hat{S}_m$ isn't the signature for m. So if a signer cannot execute this protocol, he cannot assert that $\hat{S}_m$ is the invalid signature of a message m. Figure 4 shows our denial protocol.

Signature confirmation protocol
using the parameters $\{i_1, j_1, Q_1, a_1, A_1, x_1, C_1\}$
as $\{i, j, Q, a, A, x, C\}$ in Section 5.4

Verifier V checks that the signature is invalid on these parameters as follows.

1-1, **Check** If $C_1 \overset{?}{\not\equiv} x_1{}^{A_1} \pmod{n_v}$,

then A_1, x_1 are **incorrect** and **Stop.**

1-2, **Check** If $A_1{}^2 \overset{?}{\equiv} (m^{i_1} w_p{}^{j_1})^2 \pmod{n_p}$,

then the signature is **valid** and **Stop.**

Repeat of signature confirmation protocol
using the parameters $\{i_2, j_2, Q_2, a_2, A_2, x_2, C_2\}$
as $\{i, j, Q, a, A, x, C\}$ in Section 5.4

V checks that the signature is invalid on these parameters as follows.

2-1, **Check** If $C_2 \overset{?}{\not\equiv} x_2{}^{A_2} \pmod{n_v}$,

then A_2, x_2 are **incorrect** and **Stop.**

2-2, **Check** If $A_2{}^2 \overset{?}{\equiv} (m^{i_2} w_p{}^{j_2})^2 \pmod{n_p}$,

then the signature is **valid** and **Stop.**

Final Check

V checks that these confirmation protocols are executed correctly as follows.

3, **Check** If $(A_1 w_p{}^{-j_1})^{2i_2} \overset{?}{\equiv} (A_2 w_p{}^{-j_2})^{2i_1} \pmod{n_p}$,

then the signature is **invalid.**

Fig. 4. Signature Denial Protocol

This protocol is based on Chaum's undeniable signature[10]. We improve this protocol to adapt Jakobsson-Sako-Impagliazzo's one[7] to resist the hidden verifier attacks.

This protocol consists of three parts. In the first and second part, a prover P and a verifier V execute the signature confirmation protocol twice by changing random values. Then in the last part, V checks that P executes them correctly.

In this protocol, the signature confirmation protocol is used to check whether the signature is invalid or the protocol stops if the signature is valid.

If this protocol is executed correctly, the equality in phase 3 of the final check holds because

$$(A_1 w_p{}^{-j_1})^{2i_2} \equiv (\pm \hat{S}_m^{d_{2p} i_1} w_p{}^{j_1} w_p{}^{-j_1})^{2i_2} \pmod{n_p}$$
$$\equiv \hat{S}_m^{2d_{2p} i_1 i_2} \pmod{n_p}$$
$$(A_2 w_p{}^{-j_2})^{2i_1} \equiv (\pm \hat{S}_m^{d_{2p} i_2} w_p{}^{j_2} w_p{}^{-j_2})^{2i_1} \pmod{n_p}$$

$$\equiv \hat{S}_m^{2d_{2p}i_1i_2} \pmod{n_p}.$$

On the other hand, when dishonest P wants to deny the valid signature S_m, he selects one of two cases the following.

1. P ignores the values Q_1, Q_2

   In this case, P selects (A_1, C_1, x_1) and (A_2, C_2, x_2) which satisfy $C_1 \equiv x_1^{A_1}$, $C_2 \equiv x_2^{A_2}$. Then this protocol can execute checks of 1-1 to 2-2 and the equality holds in phase 3 of the final check when the equation

   $$A_2^{2i_1} \equiv (A_1 w_p^{-j_1})^{2i_2} w_p^{2j_2} \pmod{n_p}$$

   is satisfied. However, P cannot specify A_2 because he cannot know i_2 when C_2 is calculated in the second check.

2. P uses the values Q_1, Q_2

   In this case, P uses $\hat{A}_1, \hat{A}_2$ instead of correct A_1, A_2 as follows.

   $$\hat{A}_1 \equiv Z_1 Q_1^{d_{2p}} \pmod{n_p}$$
   $$\equiv Z_1 m^{i_1} w_p^{j_1} \pmod{n_p}$$
   $$\hat{A}_2 \equiv Z_2 Q_2^{d_{2p}} \pmod{n_p}$$
   $$\equiv Z_2 m^{i_2} w_p^{j_2} \pmod{n_p}$$
   $$(Z_1, Z_2 \neq 0, 1)$$

Then this protocol also can execute checks of 1-1 to 2-2 and the equality holds in phase 3 of the final check when the following equation is satisfied.

$$(\hat{A}_1 w_p^{-j_1})^{2i_2} \equiv (\hat{A}_2 w_p^{-j_2})^{2i_1} \pmod{n_p}$$
$$(Z_1 m^{i_1})^{2i_2} \equiv (Z_2 m^{i_2})^{2i_1} \pmod{n_p}$$
$$Z_1^{2i_2} \equiv Z_2^{2i_1} \pmod{n_p}$$

The order of any number on modulo n_p must be one of $\{1, 2, p_p', 2p_p', q_p', 2q_p', p_p'q_p', 2p_p'q_p'\}$ [5]. Then if the order of either Z_1 or Z_2 is bigger than 2, then it is hard to find Z_1 and Z_2 satisfying above relations. So the order of both Z_1 and Z_2 must be either 1 or 2. In this case, the protocol must stop at the check of both 1-2 and 2-2 because they check A_1^2, A_2^2 instead of A_1, A_2, and it always satisfies that both $\hat{A}_1^2 \equiv A_1^2$ and $\hat{A}_2^2 \equiv A_2^2$ on this case.

These mean that the dishonest P cannot deny his valid signatures.

5.6 Resistance to Hidden Verifier Attack

We will show that in our scheme, the verifier can resist confirming any of undeniable signature forced by hidden verifiers. We define a signer of the undeniable signature as P, a verifier whom P allows to verify it as V and a hidden verifier whom P doesn't allow to verify it as H. H forces V to execute the signature confirmation protocol of it by using the value decided by H instead of V and check C, A and x indirectly. Figure 5 shows a detail of this situation.

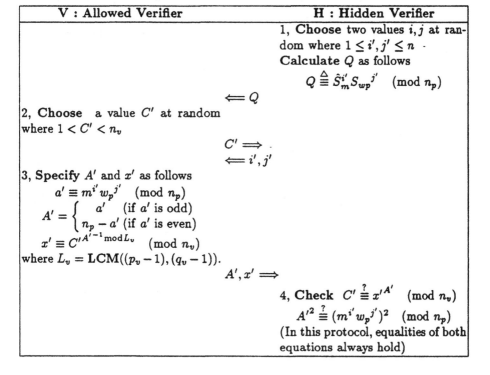

Fig. 5. Signature Confirmation by Hidden Verifier

V : Allowed Verifier	H : Hidden Verifier
	1, **Choose** two values i, j at random where $1 \le i', j' \le n$. Calculate Q as follows $$Q \triangleq \hat{S}_m^{i'} S_{wp}^{j'} \pmod{n_p}$$
	$\Longleftarrow Q$
2, **Choose** a value C' at random where $1 < C' < n_v$	
	$C' \Longrightarrow$.
	$\Longleftarrow i', j'$
3, **Specify** A' and x' as follows $$a' \equiv m^{i'} w_p^{j'} \pmod{n_p}$$ $$A' = \begin{cases} a' & \text{(if } a' \text{ is odd)} \\ n_p - a' & \text{(if } a' \text{ is even)} \end{cases}$$ $$x' \equiv C'^{A'^{-1} \bmod L_v} \pmod{n_v}$$ where $L_v = \text{LCM}((p_v - 1), (q_v - 1))$.	
	$A', x' \Longrightarrow$
	4, **Check** $C' \stackrel{?}{\equiv} x'^{A'} \pmod{n_v}$ $A'^2 \stackrel{?}{\equiv} (m^{i'} w_p^{j'})^2 \pmod{n_p}$ (In this protocol, equalities of both equations always hold)

Fig. 6. Resistance Protocol to Hidden Verifier Attack

However, V can execute the resistance protocol in Figure 6.

In this protocol, two equations H checking is always hold without regard to the signature is either valid or invalid. V specifies x' as

$$x'^{A'} \equiv C' \pmod{n_v}$$
$$x' \equiv C'^{A'^{-1}} \pmod{n_v}.$$

This is equivalent to specifying the private key corresponding to the public key

A'. Because A' is odd number, $A'^{-1} \bmod L_v$ always exists except the case where A' is a multiple of either p' or q'.

From existence of this protocol, H cannot trust that V executes the signature confirmation protocol honestly as forced by H and confirm P's signature.

This means that this protocol can resist hidden verifier attacks.

5.7 Conversion of the Undeniable Signature

Here, we present the way to convert the our undeniable signature proposed in section 5.3 to a usual RSA signature presented in section 5.2.

There exists an undeniable signature S_m corresponding to a message m. If this signature is valid, then we have

$$S_M \equiv M^{e_p d_{1p}} \pmod{n_p}.$$

When a signer wants to convert only the signature S_M into a usual RSA signature, he shows the following converting information C_M as

$$C_M \equiv M^{d_{1p}(d_{2p}-e_p)} \equiv M^{d_p - e_p d_{1p}} \pmod{n_p}. \tag{5}$$

Anyone who gets the message M, the undeniable signature S_M corresponding to M and the converting information C_M corresponding to S_M can verify them by

$$(C_M S_M)^{e_p} \overset{?}{\equiv} M \pmod{n_p}.$$

When a signer wants to convert all of signatures he signed previously, he can select two ways.

1. Open d_{2p} to the public
   If d_{2p} is opened to the public, then anyone can verify all of the undeniable signatures S_m as

   $$S_m{}^{d_{2p}} \overset{?}{\equiv} m \pmod{n_p}.$$

2. Open C_* to the public
   We define the converting value C_* as

   $$C_* \equiv d_p - e_p d_{1p} \pmod{L_p}$$

   where $L_p = \mathbf{LCM}((p_p - 1), (q_p - 1))$. If the converting value C_* is opened to the public, then anyone can verify all of the undeniable signatures S_m as follows,

   $$(m^{C_*} S_m)^{e_p} \overset{?}{\equiv} m \pmod{n_p}.$$

6 Conclusion

In this paper, we pointed out some problems in Gennaro-Krawczyk-Rabin's undeniable signature scheme based on RSA. These were some restrictions on the signature's convertibility, existence of hidden verifier and probabilistic denial protocol.

We proposed new undeniable signature scheme by modifying the original one to solve these problems.

The key generation of our scheme made each signature convertible to a usual RSA signature individually. Both the signature confirmation protocol and the denial protocol were improved to resist to the hidden verifier by using Jakobsson-Sako-Impagliazzo's protocol.

The denial protocol was also improved to be deterministic one by using Chaum's decisive denial protocol.

Then we also showed the way to convert the undeniable signature to a usual RSA signature individually.

References

1. D. Chaum and H. V. Antwerpen, "Undeniable Signatures", Proc. *CRYPTO'89* pp.212-217, 1990.
2. D. Chaum, "Zero-Knowledge Undeniable Signatures", Proc. *EUROCRYPTO'90* pp.458-464, 1990.
3. J. Boyar, D. Chaum, I. Damgård and T. Pedersen, "Convertible Undeniable Signatures", Proc. *CRYPTO'90* pp.189-205, 1991.
4. D. Chaum, "Designated Confirmer Signatures", Proc. *EUROCRYPTO'94* pp.86-91, 1994.
5. R. Gennaro, H. Krawczyk and T. Rabin, "RSA-Based Undeniable Signatures", Proc. *CRYPTO'97* pp.132-149, 1997.
   http://www.research.ibm.com/security/RSAunden.ps
6. Y. Desmedt and M. Yung, "Weaknesses of Undeniable Signature Schemes", Advances in Cryptology Proceedings of EUROCRYPTO'91
7. M. Jakobsson, K. Sako and R. Impagliazzo, "Designated Verifier Proofs and Their Applications", Advances in Cryptology-EUROCRYPT'96, Lecture Notes in Computer Science 1070, Springer-Verlag, pp.143-154, 1996.
8. R. Housley, SPYRUS, W. Ford, VeriSign, W. Polk, NIST, D. Solo, and Citicorp "Internet X.509 Public Key Infrastructure Certificate and CRL Profile", RFC2459, 1999.
9. R. Gennaro, D. Micciancio and T. Rabin, "An Efficient Non-Interactive Statistical Zero-Knowledge Proof System for Quasi-Safe Prime Products", Proc. of the Fifth ACM Conference on Computer and Communications Security, 1998.
   http://www.research.ibm.com/security/safeprime.ps
10. D. R. Stinson, CRYPTOGRAPHY, pp.218-225, CRC Press, 1996.

Efficient and Secure Member Deletion in Group Signature Schemes

Hyun-Jeong Kim[1], Jong In Lim[2], and Dong Hoon Lee[3]

Center for Information Security Technologies(CIST),
Korea University, Seoul, KOREA,
khj@cist.korea.ac.kr[1]
jilim@tiger.korea.ac.kr[2]
donghlee@tiger.korea.ac.kr[3]

Abstract. Group signature schemes allow a group member to sign messages anonymously on behalf of the group. In case of dispute, only a designated group manager can reveal the identity of the member. During last decade, group signature schemes have been intensively investigated in the literature and applied to various applications. However, there has been no scheme properly handling the situation that a group member wants to leave a group or is excluded by a group manager. As noted in [2], the complexity of member deletion stands in the way of real world applications of group signatures and the member deletion problem has been a pressing open problem. In this paper, we propose an efficient group signature scheme that allows member deletion. The length of the group public key and the size of signatures are independent of the size of the group and the security of the scheme relies on the RSA assumption. In addition, the method of tracing all signatures of a specific member is introduced.

1 Introduction

The concept of a group signature was introduced by D. Chaum and van Heyst [7]. It allows members of a group to sign messages anonymously on behalf of the group. The signed messages are then verified by a group public key. In case of dispute, only a designated group manager can reveal the identity of the member. Various group signature schemes have been investigated to develop the efficient scheme of which the length of signatures and the size of the group public key are independent of the size of the group. Group signature schemes in [3,7,8,10] do not satisfy the length requirement and/or the size requirement. Only schemes proposed in [1,4–6] satisfy both requirements.

Group signature schemes should be coalition resistant. In other words, no subset of group members including the group manager is able to generate valid group signatures that are untraceable or from which the trustee revokes the identity of another group member. The schemes in [1,4] are the provable coalition resistant group signature schemes.

With the improvement in both efficiency and security of group signature schemes, the entire concept of a group signature scheme is brought to various applications such as an electronic cash system, a bidding, and a voting. However, for group signature schemes to be adapted to real applications, a few problems need to be solved. Among them one of the most important things is the efficiency in member deletion.

In practical applications, a group is dynamic, i.e., membership changes frequently. A group member may voluntarily leave the group by various reasons such as promotion in position. If a group member does a dishonest thing, a group manager must exclude him from the group. In this case, it may need to trace all of signatures generated by him while the anonymity of the others must be preserved. As stated in [2], no proposed group signature scheme adequately addresses the member deletion problem and efficient and secure member deletion has remained a pressing and interesting open problem.

In this paper, we propose a new group signature scheme which allows member deletion and sign-tracing generated by a specific member. Our scheme is based on Camenisch and Michels' Group Signature Scheme[4] that adds a member deletion procedure. Whenever a member joins or leaves the group, public information and each member's secret are modified without re-issuing membership certificates. Each modification requires only one modular multiplication. Hence our model is an acceptable solution for a large group where membership changes frequently. The group public keys, member's secret key, and the signature are all of constant size. The computational complexity of registration and deletion of a member is linear in group size, but the computational burden is decentralized, i.e., each member updates his own key. The computational complexity of other procedure is independent of group size.

This paper is organized as follows. Section 2 presents the model that permits deletion/addition of a member in group signature schemes and the security requirements. Section 3 describes the assumptions on which the security of the proposed group signature schemes is based. We develop our scheme in Section 4 and analyze the security of our scheme in Section 5. Finally, we conclude in Section 6.

2 The Model and our Approach

This section describes the model that allows member deletion in a group signature scheme, the security requirements, and the approach of our proposal. The main difference between our model and other models previously proposed is that our model has a member deletion procedure and sign-tracing capability.

2.1 The Model

A group signature scheme consists of the following procedures:

Setup : An interactive protocol between the membership manager and the revocation manager. The outputs are the membership manager's secret key

x_M and public key y_M , and the revocation manager's secret key x_R and public key y_R.

Join : An interactive protocol between the membership manager and a user that results in the user becoming a new group member. The outputs are a group member's secret key x_G, a group member's public key y_G, a group member's secret property key U_G, the group's public property key U_M and the group's public renewal property key U_N.

Delete : A member deletion algorithm that on input a member's public key y_G outputs the group's public property key U_M and the renewal property key U_N.

Sign : A signature generation algorithm that on input a message m, x_G, y_G, y_M, y_R, and U_G outputs a signature σ.

Verify : A verification algorithm that on input a message m, a signature σ, y_M, y_R, and U_M return 1 if and only if σ was generated by a proper group member using **Sign** on input m, x_G, y_G, y_M, y_R and U_G.

User-Tracing : A user tracing algorithm that on input a signature σ, a message m, x_R, and y_R outputs the identity of the group member who generated the signature σ.

Sign-Tracing : A sign tracing algorithm that on input a part of a signature σ, y_G, and x_R outputs 1 if and only if the signature was generated by a specific member.

The followings are security requirements:

Unforgeability of signatures : Only current group members are able to generate valid signatures. Furthermore, the signature can be user-traced and sign-traced by the revocation manager in need. In particular, if a group member leaves a group, he cannot generate a valid signature any more.

Anonymity : It is infeasible to find out a member who generated a given signature except the revocation manager.

Unlinkability of signatures : Given two signatures, no one except the revocation manager can decide whether the signatures have been computed by a same group member.

No framing : Any coalition of group members, the membership manager, and the revocation manager cannot compute a signature on behalf of non-involved group member. Futhermore, they can not sign message on behalf of a deleted group member.

Unforgeability of user-tracing verification : Given a signature, the revocation manager cannot falsely blame a signer for having produced the signature.

Unforgeablility of sign-tracing verification : The revocation manager cannot falsely insist that a signature was generated by a designated member.

2.2   The Approach of Our Proposal

The core idea to handle membership changes is as follows. For membership changes in the group, a membership manager maintains a group *property* key

U_M and group *renewal property* key U_N. When a new member joins the group, the membership manager issues a *secret property* key U_G to the new member.

For each registration or deletion, the membership manager updates the group public property key U_M and the group public renewal property key U_N, and publishes U_M and U_N. Each remaining member renews his secret property key U_G using the updated group renewal property key U_N and checks the validity of his new key using the updated group property key U_M. And a group member uses his secret property key U_G to sign a message. Someone who wants to verify a signature must make use of U_M. It is computationally infeasible for a deleted member to generate a valid signature using the group renewal property key and his old secret property key.

As explained above, the membership manager is only involved in issuing a new member's secret key and the secret key of each current member is re-generated by himself. Hence computations needed in membership changes are distributed to each member.

3 Preliminaries

In this section we describe the cryptographic assumptions necessary in the subsequent construction of the proposed group signature scheme. Our scheme is based on the group signature scheme by Camenisch and Michels[4] with member deletion capability. We briefly describe their scheme in this section, but do not explain the building blocks for the group signature scheme described in Section 4 of [4]. Those building blocks are proof systems in which one party can convince other parties that he knows certain values without leaking useful informations. Since they are used for the purpose of both singing a message and providing knowledge of a secret, they are called "signatures based on a proof of knowledge", SPK for short. Whenever SPK is used in the rest of the paper, what it proves is explained without details. For more details, refer to [4] and [9].

Let l_g be a security parameter and G be the group of order with length l_g is factored into two primes of length $(l_g - 2)/2$.

Problem 1. (**RSA Problem**). Given G and $(z, e) \in G \setminus \{\pm 1\} \times \mathbf{Z}$, find $u \in G$ such that $u^e = z$.

Let $\mathcal{T}$ denote a key-generator that on input 1^{l_g} outputs a G and a $z \in G \backslash \{\pm 1\}$.

Assumption 1 (RSA Assumption). *There exists a probabilistic algorithm $\mathcal{T}$ such that for all probabilistic polynomial-time algorithms $\mathcal{A}$, all polynomials $p(\cdot)$, and all sufficiently large l_g,*

$$Pr\left[\, z = u^e \mid (G, z, e) := \mathcal{T}(1^{l_g}), u := \mathcal{A}(G, z, e) \,\right] < \frac{1}{p(l_g)}.$$

The following two assumptions are due to [4]. They proposed the Modified Strong RSA assumption which is the modification of the strong RSA assumption by Fujisaki and Okamoto [9]. Let k, l_1, $l_2 < l_g$ and $\varepsilon > 1$ be security parameters. Denote $\bar{l} := \varepsilon(l_2 + k) + 1$. Let be $\mathcal{M}(G, z) = \{ (u, e) \mid x = u^e, u \in G, e \in \{2^{l_1} - 2^{l_2}, \cdots, 2^{l_1} + 2^{l_2}\}, e : \text{prime} \}$ where $z \in G$.

Problem 2. **(Modified Strong RSA Problem).** Given G, $z \in G$, and $M \subset \mathcal{M}(G, z)$ with $|M| = O(l_g)$, find a pair $(u, e) \in G \times \mathbf{Z}$ such that $u^e = z$, $e \in \{2^{l_1} - 2^{\bar{l}}, \cdots, 2^{l_1} + 2^{\bar{l}}\}$, and $(u, e) \notin M$.

Assumption 2 (Modified Strong RSA Assumption). *There exists a probabilistic algorithm $\mathcal{T}$ such that for all probabilistic polynomial-time algorithms $\mathcal{A}$, all polynomials $p(\cdot)$, all sufficiently large l_g, all $M \subset \mathcal{M}(G, z)$ with $|M| = O(l_g)$ and suitably chosen l_1, l_2, k and ε,*

$$Pr[\, z = u^e \,\wedge\, e \in \{2^{l_1} - 2^{\bar{l}}, \cdots, 2^{l_1} + 2^{\bar{l}}\} \,\wedge\, (u, e) \notin M \mid (G, z) := \mathcal{T}(1^{l_g}),$$
$$(u, e) := \mathcal{A}(G, z, M)\,] < \tfrac{1}{p(l_g)}.$$

As noted in [4], if two pairs (u, e), (u', e') with $z = u^e = u'^{e'}$, and $\gcd(e, e') = 1$ are known, it is easy to find an element $\hat{u}$ satisfying $z = \hat{u}^{ee'}$ using the extended Euclidean algorithm. But ee' does not satisfy the range constraint since $ee' \notin \{2^{l_1} - 2^{\bar{l}}, \cdots, 2^{l_1} + 2^{\bar{l}}\}$ for suitable chosen parameters l_g, l_1, l_2, ϵ, and k. Therefore, group signature schemes based on the modified strong RSA assumption are coalition resistant. The Modified Strong RSA Problem is at least as hard as Strong RSA Problem from [4] and [11].

Besides the Modified Strong RSA Assumption, Camenish and Michels' group signature scheme relies on the Diffie-Hellman Decision(DHD) assumption. To state this assumption, We define the two sets

$$DH(G) := \{ (g_1, y_1, g_2, y_2) \in G^4 \mid ord(g_1) = ord(g_2) = n', \; log_{g_1} y_1 = log_{g_2} y_2 \},$$
$$Q(G) \;\; := \{ (g_1, y_1, g_2, y_2) \in G^4 \mid ord(g_1) = ord(g_2) = ord(y_1) = ord(y_2) = n' \}$$
with $n' \mid O(G)$ and $|n'| = l_g - 2$.

Assumption 3 (Diffie-Hellman Decision Assumption). *There exists a probabilistic algorithm $\mathcal{T}$ such that for all probabilistic polynomial-time algorithms $\mathcal{A}$, and all sufficiently large l_g, the two probability distributions*

$$Pr[\, a = 1 \mid G := \mathcal{T}(1^{l_g}), \; K \in_R DH(G), \; a := \mathcal{A}(K)\,]$$

and

$$Pr[\, a = 1 \mid G := \mathcal{T}(1^{l_g}), \; K \in_R Q(G), \; a := \mathcal{A}(K)\,]$$

are computationally indistinguishable.

Consider the DHD assumption in the case $G = Z_n^*$ where n is an RSA-modulus. In this case the DHD assumption does not hold in general since the Jacobi-Symbol leaks information about $log_{g_1} y_1$ and $log_{g_2} y_2$ for some $g_1, g_2, y_1, y_2 \in G$. For example, if $(g_1|n) = (g_2|n) = (y_2|n) = -1$ and $(y_1|n) = 1$, then $log_{g_1} y_1 \neq log_{g_2} y_2$.

Note 1. In the case $G = Z_n^*$ where n is an RSA-modulus, if $G = < g >$ is defined to be a subgroup of Z_n^* with $(g|n) = 1$, then the DHD assumption holds.

4 The Proposed Scheme

This section describes our new group signature scheme that allows member deletion and sign tracing. The security of the proposed scheme is based on Assumption 1, 2 and 3. In particular, the security of property keys replies on Assumption 1. The scheme is especially described in the viewpoint of addition/deletion of a member.

4.1 System Setup

In our scheme, the membership manager supervises the group members and the revocation manager as a trustee performs tracing protocols while guarantees anonymity of legitimate members. They first set up the system with generating the group's public keys and choosing their secret keys.

The membership manager executes the setup procedure as follows :

1. Choose a group $G = < g >$ and two random elements $z, h \in G$ with the same (large) order ($\approx 2^{l_g}$) such that Assumption 2 and 3 hold. Computing discrete logarithms in G to the bases g, h, or z must be infeasible and only the membership manager can easily compute these roots. That is, he should keep the order of G secretly.
2. Choose two large random primes p and q ($\approx 2^{l_g/2}$) of the form $p = 2p'+1$, $q = 2q'+1$ where p', q' are primes, such that $p, q \neq 1 \pmod 8$ and $p \neq q \pmod 8$.
3. Keep p and q secret and publish $n := pq$.
4. Choose a public key e_N and a secret key d_N such that $e_N d_N \equiv 1 \pmod{\phi(n)}$ where n is a RSA-modulus.
5. Publish z, g, h, G, e_N and l_g and prove that g, h and z have the same order.

The revocation manager executes the setup procedure as follows :

1. Choose his secret key x_R randomly in $\{0, \cdots, 2^{l_g} - 1\}$.
2. Publish $y_R = g^{x_R}$ as his public key.

Also, the membership manager sets up a hash function $H : \{0,1\}^* \rightarrow \{0,1\}^k$ and security parameters $\tilde{l}, l_1, l_2$ and ϵ.

4.2 Join

This is an interactive protocol between the membership manager and Alice who wants to become a new group member. Alice chooses a prime x_G randomly in a appropriate range. She keeps x_G secretly. The membership manager extracts a element $y_G \in G$ such that $y_G^{x_G} = z$ holds. This pair (x_G, y_G) is the membership key of Alice. Also the membership manager regenerates group's public property key U_M and renewal property key U_N using y_G and generates Alice's secret property key U_G. Each regeneration-value of the group property key and renewal property key is published. Before generating any signature, current members check whether the group renewal property key has been updated or not.

Alice does the following :

1. Choose random primes $\hat{x}_G, x_G \in_R \{2^{\bar{l}-1}, \cdots, 2^{\bar{l}} - 1\}$ such that $\hat{x}_G x_G \neq 1 \pmod 8$ and $\hat{x}_G \neq x_G \pmod 8$.
2. Compute $\tilde{x}_G := x_G \hat{x}_G$ and $\tilde{z} := z^{\hat{x}_G}$.
3. Commit to $\tilde{x}_G$ and $\tilde{z}$.
4. Send $\tilde{x}_G$, $\tilde{z}$ and their commitments to the membership manager.
5. Execute the interactive protocols corresponding to
$$\mathcal{W} := SPK\{ (\tau, \varrho) \mid z^{\hat{x}_G} = \tilde{z}^\tau \wedge \tilde{z} = z^\varrho \wedge$$
$$(2^{l_1} - 2^{\epsilon(l_2+k)+1} < \tau < 2^{l_1} + 2^{\epsilon(l_2+k)+1}) \}(\tilde{z})$$
with the membership manager.

$\mathcal{W}$ is a statistical zero-knowledge proof of knowledge of the discrete logarithm of $\tilde{z}(= z^{\hat{x}_G})$ and an integer x_G such that $x_G \in \{2^{l_1} - 2^{\epsilon(l_2+k)+1}, \cdots, 2^{l_1} + 2^{\epsilon(l_2+k)+1}\}$ and $\tilde{z}^{x_G} = z^{\tilde{x}_G}$. Therefore the membership manager trusts Alice to have chosen $\tilde{x}_G$ and $\tilde{z}$ correctly by the proof $\mathcal{W}$.

Let $\mathcal{C} := \{G_1, G_2, \cdots, G_{m-1}\}$ be the set of current group members and G_m be a new member, Alice. Let y_{G_i} denote each member's public key. Before adding Alice to the group, the group's public property key is $U_M := y_{G_1} \cdots y_{G_{m-1}} y'$ with a random number $y' \in_R G$. The membership manager does the followings :

1. Generate Alice's public key $y_{G_m} := \tilde{z}^{1/\tilde{x}_G}$.
2. Compute a new group's public property key $U_M := y_{G_1} \cdots y_{G_{m-1}} y_{G_m} y''$ where $y'' \in_R G$.
3. Compute the new group's public renewal property key $U_N := \left(\frac{y_{G_m} y''}{y'}\right)^{d_N}$.
4. Generate the member G_m's secret property key $U_{G_m} := (y_{G_1} y_{G_2} \cdots y_{G_{m-1}} y'')^{d_N}$.

The membership manager publishes U_M and U_N, and sends a member's public key y_{G_m} and a secret property key U_{G_m} to Alice. The pair (x_{G_m}, y_{G_m}) becomes the membership key of Alice. A new member G_m, Alice, verifies her public key y_{G_m} and secret property key U_{G_m} by checking $y_{G_m}^{x_{G_m}} = z$ and $(U_{G_m})^{e_N} y_{G_m} = U_M$ respectively. Each valid group member $G_i (1 \leq i \leq m - 1)$ except a new member

G_m changes his secret property key $U_{G_i} := (y_{G_1} \cdots y_{G_{i-1}} y_{G_{i+1}} \cdots y_{G_{m-1}} y')^{d_N}$ into $U_{G_i} = U_{G_i} \cdot U_N$, that is,

$$U_{G_i} = (y_{G_1} \cdots y_{G_{i-1}} y_{G_{i+1}} \cdots y_{G_{m-1}} y')^{d_N} \cdot (y_{G_m} y'' / y')^{d_N}$$
$$= (y_{G_1} \cdots y_{G_{i-1}} y_{G_{i+1}} \cdots y_{G_{m-1}} y_{G_m} y'')^{d_N}.$$

Each group member can check new value U_{G_i} by computing $U_M = (U_{G_i})^{e_N} y_{G_i}$.

4.3 Delete

This protocol is similar to the addition of a group member. To delete the group member G_j the membership manager eliminates public key y_{G_j} from the group public property key U_M and changes a random number. The remaining group members change their secret property keys to generate a valid signature.

Let the current group's public property key be $U_M := y_{G_1} \cdots y_{G_m} y'$ where $y' \in_R$ G. The membership manager performs the deleting protocol as the followings :

1. Compute $U_M := U_M \cdot \frac{y''}{y_{G_j} y'}$ where $y'' \in_R$ G,
   i.e., $U_M = y_{G_1} \cdots y_{G_{j-1}} y_{G_{j+1}} \cdots y_{G_m} y''$.
2. Compute $U_N := \left(\frac{y''}{y_{G_j} y'} \right)^{d_N}$.
3. Publish (U_M, U_N).

Each group member G_i changes his secret property key U_{G_i} into $U_{G_i} = U_{G_i} \cdot U_N$, for example,

$$U_{G_i} := (y_{G_1} \cdots y_{G_{i-1}} y_{G_{i+1}} \cdots y_{G_m} y')^{d_N} \cdot (y'' / y_{G_j} y')^{d_N}$$
$$:= (y_{G_1} \cdots y_{G_{i-1}} y_{G_{i+1}} \cdots y_{G_{j-1}} y_{G_{j+1}} \cdots y_{G_m} y'')^{d_N}, \text{ for } i < j.$$

Each group member verifies new value U_G by checking if $(U_G)^{e_N} y_G = U_M$.

4.4 Sign

First, We define a group signature.

Definition 1. Let ϵ, l_1 and l_2 be security parameters such that $\epsilon > 1$, $l_2 < l_1 < l_g$, and $l_2 < \frac{l_g - 2}{\epsilon} - k$ holds. A group-signature of a message $m \in \{0,1\}^*$ is a tuple $(c, s_1, s_2, s_3, a, b, d, \alpha, \beta) \in \{0,1\}^k \times \{-2^{l_2+k}, \cdots, 2^{\epsilon(l_2+k)}\} \times$
$$\{-2^{l_g+l_1+k}, \cdots, 2^{\epsilon(l_g+l_1+k)}\} \times \{-2^{l_g+k}, \cdots, 2^{\epsilon(l_g+k)}\} \times G^5$$
satisfying
$$c = H(g\|h\|y_R\|z\|a\|b\|d\|\beta\|z^c b^{s_1-c2^{l_1}} / y_R^{s_2}\|a^{s_1-c2^{l_1}} / g^{s_2}\|a^c g^{s_3}\|d^c g^{s_1-c2^{l_1}} h^{s_3}\|$$
$$\beta^c y_R^{s_3} h^{s_3 e_N}\|m).$$

Remark 1. Such a group-signature would be denoted

$$\mathcal{L} = SPK\{ (\theta, \lambda, \mu) : z = b^\theta / y_R^\lambda \wedge 1 = a^\theta / g^\lambda \wedge a = g^\mu \wedge d = g^\theta h^\mu$$
$$\wedge \beta = y_R^\mu h^{\mu e_N} \wedge (2^{l_1} - 2^{\epsilon(l_2+k)+1} < \theta < 2^{l_1} + 2^{\epsilon(l_2+k)+1}) \}(m).$$

The non-interactive protocol corresponding to $\mathcal{L}$ is a statistical zero-knowledge proof of knowledge of the discrete logarithm of a and an integer x_G such that $x_G \in \{2^{l_1} - 2^{\epsilon(l_2+k)+1}, \cdots, 2^{l_1} + 2^{\epsilon(l_2+k)+1}\}$ and $y_G = g^{x_G}$.

To sign a message $m \in \{0,1\}^*$, a group member does the followings :

1. Choose an integer $w \in_R \{0,1\}^{l_g}$, compute $a := g^w$, $b := y_G y_R^w$, $d := g^{x_G} h^w$, $\alpha := U_G h^w$, and $\beta := y_R^w h^{we_N}$.
2. Choose $r_1 \in_R \{0,1\}^{\epsilon(l_2+k)}$, $r_2 \in_R \{0,1\}^{\epsilon(l_g+l_1+k)}$ and $r_3 \in_R \{0,1\}^{\epsilon(l_g+k)}$.
3. Compute $t_1 := b^{r_1}(1/y_R)^{r_2}$, $t_2 := a^{r_1}(1/g)^{r_2}$, $t_3 := g^{r_3}$, $t_4 := g^{r_1} h^{r_3}$ and $t_5 := y_R^{r_3} h^{r_3 e_N}$.
4. Compute $c := H(g||h||y_R||z||a||b||d||\beta||t_1||t_2||t_3||t_4||t_5||m)$.
5. $s_1 := r_1 - c(x_G - 2^{l_1})$ (in $\mathbf{Z}$), $s_2 := r_2 - cw x_G$ (in $\mathbf{Z}$), and $s_3 := r_3 - cw$ (in $\mathbf{Z}$).

The resulting signature on the message m is $(c, s_1, s_2, s_3, a, b, d, \alpha, \beta)$.

4.5 Verifying Signatures, User-Tracing, and Sign-Tracing

The verification procedure is an extension of the verification procedure in [4] that adds the check that the signature is verified by the group property key used at the date when the signature was generated. User-tracing procedure in [4] is unchanged and sign-tracing is newly added in our scheme. For a given signature, sign-tracing decides whether a designated user generated the signature. It may be viewed as the concept which is similar to coin-tracing in electronic cash systems. Note that for both verification and sign-tracing, the history of old group property keys with the dates updated and the date when a signature was generated should be available. This can be resolved by keeping the history of group property keys and the updated dates in a table and embedding the generated date in a signature. The size of signatures is still constant and a table look-up takes $O(\log m)$ only for the table size m.

Verifying Signature : Given a signature, it is verified that the signature satisfies the verification condition given in Definition 1. If it is satisfied, a verifier trusts that the signer had a valid membership key, chose a random number w honestly, and formed $\beta = y_R^w h^{we_N}$. Then he checks if $\left(\frac{U_M}{\alpha^{e_N}}\right)\beta = b$ holds. This equality holds if and only if the signature was generated by a valid group member of the group.

User-Tracing : To reveal the originator of a given signature $\sigma = (c, s_1, s_2, s_3, a, b, d, \alpha, \beta)$ of message m, the revocation manager first checks its correctness and then computes $y'_G := b/a^{x_R}$. For the proof of unforgeability of user-tracing, he issues a signature $P := SPK\{ (\rho) : y_R = g^\rho \wedge b/y'_G = a^\rho \}(y'_G||\sigma||m)$ and reveals $arg := y'_G||P$. This SPK shows the equality of two discrete logarithms y_R and b/y'_G, and it is a statistical zero-knowledge proof of knowledge of the discrete logarithm of $y_R(= g^{x_R})$. He looks up y'_G in the group-member list and finds the corresponding y_G.

Sign-Tracing : To find whether a signature $\sigma = (c, s_1, s_2, s_3, a, b, d, \alpha, \beta)$ was generated by a specific (illegal) member, the membership manager sends $(a, y_G^{d_N}\alpha, \beta)$ to the revocation manager where y_G is a specific member's public key. The revocation manager computes $(y_G^{d_N} \cdot \alpha)^{e_N}/(\beta/a^{x_R})$ and checks if the result equals to U_M. If the signature was generated by the member, he sends 1 to the membership manager. In case that the signature was not generated by the member, the revocation manager cannot acquire any information except that the member didn't generate it.

5 Security Analysis

We discuss the security of the proposed scheme. The following theorem implies that non-group member or a deleted group member with his obsolete secret key cannot generate any valid signature by showing that forging a valid signature is equivalent to solving the RSA problem.

Theorem 1. *There exists a probabilistic polynomial algorithm that on input y_R, y_G, h, U_M and e_N outputs (w, α) satisfying $\left(\frac{U_M}{\alpha^{e_N}}\right)\beta = b$ where $\beta = y_R^w h^{w e_N}$ and $b = y_G y_R^w$ if and only if it is able to solve the RSA problem.*

(Sketch of Proof) Suppose that given y_R, y_G, h, U_M and e_N, a probabilistic polynomial-time algorithm $\mathcal{A}$ can find a valid (w, α) such that $\left(\frac{U_M}{\alpha^{e_N}}\right)\beta = b$. Then we have the following.

$$U_M \beta = b\alpha^{e_N}. \tag{1}$$

From (1), we have the following equation.

$$(U_M/y_G)^{1/e_N} = \alpha/h^w. \tag{2}$$

This implies that $\mathcal{A}$ finds a value (w, α) satisfying (2). (Since $U_M := y_1 \cdots y_m y'$ where $y' \in_R G$, U_M is able to be regarded as a random value in G and hence U_M/y_G is random in G.)
Therefore, given a value m^{e_N} with $m \in$ G substituting m^{e_N} for (U_M/y_G) in equation (2) results in

$$(m^{e_N})^{1/e_N} = \alpha/h^w. \tag{3}$$

Thus we find $m = \alpha/h^w$ and can solve the RSA problem.

Conversely, if there exists an algorithm $\mathcal{A}'$ which can solve the RSA problem, for $w \in_R G$ $\mathcal{A}'$ outputs α on a input pair $(\frac{U_M \beta}{b}, e_N)$ such that $\frac{U_M \beta}{b} = \alpha^{e_N}$, that is $(\frac{U_M}{\alpha^{e_N}})\beta = b$. $\square$

In the rest of this section we only discuss unforgeability of signatures and sign-tracing verification. It is straightforward to check that other security properties are satisfied in our scheme as well as in the scheme in [4] which our scheme is based on. Hence those are omitted here.

Unforgeability of Signatures : Only the valid group members can generate valid signatures which will be able to be user-traced and sign-traced by the revocation manager. (Under the Assumption 2, it is infeasible that anyone without the information of the group's order computes the valid membership key. Therefore only the membership manager can generate the membership key by executing the join protocol. So only the group members who execute join protocol with the group manager can generate the valid group signs. Furthermore, the revocation manager is able to find the public key of a signer by decrypting (a, b) of a signature using his private key and computing $y'_G = b/a^{x_R}$ [4].) Due to Theorem 1, it is infeasible that the group member who left the group generates the valid signatures.

Unforgeability of sign-tracing verification: Given $(a, y_G^{d_N} \alpha, \beta)$, if the revocation manager returns 1 to the membership manager, the membership manager computes $\alpha^{e_N} b/U_M \beta$. This value is 1 if and only if the revocation manager has executed the sign-tracing correctly.

6 Conclusion

The complexity of member deletion of group signature schemes has been an obstacle in applying the concept of group signatures to real applications. In this paper, we proposed the first efficient group signature scheme that allows membership deletion. This scheme can be viewed as an extension of the scheme proposed in [4] that adds a member deletion procedure. For each deletion, our scheme requires two modular exponentiations, two modular inversions and few modular multiplications by the membership manager and only one modular multiplication by a member without re-issuing membership keys. For the verification of signatures or sign-tracing, besides the computations needed in [4], one table look-up is performed. This takes $O(\log m)$ for table size m. And the length of the group public key and the size of signatures are constant.

Our scheme is based on the specific scheme proposed in [4] and not applicable to other group signature schemes. Furthermore each group is time-stamped and the history of old group public keys should be available. Further research is required for more efficient and generic group signature schemes.

References

1. G. Ateniese, J. Camenisch, M. Joye, and G. Tsudik. A Practical and Provably Secure Coalition-Resistant Group Signature Scheme. In *Advances in Cryptology - CRYPTO 2000*, vol.1880 of LNCS, pp.255-270. Springer Verlag, 2000.

2. G. Ateniese and G. Tsudik. Group signatures a là carte. In *ACM Symposium on Discrete Algorithms*, 1999.

3. J. Camenisch. Efficient and generalized group signatures. In W.Fumy, editor, *Advances in Cryptology-EUROCRYPT '97*, vol. 1233 of Lecture Notes in Computer Science, pp.465-479. Springer Verlag, 1997.

4. J. Camenisch and M. Michels. A group signature scheme based on an RSA-variant. Tech. Rep. RS-98-27, BRICS, Dept. of Comp. Sci., University of Arhus, preliminary version in *Advances in Cryptology-ASIACRYPT '98*, vol.1514 of LNCS.

5. J. Camenisch and M. Michels. Proving in zero-knowledge that a number is the product of two safe primes. In *Advances in Cryptology-EUROCRYPT '99*, vol.1592 of LNCS, pp.107-122.

6. J. Camenisch and M. Stadler. Efficient group signature schemes for large groups. In B.Kaliski, editor, *Advances in Cryptology-CRYPTO '97*, vol.1296 of Lecture Notes in Computer Science, pp.410-424. Springer Verlag, 1997.

7. D. Chaum, and E. van Heyst, Group signatures. In D. W. Davies, editor, *Advances in Cryptology-EUROCRYPT '91*, vol.547 of Lecture Notes in Computer Science, pp.257-265. Springer-Verlag, 1991.

8. L. Chen and T. P. Pedersen. New group signature schemes. In A. De Santis, editor, *Advances in Cryptology-EUROCRYPT '94*, vol.950 of Lecture Notes in Computer Science, pp.171-181. Springer-Verlag, 1995.

9. E. Fujisaki and T. Okamoto. Statistical zero knowledge protocols to prove modular polynomial relations. In B. Kaliski, editor, *Advances in Cryptology-CRYPTO '97*, vol.1294 of Lecture Notes in Computer Science, pp.16-30. Springer Verlag, 1997.

10. H. Petersen. How to convert any digital signature scheme into a group signature scheme. In M. Lomas and S. Vaudenay, editors, *Security Protocols Workshop*, Paris, 1997.

11. A. Shamir. On the generation of cryptographically strong pseudorandom sequences. In *ACM Transaction on Computer Systems*, vol.1, pp.38-44, 1983.

An Efficient and Practical Scheme for Privacy Protection in the E-Commerce of Digital Goods

Feng Bao, Robert H. Deng, Peirong Feng

Kent Ridge Digital Labs
21 Heng Mui Keng Terrace
Singapore 119613
{baofeng, deng, pfeng}@krdl.org.sg

Abstract. It is commonly acknowledged that customers' privacy in electronic commerce should be well protected. The solutions may come not only from the ethics education and legislation, but also from cryptographic technologies. In this paper we propose and analyze a privacy protection scheme for e-commerce of digital goods. The scheme takes cryptography as its technical means to realize privacy protection for online customers. It is efficient in both computational cost and communication cost. It is very practical for real e-commerce systems compared with previous solutions. The cryptographic technique presented in this paper is rather simple. But the scheme has great application potential in reality. We give careful security analysis to the scheme.

1   Introduction

Electronic commerce is growing with a surprising speed. We have seen various predictions on the amount of transaction revenues coming from e-commerce by the year 200X. The figures are really impressive. It seems that e-commerce will be a hot topic for at least the next decade and will penetrate into everyone's daily life.

Privacy protection in e-commerce is emerging as a commonly concerned issue.

Privacy protection may refer to different problems in different backgrounds or in different applications. Currently the problem talked most often is that some e-merchants may collect and their customers' personal information and deliver it to others for business purpose. The solutions for this problem may come from some kind of legislation, as done or being done in some countries, or from certain ethics education as suggested in [7].

Another problem related to privacy is about anonymity, which interests both industry and academic. We have noticed that industry goes pretty fast in this area. One example is that Zero-Knowledge System Inc, a Canadian company, presents various solutions for anonymity in online activities such as web surfing, FTP and online chatting etc [8]. There are several new companies, the so-called anonymizers, providing such services. Chaum's anonymous digital cash [6] is an effort from academic. It is a nice mathematical work and it has been followed by a large number of good research papers in the past decade. Unfortunately, it seems that anonymous digital cash did not take fly in practice.

In this paper, we consider a different privacy protection problem. The scenario of our privacy problem is the e-commerce of digital goods. Consider the situation where an online customer buys an e-book or e-journal. The customer may not wish to disclose which book or which journal he is buying since it may, in some sense, reveal his/her favor or habit. Such a privacy protection is extremely hard for physical goods. But for digital goods, there do exist techniques to fulfill such a requirement. As long as the techniques are efficient and add no big additional cost, the e-merchants may be willing to exploit them in order to attract customers. Let us look at it from another angle. If two merchants sell the same e-goods (at the same prices) but one provides the privacy protection while the other does not, the customers definitely would like to buy from the first merchant. This is especially true if the e-goods are more or less sensitive. Examples of such goods may include online video, online music, digital pictures, digital maps(or travel guidance packages of different cities), patent drafts, e-books, e-journals, e-news, game programs etc. Buying patent drafts or other IP(intelligent property)-related information may be in B-to-B category, for which this privacy protection problem is even more important since it may directly relate to business secret.

In this paper we present and analyze an efficient and practical scheme for the above addressed privacy protection. The paper is organized as follows. In Section 2, we describe the system architecture of our privacy protection scheme and display its merit features. In Section 3 we discuss some other issues related to the scheme such as authentication, payment, non-repudiation and copyright protection. In Section 4 we present the cryptographic technique exploited in the scheme and address some previously related work. In Section 5 we give security analysis to our scheme.

2   Description and Features of the System

In our scheme we exploit two symmetric key cryptosystems, denoted by E and CE, respectively. E is a traditional symmetric key cryptosystem such as DES or AES. We denote the ciphertext of m with key k by $E(k, m)$, and the decryption of ciphertext c with key k by $E^{-1}(k, c)$.

CE is a *commutative* encryption algorithm that satisfies the property: for any two keys k_1 and k_2 and any message m,

$$CE(k_1, CE(k_2, m)) = CE(k_2, CE(k_1, m)).$$

We will present a concrete CE in Section 4. The decryption of c with key k is denoted by $CE^{-1}(k, c)$.

For simplicity, we consider an abstract model where the merchant has n digital goods for sale. Denote them by $M_1, M_2, ..., M_n$. Suppose a customer want to get M_i without the merchant knowing what i is. The procedure is as follows.

The merchant randomly choose n secret keys of cryptosystem E, denoted by K_1, $K_2, ..., K_n$, and a secret key S of cryptosystem CE. All these secret keys must be very carefully kept by the merchant. Especially the secret key S is something like a master key, which must not be compromised.

Denote

$$C_1=\mathbf{E}(K_1, M_1), D_1=\mathbf{CE}(S, K_1)$$
$$C_2=\mathbf{E}(K_2, M_2), D_2=\mathbf{CE}(S, K_2)$$
$$\cdots\cdots$$
$$C_n=\mathbf{E}(K_n, M_n), D_n=\mathbf{CE}(S, K_n)$$

Then the merchant puts $<C_1, D_1>$, $<C_2, D_2>$, ..., $<C_n, D_n>$ onto a publicly accessible directory. Anyone is allowed to download anything freely and anonymously from the directory.

The customer downloads $<C_i, D_i>$. Then he randomly chooses a secret key R of crytposystem **CE** and encrypts D_i with R. Then he sends $U=\mathbf{CE}(R, D_i)$ to the merchant.

The merchant decrypts U with S and sends $W=\mathbf{CE}^{-1}(S, U)$ to the customer. The customer obtains K_i by decrypting W with R, $K_i=\mathbf{CE}^{-1}(R, W)$. Now he obtains M_i by a further step of decryption $M_i=\mathbf{E}^{-1}(K_i, C_i)$.

The whole procedure is shown in the following figure.

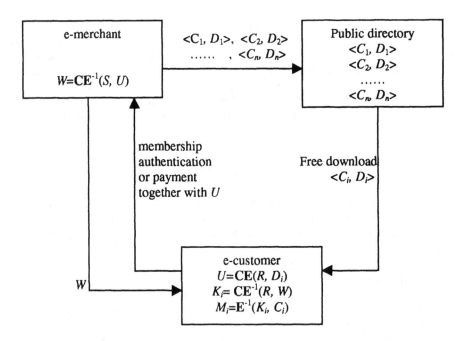

Here the public directory may be either from an Internet application service provider run by other parties, or the merchant's own public directory. We assume that the download of $<C_i, D_i>$ can be anonymous. The assumption is realistic since when the customer downloads $<C_i, D_i>$, he need not show personal information such as membership or credit card number etc. If the download is through some specific proxy, the customer's IP address can be hidden. Or if the download is through dial-up, the IP address changes every time. In addition, Zero-Knowledge System Inc. [8] provides some interesting technologies for anonymous download.

The interaction between the customer and the merchant cannot be anonymous since the merchant must know whom he is dealing with. When the merchant decrypts U for the customer, the service is a charged service. Either the membership authentication or a payment is needed, which would disclose some information about the customer. However, by our scheme the merchant completely has no idea which digital good was bought by the customer.

Our scheme has the following merit features.

1 The merchant can never know what the i is, no matter how malicious the merchant performs. Of course the merchant can make the denial of the service by not returning W or returning a fake W. But he can never get the intention of the customer. Customer's privacy is perfectly protected.
2 The customer can get at most one digital good in the implementation of the scheme once. This is very important for charged service. Also this requirement makes the scheme meaningful. Otherwise, let the customer get everything, 1 is always satisfied.
3 Statistics is still possible. Although the merchant cannot know who gets what, he can still know which digital good has been download (not bought) how many times. Such kind of data collection is realistically demanded in e-business.
4 The system is reliable. Even if some key K_i is compromised or published by some malicious customer, the master secret key S and other secret keys are not affected at all. The merchant just needs to replace $<C_i, D_i>$ with a new K_i.

The above features will be discussed in more details in the section of security analysis, after we present the concrete **CE** in Section 4.

Note. We do not consider the situation where the merchant displays wrong C_i so that the customer cannot obtain correct M_i. There is no solution at all if the merchant decides not to provide service to the customer.

3 Some Other Issues

Authentication

We skipped the authentication details when we presented our scheme in Section 2. Authentication is necessary in some situations since the merchant is not likely to do the decryption operation $W=\mathbf{CE}^{-1}(S, U)$ for everyone. Authentication is especially necessary when the customer subscribes to the merchant for a membership. There are many ways to do authentication: public key solutions, secret key solutions, password-based solutions, and so on. Actually the crypto research community spent so much effort on this topic in the past two decades. It depends on the concrete application situation to decide what kind of authentication scheme would be used in our scheme.

Payment and Prices

Payment can be in two ways. The first one is the membership. Owning a membership, a customer is allowed to have limited or unlimited access to the digital goods. In our scheme, the merchant will do a limited number of decrypting operations for the customer in the limited case. In the unlimited case, the merchant will do it whenever the customer requests. However, we recommend the limited model for our scheme since the merchant incures computation cost.

The second way is pay-per-piece. For this kind of payment, the privacy can only be guaranteed among all the digital goods with the same price. That is, the merchant cannot know which good of that price the customer is buying. However, uniform price for digital goods is much more likely than for physical goods. For example, it is not impossible to have a same price for all online videos in an online video shop.

Non-repudiation

It is possible that a customer or a merchant is dishonest sometimes. When a dispute occurs between the customer and the merchant, there may need some other parties to judge who is cheating. The solution to this problem is just asking the customer sign the U and the merchant sign the W. In the process of resolving dispute, the merchant proves to the judge that $W=CE^{-1}(S, U)$ without disclosing S. In that case, the master secret key S is not given to the judge, which is a necessary protection to the merchant. Later we will show how to prove $W=CE^{-1}(S, U)$ without disclosing S after we present CE in Section 4.

Copyright Protection

Copyright protection is a deadly demanded feature in e-commerce of digital goods. However, it is a very difficult issue from technology viewpoint. So far there is no satisfactory solution. Watermarking technology is studied for the copyright protection of multimedia digital goods, which aims at *catching* the illegal copies instead of *preventing* them. The digital goods in the form of texts such as e-book and e-journal are hardly protected by watermarking.

Another means is the so-called tamper-resistant software, which is still at its early stage. In tamper-resistant software technology, all the digital goods are encrypted and the decryption is done in the tamper-resistant software. The frame of our system meets the requirement of the tamper-resistant software protection very well. Tamper-resistant software technology can be well combined with our privacy protection scheme.

Flexible Distribution Means

The exchange or distribution of the encrypted digital goods among the customers is encouraged by the merchants. Another distribution means is by CD. It is estimated that CD with dozens GB and low cost will emerge in the near future. It is possible to store a big number of encrypted digital goods into a CD and sell it at a very low price (the price of the storage media). The decryption has to be paid when the contents are wanted.

4 Related Cryptography

Private Information Retrieval

In cryptography, the topic of retrieving information from a database without disclosing what the information is has been studied under the terminology of PIR (private information retrieval). The private information retrieval problem was first formalized and studied in [1]. The solutions provided in [1] are based on multiple databases and toward information-theoretical security. However, the assumption that the multiple databases would not communicate with one another is too strong for practical implementation. Later in [2], [3] and [5], private information retrieval schemes with single database were proposed. These solutions are based on computational assumptions, such as hardness of factoring $n=pq$. However, the computational costs of these solutions are very large due to their bit-by-bit processing manner. For example, the scheme in [3] needs a computation of $O(N)$ multiplications modulo a 1024-bit number for retrieval of only one bit, where N is the number of bits of the whole database. Such schemes can never be accepted for practical use. Any private information retrieval scheme for practical use should process messages file-by-file instead of bit-by-bit.

The security for database is studied in [4], i.e., a customer should not be able to retrieve more messages than what he should retrieve according to the scheme. This issue was ignored in [2] and [3]. This issue is important in e-commerce of digital goods.

All the previous PIR schemes aim at reducing communication cost. The scheme in [5] can even achieve a communication cost of poly(log$N$) while those in [2], [3] and [4] have communication cost $O(N^\varepsilon)$ for $\varepsilon<1$. From mathematics viewpoint, those schemes are beautiful research jobs. But from implementation viewpoint, those schemes are completely unpractical since they all require computation complexity at least $O(N)$. This make them infeasible even for a small database. Realistically, they are not necessary either. They ignore the different attributes of various Internet communication protocols as explained in Section 2.

Using Commutative Cipher

In this paper we use commutative encryption algorithm **CE** to achieve similar property of PIR. But our scheme is not a PIR scheme since we do not aim at reducing communication complexity. Instead, we build up our scheme based on the assumption that anonymous download is available.

It is well known that not every symmetric key cryptosystem has commutative property. Actually, all the noted symmetric key ciphers are not commutative. Stream ciphers are commutative (just XOR), but they cannot be applied to our scheme. When more than one messages are encrypted with a same key by a stream cipher, there must be a unique number assigned to each message to indicate which session of the key stream is being used(or the number is integrated into the key so that different key streams are generated for different messages). That number must be accompanied with the ciphertext; otherwise, decryption cannot be done. In our scheme, such a number would definitely disclose which digital good the customer is trying to buy.

We use a commutative cipher as follows.

Let $p=2q+1$ where primes p and q are public parameters of the system. The size of p is recommended to be 1024-bit. Let key $S \in Z_{2q}$ be an odd integer other than q. For any message $M \in Z_p$, the encryption of M with key S is

$$C=M^S \bmod p$$

Such an encryption is apparently commutative. But it may have security problems as to be explained in the next section. A modification is needed to make the cipher secure. The key point to secure the cipher is to add a padding format, i.e., instead of letting $C=M^S \bmod p$, we let $C=(\mathbf{P}(M))^S \bmod p$, where $\mathbf{P}$ is a padding format. It is easy to find M from $\mathbf{P}(M)$. We will discuss $\mathbf{P}$ in the next section.

The scheme presented in Section 2 can be expressed in more detail as follows.

Merchant   **Customer**

Randomly choose $K_1, \ldots, K_n \in \{0,1\}^{128}$
and a 160-bit odd integer S.   Randomly choose a
$C_1=\mathbf{E}(K_1, M_1)$, $D_1=(\mathbf{P}(K_1))^S \bmod p$ 160-bit odd integer R.
$C_2=\mathbf{E}(K_2, M_2)$, $D_2=(\mathbf{P}(K_2))^S \bmod p$
...... $Q=1/R \bmod 2q$
$C_n=\mathbf{E}(K_n, M_n)$, $D_n=(\mathbf{P}(K_n))^S \bmod p$
$T=1/S \bmod 2q$

$$\xrightarrow{\text{Anonymous download}}$$
$$<C_i, D_i>$$

$$U=D_i^R \bmod p$$

$$\xleftarrow{\hspace{3cm} U \hspace{3cm}}$$

$W=U^T \bmod p$

$$\xrightarrow{\hspace{3cm} W \hspace{3cm}}$$
$$\mathbf{P}(K_i)=W^Q \bmod p$$

Proof of Legal W without Disclosing S

In Section 3 we addressed the issue of non-repudiation. When a dispute occurs between the merchant and the customer, they can go to a third party for resolution. This requires the customer sign U and the merchant sign W in their interaction. Now the third party is convinced that U and W are indeed originated from the customer and the merchant, respectively. Still he should be convinced that U and W have the relationship $W=U^T \bmod p$ for $T=1/S \bmod 2q$. This can typically be done by the so-called *proof of equivalence of discrete logarithms*, where the merchant can prove to everyone that

$$W=U^T \bmod p \text{ and } Y=g^T \bmod p$$

hold for a same T without disclosing T. Here Y and g are values published by the merchant in advance.

5 Security Analysis

First, it is easy to see that the customer can obtain the digital good by decrypting ciphertext C_i since he can get K_i from $P(K_i)$. In other words, if the customer and the merchant follow the scheme properly, the customer can obtain what he wants. We next show that the merchant cannot know which message the customer intends to get even if the merchant is trying to perform incorrectly.

The merchant must figure out which D_i the customer has obtained. Our assumption is that the merchant cannot do it from the customer's download operation. Hence the only available information from the customer is $U(=D_i^R \bmod p)$. Since R is randomly chosen and kept secret by the customer, the i remains information-theoretically secure (all i's are equally probable). So the privacy for the customer is perfectly protected, without any computational assumption.

On the merchant side, it is required that the customer cannot get any extra messages. Without loss of generality, let's consider the following simplified situation.

The customer has already retrieved K_1, K_2, ..., K_h. Now the customer tries to recover $M_i(i \neq 1,2,...,h)$, i.e., to find K_i, without the decryption help from the merchant.

The problem is equivalent to this:

Having G_1, g_1, G_2, g_2, ..., G_h, g_h such that $G_1=(P(g_1))^r$, $G_2=(P(g_2))^r$, ..., $G_h=(P(g_h))^r$ (mod p is omitted here), finding g_i such that $G_i =(P(g_i))^r$ for some unknown r.

There are two approaches to solve the problem The first one is to find r through $G_j=(P(g_j))^r$ for $j=1,2,...,h$. But this is equivalent to computing discrete logarithm. Since the modulo p has 1024 bits, computing discrete logarithm is infeasible.

The second one is to compute G_i from G_1, G_2, ..., G_h through some arithmetic operations. This is dependent on how the padding format P is chosen.

A bad P, such as $P(K)=K$, may lead to a flaw under an attack like that in [9]. This is because those K_j's have only 128-bit size. Some of them may have a simple factorization. Even a simple attack may success with a tolerable computation complexity: a greedy customer gives $U=(D_1 D_2)^R$ to the merchant for decryption. Then the customer obtains $K_1 K_2$ since $P(K)=K$. Now K_1 and K_2 are all 128-bit integers. It is not very hard to find them by exhaustive tries following factorization.

We conjecture that a simple padding format as follows should be safe. Let H be a one-way hash function $H: \{0,1\}^* \rightarrow \{0,1\}^{160}$.

$$P(K)=H(K)\|H^2(K)\|H^3(K)\|K \oplus H^4(K)\|H^5(K)\|H^6(K)H^7(K)$$

Another possible choice is to take the padding format the same as in OAEP [10], i.e., $P(g)=g \oplus Ger(Rand) \| Rand \oplus H(g \oplus Ger(Rand))$ where Ger is a pseudo random generator and *Rand* is a random number.

Note. The padding **P** is applied only to K_i but never to U and W. Otherwise, the key K_i cannot be recovered.

6 Concluding Remarks

In this paper we propose an efficient and practical scheme for privacy protection in e-commerce of digital goods, which is based on a cryptographic technique, i.e., a commutative symmetric key cipher. However, it is not a general cipher from the cryptographic viewpoint. It is proposed specifically for our scheme. If it is used in other applications, security must be very carefully addressed. Also it is not as efficient as usual symmetric key ciphers. Although the exponential computation is involved in the cipher, the whole scheme is much more efficient than the previous PIR schemes. Another merit feature of our scheme is that is matches the overall structure of copyright protection where digital goods are encrypted with different keys and decrypted in tamper-resistant hardware/software.

References

1. B. Chor, O. Goldreich, E. Kushilevita, and M. Sudan, "Private Information Retrieval", Proc. Of 36th FOCS, pp. 41-50, 1995.
2. B. Chor and N. Gilboa, "Computational Private Information Retrieval", Proc. Of 29th STOC, pp. 304-313, 1997.
3. E. Kushilevita and R. Ostrovsky, "Singal-database Computationally Private Information Retrieval", Proc. Of 38th FOCS, 1997.
4. Y. Gertner, Y. Ishai, E. Kushilevita and T. Malkin, "Protecting Data Privacy in Private Information Retrieval Schemes", Proc. of 30th STOC, 1998.
5. C. Cachin, S. Micali, and M. Stadler, "Computationally Private Information Retrieval with Polylogrithmic Communication", in Proceedings of Eurocrypt'99, LNCS, Springer-Verlag, pp. 402-414, 1999.
6. D. Chaum, A. Fiat, and M. Naor, "Untraceable electronic cash", in Proceedings of Crypto'88, LNCS, Springer-Verlag, pp. 319-327, 1990.
7. A. Underwood, "Professional Ethics in a Security and Privacy Context - the Perspective of a National Computing Society", in Proceedings of ACISP'2000, LNCS 1841, Springer-Verlag, pp. 477-486, 2000.
8. Zero-Knowledge System Inc., http://www.zeroknowledge.com/
9. J. Coron, D. Naccache and J. Stern, "On the security of RSA padding", Crypto'99, pp. 1-18, Springer-Verlag, 1999.
   10. M. Bellare and P. Rogaway, "Optimal asymmetric encryption", Eurocrypt'94, LNCS, Springer-Verlag, 1995.

An Internet Anonymous Auction Scheme

Yi Mu and Vijay Varadharajan

School of Computing and IT, University of Western Sydney, Nepean,
P.O.Box 10, Kingswood, NSW 2747, Australia
Email: {yimu,vijay}@cit.nepean.uws.edu.au

Abstract. This paper proposes a new Internet bidding system that offers anonymity of bidders and fairness to both bidders and the auction server. Our scheme satisfies all the basic security requirements for a sealed-bid auction system, without requiring multiple servers.

1 Introduction

An Internet auction system is somewhat similar to a normal non-Internet auction scheme, but differs from it in certain aspects. One major difference is that an Internet auction system requires a sealed bidding process over the network. A sealed-bid auction is one in which secret bids are protected from disclosure before the bidding deadline. After the deadline has passed, all the bids are opened and the winning bid is determined according to some deterministic rules. Typically, it is used in the auctioning of artwork, real estate and government contracts.

A major challenge in developing a secure Internet auction is concerned with fairness between bidders and auction server. This includes the means of securing information exchange as well as achieving electronic payments. Secure information exchange provides secure and fair means for different, usually untrusted parties to exchange information via the Internet. These parties may never meet physically and can be geographically apart. Secure electronic payments need to ensure the operation and provide convenience for all the parties involved just like the conventional payment systems. To-date, there have been several research work on both fair exchange and secure payment issues [1,5,3,6,7].

Some recent work in securing sealed-bid auction have been done in [11,14]. These systems consider the auction service to be distributed over multiple non-related servers. Security, secrecy and fairness of bids are achieved, provided that no more that a certain number of the servers are faulty. However in these systems, the bidders, while they may not necessarily trust any individual auction server, they must trust the auction service as a whole. A sealed-bid auction is a two-sided transaction. On one side, we have the bidders and on the other side, it is the auction service. These two entities usually have contrasting commercial interest(s), thereby making the requirement that one entity (bidders) to trust the other (the auction service) somewhat unrealistic in practice.

In this paper, we consider a secure anonymous auction service where we place a minimal trust assumption between bidders and the auction server. In our system, a bidder can make secret bids to the server for a bidding instance advertised in the Internet. In the bidding period the server has no idea on the bid made by the bidder before the end of a bidding period. After the deadline, only the anonymous identity associated with the highest bid is revealed, while all others remain unknown. It therefore ensures maximum bidder privacy. Our scheme also prevents the winning bidder from unauthorised withdrawal. The trusted party remains off-line until it receives a request from the auction server for decrypting a cheating-winner's escrowed bid opener. Our scheme works with any on-line anonymous electronic payment systems. It does not require multiple auction servers that is deemed to be necessary for a sealed-bid auction [1,2]. The basic idea behind our protocols is similar to that for the digital cash protocol given in [8].

The rest of this paper is organized as follows. In section 2, we describe the security requirements for a sealed-bid auction. Section 3 reviews some preliminaries that are needed in our auction service scheme. Section 4 presents our new auction service scheme and describe the setup the bidding server and the bidder systems. Section 5 gives the bid casting protocols. Section 6 describes the determination of the winning bid and Section 7 concludes our work.

2 Secure Sealed-Bid Auction

Informally, a sealed-bid auction is a service in which services and goods are auctioned by an auction server. The goods and services are usually supplied by sellers. Each item is sold to a bidder through an auction. The auction consists of two phases. The first is the bidding period in which bidders submit sealed bids to the server. Once the bidding is closed, the second phase starts. In the second phase, all the bids are opened and a winner is determined and possibly announced. The choice of the winner is done using a publicly known deterministic process agreed between the seller and the auction server. For convenience, we assume that the winning bid is the highest bid.

Secure an auction system involves several issues that are conerned with the relationship between the auction server and the bidders. The auction server may involve a large number of bidders whose behaviors may not be trusted. Also, as all the bidders compete against one another with a common objective of obtaining the goods/services, the system must ensure fairness for all parties. To achieve these objectives, the auction system must satisfy the following requirements:

- **Auction server**
  - **Warranty of Funds/Elimination of Faulty Bids/Defaults Prevention:** Auction server should be able to eliminate the bids that are not valid or have insufficient funds prior to the selection of the winning bid. Once a bid is awarded, it is infeasible for the bidder to default the bid. This prevents the winning bidder to default his/her bid as evidenced by the FCC auction in which 13 winning bidders defaulted on their bids[10]. This is crucial to the success of the auction as our system is

anonymous, i.e., the identities of the bidders are not known at any stage and it is infeasible and expensive to hold faulty bidders accountable.

- **Fairness:** When the auction server delivers the goods/services to the winning bidder, it must be able to receive the payments regardless of how malicious the bidder may be.

- **Bidders**
  - **Privacy:** The identities of bidders(winners and losers) must be protected.
  - **No Misuse of Bids:** Once a bid is submitted, it should be used solely for the bidding purpose. It should be infeasible for anyone even the auction server to misuse the bid in order to disadvantage some bidders. Such cases occur when the server opens submitted bids and informs a collaborator of their amounts so that the collaborator can submit a higher bid.
  - **Fairness:** The bidders should be ensured that the winning bid is chosen properly. It is infeasible for the server to incorrectly award the goods/services to someone other than the winning bidder without being detected. The winning bidder should have sufficient evidence to prove the misbehavior of the auction server. The winning bidder must be able to receive the goods/services once the payment is made. Other bidders should not loose any funds in the auction.

It is worth noting that from time to time, disputes may occur during a particular auction. Therefore it is necessary to assume the existence of a trusted authority to resolve these disputes. However, our model is optimistic[1,2] in that the trusted authority is not involved unless there is a fault. In most disputes that occur between the auction server and the bidders, in which the action does not benefit the seller, the seller could assume the role of the trusted authority. This holds for many disputes that we are aware of, except that the server (in collusion with the sellers) does not deliver the goods/service while still being able to obtain the payment from the winning bidder.

In Franklin and Reiter's system[11], there are multiple servers, and each of these can receive bids from bidders. The information hiding is based on a (t, n) secret sharing scheme. That is, the information about a bid consists of a number of shares; each server can at most obtain one share for one bidder. Therefore, servers have no idea about the committed bid until the end of the bidding period. The discovery of the bid can be done by computing the secret bid by t out of n servers.

3 Preliminary

This section reviews some cryptographic primitives that are used in our auction protocol.

Some common notations will be used throughout this paper:

- p_1, p_2: two large prime numbers such that $p = 2q + 1$ is a prime number as well, where $q = p_1 p_2$.

- $\mathcal{Z}_p^*$: a primitive group of order $p - 1$.
- $\mathcal{Z}_q^*$: a primitive group of order $\phi(q)$ or $(p_1 - 1)(p_2 - 1)$.
- $\mathcal{H}$: a strong one-way hash function.

3.1 Optimistic Fair Exchange

Optimistic fair exchange[2,1,3] is a protocol that allows two untrusted parties to exchange their information in a fair manner. After the exchange, either both parties get the other's information or neither party gets anything. The information held by the two parties can be of any format. It can be a file, a document or even a signature. Optimistic fair exchange protocol assumes the existence of an off-line trusted authority. However, the trusted authority is invisible and only becomes apparent in the case of disputes.

Optimistic fair exchange makes use of a cryptographic primitive called verifiable encryption. Verifiable encryption consists of a ciphertext under the trusted authority's public key and a non-interactive proof that the plaintext corresponding to the ciphertext is indeed the required information. Verifiable encryption can be constructed for any type of information. For simplicity, we leave out the construction of verifiable encryption. The reader is referred to [1,4] for details.

Let $VE(m)$ be the verifiable encryption of the message m using the public key of the trusted authority (TA). The optimistic fair exchange between the two parties A and B, each holding the information α, β wanted by the other party respectively, works as follows:

Fair Exchange Protocol (run in a secure channel)

1: First A and B agree on the condition of the exchange. Let us assume that A initiates the exchange.
2: $A \rightarrow B$: $VE(\alpha)$
3: $B \rightarrow A$: $VE(\beta)$
4: $A \rightarrow B$: α
5: $B \rightarrow A$: β

Then the exchange is completed. However, if the exchange is not successfully terminated, any party, who has received the verifiable encryption from the other party, runs the following recovery protocol with the trusted authority. Without the loss of generality, let us assume that A is the party involved.

Recovery Protocol (run in a secure channel)

1: $A \rightarrow TA$: $\alpha, VE(\beta)$ and information about the exchange.
2: TA decrypts $VE(\beta)$ to get β. Then $TA \rightarrow A : \beta$ and $TA \rightarrow B : \alpha$. Here TA should verify all the information and send α, β to the respective parties if and only if TA is satisfied with all the checks.

Note that the recovery protocol is always fair. Upon completing the protocol, both parties will get the other's information. In fact, in the exchange protocol,

B is never disadvantaged. The only possible unfair state is when B gets α, while A does not receive β. However A will never send α to B unless A has received $VE(\beta)$. With $VE(\beta)$, A can run the recovery protocol with TA to obtain β. Hence, at no stage A is disadvantaged, i.e. the protocol is fair for both A and B.

For our auction service, we are mainly interested in the fair exchange of digital signature knowledge[16]. In [1], an optimistic fair exchange protocol of digital signatures was proposed for several digital signature schemes. It can be used to exchange signature knowledge as well. For convenience, we denote by $OFE(a, b)$ an Optimistic Fair Exchange of a and b.

3.2 Blind Nyberg-Rueppel digital signature

Assume that $x \in \mathcal{Z}_q$ is the secret key of the signer. $h = g^x \bmod p$ is then the public key of the signer, where $g \in \mathcal{Z}_p^*$. g, q, p are public.

To obtain a blind Nyberg-Rueppel digital signature on a message m from the signer, the verifier or the signature receipient needs to get a pair (r, s) in the form:

$$r = mg^k \bmod p, \quad (k \in_R \mathcal{Z}_q)$$
$$s = xr + k \bmod q,$$

in such a way that the signer does not learn anything about either r or s. This can be achieved using the following process:

1: The signer selects $\tilde{k} \in \mathcal{Z}_q$, computes $\tilde{r} = g^{\tilde{k}} \bmod p$, and sends $\tilde{r}$ to the verifier.
2: The verifier selects $\alpha, \beta \in \mathcal{Z}_q$, computes $r = mg^\alpha \tilde{r}^\beta \bmod p$, $\tilde{m} = r\beta^{-1}$ and sends $\tilde{m}$ to the signer.
3: The signer computes $\tilde{s} = \tilde{m}x + \tilde{k}$ and then forwards $\tilde{s}$ to the verifier.
4: The verifier computes $s = \tilde{s}\beta + \alpha \bmod q$.

The pair (r, s) is then a blind signature of the signer on message m. The verification of the signature (r, s) for message m is done by verifying

$$g^{-s}h^r r = mg^{-\tilde{s}\beta + xr + \tilde{k}\beta + \alpha} = mg^{-\tilde{m}x\beta - \tilde{k}\beta + xr + \tilde{k}\beta} = m \bmod p.$$

Furthermore, as α and β are randomly chosen, the signer does not learn anything about (r, s). For a given signature (r, s), there exists an unique pair of α and β. Thus for each signature from the signer, the verifier can generate only one blind signature. Detailed discussion on the security of this scheme can be found in [15].

3.3 Proof of Discrete Logarithm

We now take a look at the scheme of proving knowledge without revealing anything about the content that is being proved. This proof scheme was initially proposed in [12,13]. We briefly summarise here the method of discrete logarithm proof. The objective of the proof is that, given a primitive $g \in_R \mathcal{Z}_p^*$ and

$\varrho = g^o \bmod p$ where $o \in \mathcal{Z}_q$ is the secret known to the prover only and ϱ is public, the prover proves her knowledge on o but not revealing the secret to the verifier. The proof protocol can be either interactive or non-interactive. Here we consider the interactive one only:

1: The Verifier: sends a challenge c to the Prover.
2: The Prover:
   - selects a secret number $\delta \in_R \mathcal{Z}_q$ and then computes $h = g^\delta \bmod p$ and $w = (co + \delta) \bmod q$.
   - sends c, g, h, ϱ to the Verifier.
3: The Verifier then checks: $g^w \stackrel{?}{=} \varrho^c h \bmod p$.

For convenience, we denote by $DLP[o : \varrho]$ the discrete log proof.

4 System Setup

Our system consists of an auction server $\mathcal{S}$, a/several financial institution(s), several bidders. We denote by $\mathcal{B}$ a bidder and by $\mathcal{F}$ a financial institution. There are two possible design choices regarding Trusted Third Party (TTP): (1) There exists an independent TTP, who acts as a key escrow agent, and (2) The financial institution also acts as a TTP. For simplicity, we assume that the financial institution is also the TTP in our system.

4.1 The Server Setup

The auction server should not know the bids of bidders before the bidding deadline. This is ensured by the bid structure itself and a TTP rather than secret sharing. Bidders are confident that their bids are safe, since the winning bidder is the only party who sends its bid opener to the server once it is informed that its bid is the highest. With a bid opener escrow scheme, the server can also prevent any winning bidder from withdrawal.

The auction server has the following tasks:

- Displays auction objects on its web page.
- Receives/Validates the anonymous bids from bidders.
- Receives/Validates the escrowed bid openers
- Implements a fair exchange protocol.

4.2 Bidder Setup

Each bidder needs to establish anonymous account at the financial institution. We assume that an alias or anonymous identity has been issued to the bidder $\mathcal{B}$. Using it, $\mathcal{B}$ can obtain an anonymous account from the financial institution. Let $A \stackrel{\text{def}}{=} g^\sigma \bmod p$ be the authorised alias for bidder $\mathcal{B}$. σ is a secret number known only to $\mathcal{B}$. Let A be the legitimate account name for $\mathcal{B}$. For the bidding purpose, $\mathcal{B}$ should obtain an account certificate associated with A by the following process:

We assume that the financial institution has a web page providing on-line service. In order to obtain an account certificate, the bidder accesses the web and sends her request. The details of the process are as follows:

Let (p, q, g) be public information, where $g \in Z_p^*$ and $g \in Z_q^*$, i.e., $g^q \bmod p = 1$ and $g^{\phi(q)} \bmod q = 1$. Let x be $\mathcal{F}$'s secret key and $h = g^x \bmod p$ and $h' = g^x \bmod q$ be its public keys. $\mathcal{F}$ chooses two random numbers w_1 and w_2, computes $g_1 = g^{w_1} \bmod p$, $g_2 = g^{w_2} \bmod p$ as well as $h_1 = g_1^x \bmod p$ and $h_2 = g_2^x \bmod p$. g_1, g_2, h_1 and h_2 are then made public. Each bidder needs to pre-determine its bid and the bidding commitment $v = g_1 g_2^u \bmod p$, where u is the value of concatenation of the bid and a secret random number. Without loss of generality, we will call u the bid for $\mathcal{B}$. The bidding commitment v is registered with the financial institution as $\mathcal{B}$'s current identity is associated with A. $\mathcal{B}$ is given $w = v^x \bmod p$ as the certificate of $\mathcal{B}$'s identity. It might also be necessary for $\mathcal{F}$ to prove to $\mathcal{B}$ that $\log_v w = \log_g g^x$. This can be done using a bi-proof of discrete logarithm. Note that v, y, w are used for one bid only.

Bidding setup protocol:

0: $\mathcal{B}$ sends her discrete log proof $DLP([\sigma : A]$ to $\mathcal{F}$ for the authentication purpose.

1: $\mathcal{F}$ chooses a random number $k \in Z_q$, computes $\delta = v^k \bmod p$ and forwards δ to $\mathcal{B}$.

2: $\mathcal{B}$ generates three random numbers (y, x_1, x_2), computes $\alpha = w^y \bmod p$, $\beta = v^y \bmod p$ and $\lambda = h_1^{x_1} h_2^{x_2} \bmod p$.

3: $\mathcal{B}$ forms the message $m = \mathcal{H}(\alpha, \beta, \lambda)$, generates a random number a and a Nyberg-Rueppel blind factor b, calculates $r = m\beta^a \delta^{by} \bmod p$ and sends $m' = r/b$ to $\mathcal{F}$.

4: $\mathcal{B}$ computes its Nyberg-Rueppel signature on the blind message m' by forming $s' = m'x + k$ to $\mathcal{B}$ and sends it to $\mathcal{F}$.

5: $\mathcal{B}$ removes the blind factor b and obtains $s = s'b + a = rx + kb + a$.

$Cert_B \stackrel{\text{def}}{=} \{\alpha, \beta, \lambda, r, s\}$ represents a valid anonymous account certificate for the bid u. It can be verified using the following equation:

$$\mathcal{H}(\alpha, \beta, \lambda) = \beta^{-s}\alpha^r r \bmod p.$$

$Cert_B$ is used for one bid only.

5   Bidding Protocol

When $\mathcal{B}$ wishes to cast a bid to the auction server $\mathcal{S}$, she needs to obtain her unique bidding number N and bidding account certificate $Cert_B$ with respect to her bid u from the on-line $\mathcal{F}$. The bidding protocol is divided into two stages: casting a bid and sending an escrowed bid opener. $\mathcal{S}$ is able to verify the correctness of the bid and the corresponding escrowed opener. The following protocol is used for casting a bid:

$\mathcal{B}$		$\mathcal{S}$
Casting a bid:		
		$c \leftarrow \mathcal{H}(\mathcal{S}\|Date\|Time\|\ldots)$
	$\xleftarrow{\quad c \quad}$	
$r_1 = x_1 + cy \bmod q$		
$r_2 = x_2 + ucy \bmod q$		
	$\xrightarrow{\quad N, r_1, r_2, Cert_B \quad}$	
		$\mathcal{H}(\alpha, \beta, \lambda) \stackrel{?}{=} \beta^{-s}\alpha^r r \bmod p$
		$h_1^{r_1} h_2^{r_2} \stackrel{?}{=} \alpha^c \lambda \bmod p$
Creating escrowed bid opener:		
		$c' \leftarrow \mathcal{H}(\mathcal{S}\|Date\|Time\|\ldots)$
	$\xleftarrow{\quad c' \quad}$	
$\epsilon \in_R \mathcal{Z}_{\phi(q)}$		
$r_1' = x_1 + c'y \bmod q$		
$r_2' = x_2 + uc'y \bmod q$		
$k = g^\epsilon \bmod q, \; \bar{h}' = h'^\epsilon \bmod q$		
$R_1 = r_1' k \bmod q, \; R_2 = r_2' k \bmod q$		
$\alpha' = \alpha^k \bmod p, \; \lambda' = \lambda^k \bmod p$		
	$\xrightarrow{\quad R_1, R_2, \bar{h}', \alpha' \lambda' \quad}$	
		$h_1^{R_1} h_2^{R_2} \stackrel{?}{=} \alpha'^{c'} \lambda' \bmod p$
Proving:		
$\log_{h'} \bar{h}' = \log_g (\log_\alpha \alpha')$		
$\log_{h'} \bar{h}' = \log_g (\log_\lambda \lambda')$		
$KREP[(\epsilon, 1) : h'^\epsilon \wedge g^\epsilon r_1']$		
$KREP[(\epsilon, 1) : h'^\epsilon \wedge g^\epsilon r_2']$	$\xrightarrow{\quad Proof\ tokens \quad}$	Verification

1: S generates a random challenge c and sends it to B. This challenge should be unique for each bid. For example, it can be computed as

$$c = \mathcal{H}(S\|Date\|Time\|\dots)$$

2: Upon receiving c, B computes a response (r_1, r_2) with respect to her bid, u, where $r_1 = x_1 + cy \bmod q$ and $r_2 = x_2 + ucy \bmod q$. The bidding token (r_1, r_2) along with her bidding number N and the account certificate is then sent to S.

3: S checks the Nyberg-Rueppel signature on the message $\mathcal{H}(\alpha, \beta, \lambda)$ and verifies that (r_1, r_2) are indeed consistent with the challenge c. S accepts the account certificate for the bid if $\mathcal{H}(\alpha, \beta, \lambda) = \beta^{-s}\alpha^r r$ and $h_1^{r_1} h_2^{r_2} = \alpha^c \lambda$. S then stores $(c, r_1, r_2, Cert_B)$. It is obvious that if the bidder and the server follow the correct procedures given above, then $\mathcal{H}(\alpha, \beta, \lambda) = \beta^{-s}\alpha^r r$ and $h_1^{r_1} h_2^{r_2} = \alpha^c \lambda$ must hold. In fact, we find that

$$
\begin{aligned}
\alpha^c \lambda &= w^{yc} h_1^{x_1} h_2^{x_2} \\
&= (g_1 g_2^u)^{xyc} h_1^{x_1} h_2^{x_2} \\
&= (h_1 h_2^u)^{yc} h_1^{x_1} h_2^{x_2} \\
&= h_1^{x_1 + cy} h_2^{x_2 + cyu} \\
&= h_1^{r_1} h_2^{r_2} \bmod p.
\end{aligned}
$$

4: S sends a new challenge, c', to B for computing the escrowed bid opener.

5: B computes the bid opener (r_1', r_2'), $r_1' = x_1 + c'y \bmod q$, $r_2' = x_2 + uc'y \bmod q$, which are then encrypted with the TTP's public key h,

$$\epsilon \in_R \mathcal{Z}_\phi(q),\ k = g^\epsilon \bmod q,\ \bar{h}' = h'^\epsilon \bmod q,\ R_1 = kr_1' \bmod q,\ R_2 = kr_2' \bmod q,$$

where ϵ and k are kept secret. In order to allow S to prove that the escrowed opener is correct, B needs to compute

$$\alpha' = \alpha^k,\ \lambda' = \lambda^k\ (\bmod p).$$

R_1, R_2, $\bar{h}'$, α', and λ' are then sent to S.

6: S checks correctness of the escrowed bid opener, $h_1^{R_1} h_2^{R_2} \stackrel{?}{=} \alpha'^{c'} \lambda' \bmod p$

There are two additional proofs are needed from the bidder: (1) The knowledge proof on representations of discrete logarithms that proves r_1' and r_2' are indeed encrypted correctly, and (2) The discrete log proofs on the equalities of $\log_{h'} \bar{h}' = \log_g (\log_\alpha \alpha')$ and $\log_{h'} \bar{h}' = \log_g (\log_\lambda \lambda')$.

For the first proof, the readers are referred to [17] for details. We here use the notation: $KREP[(\epsilon, 1): h'^\epsilon \wedge g^\epsilon r_1']$, which means that the discrete logarithm $\bar{h}'$ to base h' and a representation of R_1 to bases g and r_1', and the g-part of this representation equals the discrete logarithm of $\bar{h}'$ to base h'. Similarly to that, we have $KREP[(\epsilon, 1): h'^\epsilon \wedge g^\epsilon r_2']$. The second proof is realised as follows:

1: The bidder (prover):
- Selects an integer $\delta \in_R \mathcal{Z}_{\phi(q)}$ and computes $f = g^\delta \bmod q$.
- Sends f to the server $\mathcal{S}$.

2: The server $\mathcal{S}$ sends a challenge τ to the bidder.

3: The bidder:
- Computes $t = \epsilon + \tau\delta$.
- Sends t to $\mathcal{S}$.

4: The server $\mathcal{S}$:
- Checks $h'^t \stackrel{?}{=} \bar{h}'f^\tau \bmod q$, $\alpha^{g^t} \stackrel{?}{=} (\alpha')^{f^\tau} \bmod p$, and $\lambda^{g^t} \stackrel{?}{=} (\lambda')^{f^\tau} \bmod p$.

This protocol can be easily converted into a non-interactive version.

6 Opening the Winning Bid

After the closing day of bidding, all bidders anonymously submit the values of their submitted bids along with their bidding numbers to the auction server. The auction server then broadcasts the highest bid and a new challenge, c''. The bidder responsible for the highest bid computes (r''_1, r''_2) using c'',

$$r''_1 = x_1 + c''y \bmod q, \quad r''_2 = x_2 + uc''y \bmod q,$$

$\mathcal{B}$ and $\mathcal{S}$ then carry out an Optimistic Fair Exchange on $r'' \leftarrow (N, r''_1, r''_2)$ and the auction object R, i.e, $OFE(r'', R)$, where R could be either the electronic auction object or a receipt that ensures the physical delivery of the auction goods at a later stage. As part of optimistic fair exchange process, the server must check the legitimacy of r'' as follows. The server computes the bid u and the current alias v:

$$\frac{r_1 - r''_1}{c - c''} = \frac{x_1 + cy - x_1 - c''y}{c - c''} = y,$$

$$\frac{r_2 - r''_2}{c - c''} = \frac{x_2 + cyu - x_2 - uc''y}{c - c''} = yu.$$

From y and yu, $\mathcal{S}$ can easily obtain the bid value u and check if u equals the highest bid. After obtaining u, $\mathcal{S}$ computes the bidder's anonymous bidding identity $v = g_1 g_2^u \bmod p$. v is then sent to $\mathcal{F}$, who can find the match between this value and the one in the $\mathcal{F}$'s alias list. With such a match, $\mathcal{F}$ can transfer the money from $\mathcal{B}$'s account represented by anonymous identity A to $\mathcal{S}$'s account. The evidence is *undeniable* as u is the bidder's secret information, which is not possible to determine unless the bidder sends r''.

If the winner refuses to send r'' to the server, the server can send $Cert_B$ and (R_1, R_2, c') to the TTP, who can then decrypt (R_1, R_2) and send (r'_1, r'_2) to the server. The server can compute u and v using (r_1, r_2, c) and (r'_1, r'_2, c').

7 Conclusion

We have proposed a new anonymous sealed-bid action scheme for the Internet. Our scheme satisfies all basic requirements for a sealed-bid auction, without requiring multiple servers. In our system, each bidder has an anonymous account with a financial institution. Before casting its bid, each bidder logs onto the financial institution's on-line web page to obtain an account certificate, associated with its account, for its bid. All bids submitted to the auction server are protected from disclosure, before the auction deadline. After the deadline, only the anonymous identity associated with the highest bid is revealed, while all others remain unknown. It therefore ensures the maximum bidder privacy. Our scheme also prevents the winning bidder from unauthorised withdrawal, since an escrowed bid opener is sent to the server who can prove the escrowed opener is legitimate. The TTP remains off-line until receiving a request from the server for decrypting a cheating-winner's escrowed bid opener.

Additional note: After this paper has been submitted, we noted that a similar method has been just proposed in [18].

References

1. N. Asokan, V. Shoup and M. Waidner. "Asynchronous protocols for optimistic fair exchange", In Proceedings of IEEE Symp. on Research in Security and Privacy Available on-line at www.cs.wisc.edu/~shoup.
2. N. Asokan, V. Shoup and M. Waidner. "Optimistic fair exchange of Digital Signatures", Advances in Cryptography – Eurocrypt'98 (Lecture Notes in Compute Science 1403), Springer-Verlag, 1998. Pages 591–606.
3. F. Bao, R. H. Deng and W. Mao, "Efficient and practical fair exchange protocols with off-line TTP", in Proceedings of the 1998 IEEE Symposium on Security and Privacy, IEEE Computer Press, Oakland, CA, 1998.
4. F. Bao, "An efficient verifiable encryption scheme for the encryption of discrete logarithms", in Proceedings of CARDIS'98, LNCS, Springer-Verlag, 1998.
5. S. Brands. "Untraceable off-line cash in wallets with observers". In Advances in Cryptology – Crypto'93, Springer-Verlag 1994. Pages 302–318.
6. D. Chaum. "Security without identification: Transaction systems to make big both obsolete", Communications of the ACM, 28:1030–1044, 1985.
7. D. Chaum. "Online cash checks", Advances in Cryptology - EUROCRYPT '89. Springer-Verlag, 1990. Pages 288-293.
8. K. Q. Nguyen, Y. Mu, and Vijay Varadharajan, "A new digital cash scheme based on blind Nyberg-Rueppel digital signature", in Proceedings of the Information Security Workshop, 1997, Japan, Lecture Notes in Computer Science, Springer Verlag.
9. T. ElGamal, "A public key cryptosystem and a signature scheme based on discrete logarithms", IEEE Transactions on Information Theory, IT-31(4):469-472, 1985.
10. FCC takes licenses, denies more time to 13 bidders. AP-Dow Jones News, August 10,1994.

11. M. Franklin and M. Reiter. "The design and implementation of a secure auction service", Proceedings of 16th IEEE Symposium on Security and Privacy, Oakland, California, May 1995. Pages 2–14.

12. C. P. Schnorr, "Efficient signature generation for smart cards", Proceedings Crypto'89, LNCS, Springer-Verlag, pp.225-232, 1990.

13. D. Chaum, J.-H. Evertse, and J. V. de Graaf, "An improved protocol for demonstrating possession of discrete logarithms and some generalizations," pp. 128–141, Springer-Verlag, 1988.

14. Dawn Song and Jonathan Millen, "Secure Auctions in a Publish/Subscribe System", available from *http://www.csl.sri.com/~millen*

15. J. Camenisch and J.Piveteau, M. Stadler, "Blind Signatures Based on the Discrete Logarithm Problem", *Advances of Cryptology- Eurocrypt'94 Proceedings*, Springer-Verlag, 1994, pp.428-432.

16. Jan Camenisch, "Efficient and Generalized Group Signatures", *Adances in cryptology - EUROCRYPT'97, Lecture Notes in Computer Secience 1233*, Springer-Verlag, Berlin, 1997, pp465-479.

17. J. Camenisch and M. Stadler, "Efficient group signature schemes for large groups," pp. 410–424, Springer-Verlag, Berlin, 1997.

18. K. Viswanathan, C. Boyd, E. Dawson: "A Three Phased Scheme for Sealed Bid Auction System Design" in 5th Australasian Conference on Information Security and Privacy (ACISP 2000), LNCS 1841, Springer-Verlag, pp.412-427,

Efficient Sealed-Bid Auction Using Hash Chain

Koutarou Suzuki, Kunio Kobayashi and Hikaru Morita

NTT Laboratories
1-1 Hikari-no-oka, Yokosuka, Kanagawa, 239-0847 Japan
{koutarou,kunio,morita}@isl.ntt.co.jp

Abstract. This paper proposes the first sealed-bid auction method which uses *only hash function*. We use a hash chain to commit a bidding price. By using the hash chain, we can drastically reduce the time needed for bidding and opening bids. If we use a practical hash function e.g. SHA-1, our method is 200,000 times faster than former methods that use public key cryptosystems. Accordingly, our method is capable of wide application in terms of the number of bidders and the range of bidding prices.

1 Introduction

Auctions are a basic and important method to establish the price of goods. As the importance of the Internet continues to increase, electronic auction services are also increasing. There has been a lot research on auctions. However, existing sealed-bid auction methods use public key cryptosystems and so are computationally expensive or require a lot of communication. The inefficiency of the public key cryptosystems forces the auction methods to limit the number of bidders and the range of bidding prices. Moreover, additional trusted third parties may be needed to ensure the secrecy of bidders.

This paper proposes the first sealed-bid auction method which uses *only hash function*. We use a hash chain to commit a bidding price. Though a hash chain method has been used for authentication e.g. SKEY [Lam81] or payment e.g. PAYWORD [RS96], the authors found a new application : sealing and comparing bids where the bidding prices except for the winning price is kept secret [KM99] [Suz99] [KMSH01]. The hash chain drastically reduces the time needed for bidding and opening bids. If we use a practical hash function e.g. SHA−1 [FIPS], our method is 200,000 times faster than former methods that use public key cryptosystems. Accordingly, our method is widely applicable in terms of the number of bidders and the range of bidding prices.

In Section 2, we explain sealed-bid auction and its requirements, and review two previous works. In Section 3, we introduce our sealed-bid auction method and discuss its security and efficiency. In Section 4, we conclude the paper.

1.1 Related Works

Kikuchi, Harkavy and Tygar showed the necessity of bid secrecy, and presented a sealed-bid auction method that uses a distributed decryption technique [HTK98]

[KHT98]. Kudo used a time server to realize sealed-bid auctions [Kud98]. Naor, Pinkas and Sumner realized sealed-bid auctions by combining Yao's secure computation with oblivious transfer [NPS99]. The method is capable to compute any circuit, so can realize various types of auctions (e.g. Vickrey auction). Cachin proposed a sealed-bid auction using a homomorphic encryption and oblivious third party [Cac99]. The complexity of the method is in polynomial of logarithmic of the number of possible prices. Stubblebine and Syverson proposed an open-bid auction method which uses a hash chain technique[SS99]. Sakurai and Miyazaki proposed a sealed-bid auction in which a bid is represented by the bidder's undeniable signature of his bidding price [SM99]. Sako proposed a sealed-bid auction in which a bid is represented by an encrypted message with a public key that corresponds to his bidding price [Sak00].

However, these sealed-bid auction methods use public key cryptosystems and so are computationally expensive or require a lot of communication. The inefficiency of the public key cryptosystems forces the auction methods to limit the number of bidders and the range of bidding prices.

2 Sealed-bid Auction

2.1 Sealed-bid Auction

The sealed-bid auction is a type of auction in which bids are kept secret during the bidding phase. In the bidding phase, each bidder sends his sealed bidding price. In the opening phase, the auctioneer opens the sealed bids and determines the winning bidder and winning price according to a rule : the bidder who bids the highest (or lowest) price of all bidder is the winning bidder and their bidding price is the winning price.[1] Since there is no difference essentially between the highest price case and the lowest price case, we discuss only the former hereafter.

Bidding : Let A be an auctioneer and $B_1, \cdots, B_m$ be m bidders. The auctioneer shows some goods for auction (e.g. a painting from a famous painter) and calls the bidders to bid their price for this item. Each bidder B_i then decides his price, and seals his price (e.g. by envelope) and puts his sealed price into auctioneer's ballot box.

Opening : After all bidders have input their sealed prices, the auctioneer opens his ballot box. He reveals each sealed price and finds the bidder who bid the highest price. The bidder who bid the highest price wins and he can buy the item at his bid price.

2.2 Requirements

To achieve a fair auction, the sealed-bid auction protocol must satisfy the following properties.

Secrecy of bidding price : All bidding prices except winning price must be kept secret even from the auctioneer. If the auctioneer can see some bidding

---

[1] Here we accept the case where several bidders submit identical highest bids.

prices before opening, he can tell the prices to a colluding bidder in order to cheat the auction.

Verifiability : Anyone must be able to verify the correctness of the auction. As above, all bidding prices except winning price are kept secret. So verifiability is necessary to convince all bidders.

Undeniability : No bidder must be able to deny his bidding price. If a bidder can deny his bidding price, two bidders can collude to cheat the auction as follows. Two bidders B_1 and B_2 collude and bid their prices p_1 and $p_2 = p_1 - 1$. If no bidder bid at the price higher than or equal to p_1, bidder B_1 denies his bid at price p_1 and bidder B_2 wins the auction at his price p_2. If some bidders bid at price p_1, bidder B_1 does not deny his bid at price p_1 and wins the auction at his price p_1. By this, bidders B_1 and B_2 can bid at two selectable prices p_1 and $p_2 = p_1 - 1$ and so cheat the auction.

Anonymity : The bidders must be able to bid anonymously. It is important to keep the privacy of bidders to prevent collusion between bidders and auctioneer.

2.3 Previous Work

To compare with our methods 1 and 2, we review here two former methods. These methods as well as our methods satisfy above requirements.

Former Method 1. Sakurai and Miyazaki proposed a sealed-bid auction in which a bid is represented by the bidder's undeniable signature of his bidding price [SM99]. We describe their protocol as a good reference against which we compare our method 1.

Bidding : Each bidder B chooses his bidding price p from the price list $P = \{1, 2, \cdots, n\}$. He creates an undeniable signature $Sig_B(p)$ of his bidding price, and publishes it as the commitment of his bidding price.

Opening : After all bidders commit their bidding prices, the auctioneer iterates following steps for $j = n, n - 1, \cdots$ to open bids.

If bidder B prove that his bidding commitment $Sig_B(p)$ is a valid signature of price j by the confirmation protocol of the undeniable signature, the auctioneer is convinced that the winning bidder is bidder B and winning price is j.

If all bidders prove that his bidding commitment $Sig_B(p)$ is not a valid signature of price j by the repudiation protocol of the undeniable signature, the auctioneer is convinced that no bidder bid at price j. Then auctioneer decreases j by 1 and repeat the above steps.

This protocol satisfies all the above required properties. However, all bidders have to communicate with auctioneer in opening phase.

Former Method 2. Sako proposed a sealed-bid auction in which a bid is represented by the encrypted message with the key which corresponds to the bidding price [Sak00]. We describe her protocol as a good reference against which we compare our method 2.

Let E and D be probabilistic encryption and decryption function. Auctioneer publishes keys K_j and messages M_j corresponding to each prices $j \in \{1, 2, \cdots, n\}$.

Bidding : Each bidder B chooses his bidding price p from the price list $P = \{1, 2, \cdots, n\}$. He creates his encrypted message $C_B = E_{K_p}(M_p)$ with the key K_p which corresponds to his bidding price p, and publishes it as his bid.

Opening : After all bidders send their bid, the auctioneer iterates following steps for $j = n, n - 1, \cdots$ to open bids.

If relation $D_{K_j}(C_B) = M_j$ holds, the auctioneer is convinced that the winning bidder is bidder B and winning price is j.

If relation $D_{K_j}(C_B) = M_j$ does not hold for all bidders, the auctioneer is convinced that no bidder bid at price j. Then auctioneer decreases j by 1 and repeat the above steps.

This protocol satisfies all the above required properties. Moreover, bidders need not to communicate with auctioneer in opening phase. However, malicious auctioneer can reveal all bidding prices. To prevent this, we have to use plural auctioneers and distributed decryption technique.

3 Sealed-bid Auction using Hash Chain

Our sealed-bid auction method is the first method which uses *only hash function*. It is based on a chain of a hash function. By using a hash chain we can drastically reduce the time taken for bidding and opening bids. If we use a practical hash function e.g. SHA–1, our method is 200,000 times faster than former methods that use public key cryptosystems. The method resembles SKEY in the point that the proof of the commitment plays the role of the next commitment, and PAYWORD in the point that the number of the iteration of the hash function represents the value.

3.1 Proposed Sealed-bid Auction Method 1

The first proposed sealed-bid auction method is as follows.

Preparation : Let A be an auctioneer and $B_1, B_2, \cdots, B_m$ be bidders. Auctioneer A publishes

- $P = \{1, 2, \cdots, n\}$: a price list of the auction,
- $h : \{0, 1\}^{s+t} \to \{0, 1\}^t$: a collision intractable[2] random hash function,
- $M_{no}, M_{yes} \in \{0, 1\}^s$: messages for "I do NOT bid" and "I DO bid".

Bidding : In the bidding phase, each bidder B_i decides his bidding price $P_i \in P$. He chooses his secret seed $S_i \in \{0, 1\}^t$ randomly and computes a hash chain

$$L_{P_i-1, i} = S_i, \quad L_{P_i, i} = h(M_{yes} || L_{P_i-1, i}),$$

---

[2] In this paper "collision intractable" means that to find x, y s.t. $h(x) = h(y)$ is computationally infeasible. We consider a practical hash function e.g. SHA–1.

$$L_{j,i} = h(M_{no}||L_{j-1,i}) \; (j = P_i + 1, P_i + 2, \cdots, n).$$

He then sends the commitment of his biding price with his signature

$$(L_{n,i}, Sig_{B_i}(L_{n,i})).$$

to auctioneer A. Auctioneer A receives all bids $(L_{n,i}, Sig_{B_i}(L_{n,i}))$ $(i = 1, 2, \cdots, m)$, verifies the signatures and publishes all bids.

Opening : In the opening phase, auctioneer A iterates following steps for prices $j = n, n - 1, \cdots$.

Auctioneer A receives messages $L_{j-1,i}$ $(i = 1, 2, \cdots, m)$ from all bidders B_i and checks the following equality for $i = 1, 2, \cdots, m$

$$L_{j,i} = h(M_{no}||L_{j-1,i}).$$

If the above equation holds for all i's, auctioneer A concludes that no bidder bids price j. Auctioneer A then publishes all messages $L_{j-1,i}$ $(i = 1, 2, \cdots, m)$, decreases price j by 1, and receives messages $L_{j-2,i}$ $(i = 1, 2, \cdots, m)$...

If the above equation does not hold for some i's, auctioneer A checks the following equality for these i's

$$L_{j,i} = h(M_{yes}||L_{j-1,i}),$$

and concludes that the winning bidders are these B_i's and their winning price is j. Finally, auctioneer A publishes the winning bidders, the winning price and all messages $L_{j-1,i}$ $(i = 1, 2, \cdots, m)$.

Notice that we can omit the messages M_{no} i.e. each bidder bids $L_{n,i} = h^{n-P_i+1}(M_{yes}||S_i)$ with his signature.

We can also realize a sealed-bid auction with quantity as follows. Each bidder B_i decides his bidding quantity for each price j and sets the messages

$$M_{j,i} = \text{"bidding quantity of bidder } B_i \text{ at price } j \text{ "}.$$

He then makes his hash chain $L_{j,i} = h(M_{j,i}||L_{j-1,i})$, and bids his commitment with his signature $(L_{n,i}, Sig_{B_i}(L_{n,i}))$.

3.2 Proposed Sealed-bid Auction Method 2

In our first method, each bidder has to communicate with auctioneer to open his bid. In our second method, since each bidder send the secret seed of his hash chain to auctioneer, he does not need to communicate to open his bid. To keep the secrecy of bidding price, we use plural auctioneers who cooperate to open bids. The secrecy of bidding price is then kept besides the case all auctioneers collude. The second proposed sealed-bid auction method is as follows.

Preparation : Let $A_1, A_2, \cdots, A_a$ be auctioneers and $B_1, B_2, \cdots, B_m$ be bidders. Auctioneer A_1 publishes

- $P = \{1, 2, \cdots, n\}$: a price list of the auction,
- $h : \{0, 1\}^* \to \{0, 1\}^t$: a collision intractable random hash function,

Bidding : In the bidding phase, each bidder B_i decides his bidding price $P_i \in P$. He chooses his secret seed $S_{1,i}, S_{2,i}, \cdots S_{a,i} \in \{0,1\}^t$ randomly and computes his bid

$$Bid_i = \{b_i, c_{1,i}, c_{2,i}, \cdots, c_{a,i}\}$$

where

$$b_i = h(h^{P_i}(S_{1,i})|h^{P_i}(S_{2,i})|\cdots|h^{P_i}(S_{a,i})), c_{j,i} = h^{n+1}(S_{j,i}).$$

He then sends the commitment of his biding price with his signature

$$(Bid_i, Sig_{B_i}(Bid_i)).$$

to auctioneer A_1. Auctioneer A_1 receives all bids $(Bid_i, Sig_{B_i}(Bid_i))$ $(i = 1, 2, \cdots, m)$, verifies the signatures and publishes all bids. After all bids are published, each bidder B_i sends his secret seed $S_{j,i}$ to each auctioneer A_j through a secure channel (e.g. using encryption with auctioneer A_j's key).

Opening : In the opening phase, auctioneers $A_1, A_2, \cdots, A_a$ cooperate to open the bids. First, each auctioneer A_j check the relation

$$c_{j,i} = h^{n+1}(S_{j,i})$$

to verify the validity of secret seed $S_{j,i}$. Then auctioneers $A_1, A_2, \cdots, A_a$ iterates following steps for prices $k = n, n-1, \cdots$.

Each auctioneer A_j publishes the values $h^k(S_{j,i})$ $(i = 1, 2, \cdots, m)$. Auctioneer A_1 then checks the following equality for $i = 1, 2, \cdots, m$

$$b_i = h(h^k(S_{1,i})|h^k(S_{2,i})|\cdots|h^k(S_{a,i})).$$

If the above equation does not hold for all i's, auctioneer A_1 concludes that no bidder bids price k. Auctioneer A_1 then publishes all values $h^k(S_{j,i})$ $(j = 1, 2, \cdots, a,\ i = 1, 2, \cdots, m)$, decreases price k by 1, and each auctioneer A_j publishes the values $h^{k-1}(S_{j,i})$ $(i = 1, 2, \cdots, m)$...

If the above equation hold for some i's, auctioneer A_1 concludes that the winning bidders are these B_i's and their winning price is k. Finally, auctioneer A_1 publishes the winning bidders, the winning price and all values $h^k(S_{j,i})$ $(j = 1, 2, \cdots, a,\ i = 1, 2, \cdots, m)$.

3.3 Security

The proposed method satisfies the properties listed in Section 2.

Secrecy of bidding price : Since bids are opened from the highest price to the price at which the wining bidders bid, no bids under the winning price are opened. Due to the one-wayness of the hash function, no one can compute the rest of the hash chain that has not yet been revealed. Due to the randomness of the hash function, no information about the rest of the hash chain that has not yet been revealed can be obtained. Thus all bidding prices except winning price are kept secret even from the auctioneer. In the case of method 2, since the hash chain is distributed to plural auctioneers, all bidding prices except winning price are kept secret besides the case all auctioneers collude.

Verifiability : Since values of hash chain which are opened are published, anyone can verify the correctness of the hash chains. Due to the collision intractability of the hash function, no one can change the hash chain without changing commited head of hash chain. Thus anyone can verify the correctness of the auction.

Undeniability : Since each bid has the bidder's signature, no bidder can deny his bid. Due to the collision intractability of the hash function, no one can change the hash chain without changing commited head of hash chain. So no bidder can deny his bidding price.

Anonymity : We can introduce a registration center to register bidders and issue a pseudonym (= public key of signature) for each bidder. Each bidder can participate in an auction using his pseudonym and so bid anonymously.

By these properties, our method achieves a fair sealed-bid auction.

3.4 Efficiency

We compare our method 1 to former method [SM99] and our method 2 to former method [Sak00] in terms of the volume of data communications and computational complexity. We consider only the opening phase, since the load factor of this phase is the highest of all phases. Table 1 shows the comparison result in the worst case. Here, m bidders bid and there are n bidding prices. In the case of our method 2 and former method [Sak00], the number of auctioneer is a.

Table 1. The load of the opening phase.

Method	Volume of data communication (bit)
Our method 1	$L_{hash} \times mn = 160mn$
[SM99]	$L_{exp} \times 10mn = 10240mn$
Our method 2	0
[Sak00]	0

Method	Computational complexity (μs)
Our method 1	$T_{hash} \times mn = 3.6mn$
[SM99]	$T_{exp} \times 10mn = 800000mn$
Our method 2	$T_{hash} \times (amn + mn) = 3.6(amn + mn)$
[Sak00]	$T_{exp} \times amn = 80000amn$

Here, L_{hash} and L_{exp} denote the size of the output data of the hash function and the size of modular exponentiation, respectively. Similarly, T_{hash} and T_{exp} denote the time taken for performing a single hash function evaluation and a single modular exponentiation, respectively. For the hash function SHA–1, L_{hash} is 160 bits and T_{hash} is 3.6 μs on Pentium II 233MHz (translated from the original data 55.1 Mbit/s on Pentium 90MHz [Bos]). For 1024 bits modular exponentiation, L_{exp} is 1024 bits and T_{exp} is 80000 μs on Pentium II 233MHz [Sha].

Our method is 200,000 times faster than former methods. Thus, our method supports a large number of bidders and a wide range of bidding prices.

4   Conclusion

We have proposed the first sealed-bid auction method which uses *only hash function*. In our method, the length of the hash chain represents the bidding price, and the bidder uses the head of his hash chain to commit his bid. We showed that our method satisfies all the requirements of sealed-bid auctions. The method drastically reduces the time taken for bidding and opening bids. If we use a practical hash function e.g. SHA-1, our method is 200,000 times faster than former methods that use public key cryptosystems. Accordingly, our method is widely applicable in terms of the number of bidders and the range of bidding prices.

Acknowledgments

The first author would like to thank Masayuki Abe (NTT Laboratories) for his helpful suggestions.

References

[Bos]   A.Bosselaer, *The hash function RIPEMD-160*, http://www.esat.kuleuven.ac.be/~bosselae/ripemd160.html.

[Cac99]   C.Cachin, *Efficient Private Bidding and Auctions with an Oblivious Third Party*, Proc. of 6th ACM Conference on Computer and Communications Security, pp. 120–127, 1999.

[FIPS]   FIPS Publication 180-1, *Secure Hash Standard*, Federal Information Processing Standards Publication 180-1, NIST, 1993.

[HTK98]   M. Harkavy, J. D. Tygar and H. Kikuchi, *Electronic Auctions with Private Bids*, Proc. of Third USENIX Workshop on Electronic Commerce, pp. 61–74, 1998.

[KHT98]   H. Kikuchi, M.Harkavy and J. D. Tygar, *Multi-round Anonymous Auction Protocols*, Proc. of first IEEE Workshop on Dependable and Real-Time E-Commerce Systems, pp. 62–69, 1998.

[KM99]   K. Kobayashi and H. Morita, *Efficient Sealed-bid Auction with Quantitative Competition using One-way Functions*, Tech. Report of IEICE ISEC99-30 (Japanese), 1999.

[KMSH01]   K. Kobayashi, H. Morita, K. Suzuki and M. Hakuta, *Efficient Sealed-bid Auction by using One-way Functions*, to appear in IEICE Trans. Fundamentals, Jan. 2001.

[Kud98]   M.Kudo, *Secure Electronic Sealed-Bid Auction Protocol with Public Key Cryptography*, IEICE Trans. Fundamentals, vol. E81-A, no. 1, pp. 20–27, Jan. 1998.

[NPS99]   M. Naor, B. Pinkas and R. Sumner, *Privacy Preserving Auctions and Mechanism Design*, Proc. of ACM conference on E-commerce, pp. 129–139, 1999.

[RS96]   R.L.Rivest, A.Shamir, *PayWord and MicroMint : Two simple micropayment schemes*, preprint 1996.

[Sak00]   K. Sako, *Universally verifiable auction protocol which hides losing bids*, Proc. of Public Key Cryptography 2000, pp. 35–39, 2000.

[Lam81]   L. Lamport, *Password Authentication with insecure Communication*, Communications of the ACM, vol. 24, No. 11, pp. 770–772, Nov. 1981.

[Sha]   Shamus Software Ltd., *Benchmarks*, http://indigo.ie/~mscott/.

[SM99]   K. Sakurai and S. Miyazaki, *A bulletin-board based digital auction scheme with bidding down strategy*, Proc. of 1999 International Workshop on Cryptographic Techniques and E-Commerce, pp. 180–187, 1999.

[SS99]   S. G. Stubblebine and P. F. Syverson, *Fair On-line Auctions Without Special Trusted Parties*, Proc. of Financial Cryptography 99, 1999.

[Suz99]   K. Suzuki, *Efficient Sealed-bid Auction by Sequentially Revealable Commitment*, Tech. Report of IEICE ISEC99-67 (Japanese), 1999.

Micropayments for Wireless Communications

DongGook Park [1,2], Colin Boyd [2] and Ed Dawson[2]

[1] Korea Telecom, Access Network Laboratory,
17 WooMyeon-Dong, SeoCho-Gu, 137-792, Seoul, Korea
dgpark6@kt.co.kr
[2] Queensland University of Technology, Information Security Research Centre,
2 George Street, GPO Box 2434, Brisbane, Queensland 4001, Australia
{c.boyd, e.dawson}@qut.edu.au

Abstract. Electronic payment systems for wireless devices need to take into account the limited computational and storage ability of such devices. Micropayment schemes seem well suited to this scenario since they are specifically designed for efficient operation. Most micropayment schemes require a digital signature and therefore users must support public key operations and, furthermore, a public key infrastructure must be available. Such schemes are not suitable for current wireless systems since public key technology is not supported. We examine the SVP micropayment scheme which overcomes this problem by using only symmetric key cryptography and relying on tamper resistance. Some limitations are observed in the SVP micropayment scheme and an enhanced scheme is proposed suitable for current generation wireless communications.

1 Introduction

Over recent years, there has been a significant increase in both the scale and the diversity of electronic transactions over the Internet. Electronic commerce (E-commerce) means electronic payment (E-payment) in a narrow sense but it may mean *electronic business* in a broader sense which also includes the exchange of information not directly related to the actual purchasing activities [GrFe99]. In this paper, the term *E-payment* will be used to describe purchasing activity itself between buyers and sellers. Secure electronic payments will not only make purchasing activities more flexible and convenient but also create as yet unimagined new markets [Wayn97].

As the integration of computing and communications continues, wireless computing devices are of increasing importance as devices to access the Internet and to make electronic transactions. Many E-payment schemes have been proposed, and a lot of them assume the use of today's well-established credit card business environment. The most well-agreed and dominant E-payment protocol is the SET (Secure Electronic Transaction) protocol, produced by Visa and MasterCard to be their standard for processing credit card transactions over networks like the Internet. This, and other similar schemes, use extensive cryptographic technologies, a lot of which are based on public-key cryptography, to satisfy high level security requirements such as nonrepudiation. These protocols are all appropriate for medium to large transactions (macropayments) of more than $5 or $10.

These macropayment protocols will be too expensive and time-consuming when applied to inexpensive transactions, 50 or 25 cents and less, because of the transaction charges of card companies and the computational cost of public-key signature/verification. They also place a heavy burden on the computational and storage capabilities of currently available wireless devices. Without appropriate cheap alternative schemes, the light-weight transaction market cannot be developed to its full potential. This market typically includes selling *inexpensive* information software, and services (e.g. directory search or games), usually delivered online, and is a prime market for wireless communications.

Several schemes have been proposed for micropayment. To reduce the computational and signalling burden down to a reasonable level which can be justified in micropayment environments, they try to avoid *public-key* cryptography partially or entirely. Their dependency on *on-line* access to banking/clearing systems is also small compared to macropayment schemes. In this paper, we focus on micropayment schemes because this category not only directly addresses the limited resources of mobile communications but also is the most reasonable option for applying to the light-weight E-payment by mobile users.

Figure 1 shows a general scenario in E-payment environments, adopted from [Ahuj96].

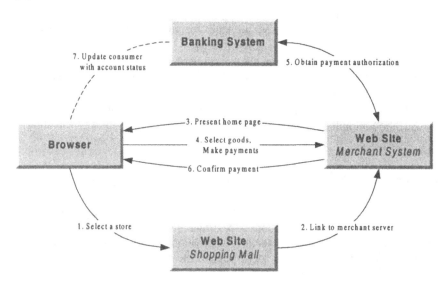

Figure 1: E-payment Environments

In this scenario, there are four elements or role players as follows.

- **A consumer** along with a Web browser uses the hyperlinks from the mall to access the merchant's home page.

- **A merchant** system residing on an online Web server with a connection to Web browsers over the Internet consists of the home page and related software to manage business.

- **An online shopping mall** may help direct consumers to the merchant server. It may pay to enlist with one or more well-known shopping malls.

- **A background banking network** supports electronic payments from consumers to the merchant. This network may include two types of banks:
  - **a merchant's bank** maintains the account for the merchant, authorizes and processes the payments. It may use on-line real-time link to the merchant so as to allow online authorization of consumer payments, and the link with the consumer's bank for verifying the transactions.
  - **a consumer's bank** manages the account for the consumer, and has an offline link to the consumer, such as via postal mail or e-mail.

These four role players take part in the following sequence of E-payment related activities.

1. The consumer accesses the shopping mall and selects a shop for purchasing certain items.
2. The shopping mall server accesses the merchant system for the selected shop.
3. The merchant system presents the store's home page to the consumer. It also includes information on the various goods available from this store.
4. The consumer selects the desired goods, interacts with the merchant's system, and makes the payments.
5. The merchant system accesses its bank for authorization of the consumer payment.
6. The merchant system informs the consumer that the payment is accepted and the transaction is completed. (At a later time, the merchant's bank obtains payment from the consumer's bank.)
7. The consumer's bank informs the consumer of the money transfer through mail such as a monthly report or online bank account.

2   E-Payment Mechanisms in Mobile Environments

The European ASPeCT project [ASPe96] has investigated security services for next generation wireless communications. The project included proposals for secure billing using hash based micropayments in combination with a digital signature scheme [ASPe97]. The digital signature serves a dual role as it is also used for authentication purposes during the establishment of a session key to secure the session data [HoPr98].

Provision of secure and trustworthy E-payment mechanisms will be the most critical factor for the success of E-commerce. Such a payment scheme must satisfy the following requirements [Ahuj96].

- Strong authentication of each party using certificate and digital signature
- Privacy of transaction using encryption
- Transaction integrity using message digest algorithms
- Nonrepudiation to handle disputes about the transaction

There are many classification methods for E-payment schemes, many of them rather orthogonal to each other. The following list shows an example of many classification criteria, most of which are described in detail in the report of the ASPeCT protocol [ASP97].

- Electronic purse/cash/credit
- On-line/off-line
- Credit-based/debit-based
- Software-based/tamper-resistant hardware
- Macropayment/micropayment

The public-key based security protocols for mobile users are not likely to be deployed in the early stage of future mobile communications such as IMT2000 [Buha97, ITU97]. They will be introduced into the systems when a fully-fledged public key infrastructure is available which may yet take a number of years. Therefore the assumption of limited use of tamper resistant devices by mobile users for E-payment in the near future is very likely and reasonable. Payment takes place using hashing mechanisms which are very cheap in computation while the broker can remain off-line. In mobile communication environments, there is already a well established infrastructure for billing users. This means that we do not need to establish extra clearing/banking infrastructure for mobile E-payment. A suggested model for billing users for micropayments is presented in Figure 2.

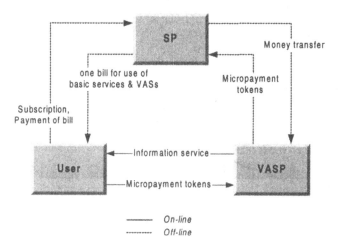

Figure 2: Billing of Micropayments

The role players in the mobile micropayment shown in Figure 2 comprise mobile users, service providers (SPs) and value added service providers (VASPs). Here, a SP plays the role of the broker in general micropayment environments. It bills the user for both basic and value-added services, and then redeems the relevant payment to the VASP. Considering the light-weight nature of most transactions to be carried out through mobile communications, the VASP-SP interface will be usually off-line.

3 Micropayments using Hash Chains

The most widely studied and promising approach involves using a public-key signature together with a hash-chain. Four similar schemes have been proposed: PayWord [RiSh96], *i*KP's micropayment [HSW96], Netcard [AMS95], and Pedersen's scheme [Pede95]. The basic idea is that a signature value generated using a public key operation is spread over many other cryptographic values derived by much more efficient one-way functions such as hash functions. In other words, the effect of a digital signature is reused many times over subsequent messages (containing preimages of a specific hash). This mechanism was originally proposed for use in an authentication scheme by Lamport [Lamp81]. The following description of the hash chain scheme is based on *PayWord* proposal of Rivest and Shamir [RiSh96].

3.1 Issuing user certificate

- The *user* U establishes an account with a *broker* B. U supplies personal details to B, such as a credit card number and delivery address, together with U's public key, K_U. U's aggregate charges will be charged to her credit-card number.
- The broker issues to U a PayWord Certificate, which is a signed statement by B containing:
  - broker's name B
  - user *name* U
  - user's IP-address
  - user's public key K_U
  - expiration date ExpDate
  - other information, possibly including user-specific information such as:
    - a certificate serial number,
    - credit limits to be applied per vendor,
    - information on how to contact the broker,
    - broker/vendor terms and condition.

The user's certificate has to be renewed by the broker regularly (e.g. monthly); the broker will do so only if the user's account is in good standing.

3.2 Typical scenario

The user's certificate authorizes the user to make Payword chains, and assures vendors that the user's paywords are redeemable by the broker. When the user U wishes to make a micropayment she clicks on a link to a vendor V's charged web page.
- The user's browser determines whether this is the first request to V that day.
  - For a first request, U computes and signs a *commitment* to a new user-specific and vendor-specific chain of payments $c_1, c_2, ..., c_N$.
    - The user creates the payword chain in reverse order by picking the last payword c_n at random, and then computing

$$c_i = h(c_{i+1}) \qquad \text{for } i = N-1, N-2, ..., 0.$$

Here c_0 is the root of the payword chain, and is not a payword itself. The *commitment* contains the root c_0, but not any payword c_i for $i > 0$.

- commitment M = {V, UCert, c_0, Date, OtherInfo}
- commitment includes both identities of the user and the vendor, and so is both user-specific and vendor-specific.
  - The user provides this commitment and her certificate to the vendor V, who verifies the signatures.
- The i-th payment (for $i = 1,2,...$) from U to V consists of the pair (c_i, i), which V can verify using c_{i-1}.
- At the end of each day, V reports to the broker B the last (highest-indexed) payment (w_l, l) received from each user that day, together with each corresponding commitment.
- The broker charges subscription and/or transaction fees.

Figure 3 shows the generation of hash-chain and commitment in the above scheme.

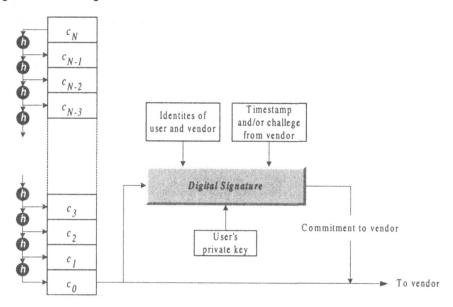

Figure 3: Hash Chain Protocol

3.3 Problems with Existing Hash Chains Schemes

There are a number of significant overheads which are implied by this model. We would like to emphasise that these are not currently present for most mobile users.

- The user must have the capability of public key operations in her mobile communications device.
- The user must have the ability to generate, or acquire, a suitable public key.
- There must be some mechanism for users to be able to revoke certificates when they are compromised, and for merchants to know that certificates have been revoked.

Wireless communications systems in the current generation use only symmetric key based cryptography [Mehr97, Moha96]. This is the main problem with the wireless application of existing micropayment schemes based on hash-chains. In Section 5 we will examine how to include hash chains into a symmetric cryptography setting so as to be able to benefit from the idea in the current wireless context.

4   SVP Scheme based on Tamper-Resistant Device

An alternative to use of a digital signature for micropayments is to employ a tamper-resistant device together with symmetric key cryptography. One such scheme called Small Value Payment (SVP) was proposed by Stern and Vaudenay [StVa97]. It aims to provide an even cheaper and more effective micropayment scheme than the approach using hash chains, by avoiding the use of asymmetric key cryptography. Instead, it requires the use of tamper-resistant devices both at the consumer and the merchant sides. We note that current mobile communications systems include use of a smart card which provides a degree of tamper resistance. The SVP scheme is illustrated in Figure 4.

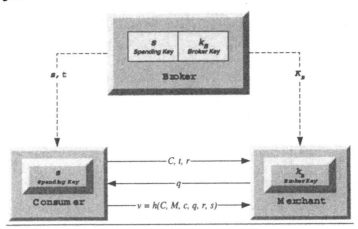

Figure 4: Small Value Payment Protocol

Initialization. The broker B generates its own secret key k_B and communicates it in a secure way to the device of each merchant, where k_B is a common value to all merchants. It also generates and computes a *random value* t and a *spending key* $s = Mac_{k_B}(C,t)$ for a consumer C, where s is unique to the consumer. In this equation *Mac* is a message authentication code which should have the property that it cannot be formed without knowledge of the key k_B and it is infeasible to find a different input (C, t) which gives the same output s. Both s and t are given by B to C as authorisation for C to spend an agreed amount of money.

4.1 Payment protocol

The payment protocol can be described as follows. In this description h is a hash function which can be regarded as a *MAC* operation with the spending key s as the *MAC* key.

C: consumer, M: merchant	
1. $C \rightarrow M$: C, t, r	(r: random number chosen by C)
2. $C \leftarrow M$: q	(q: random challenge)
3. $C \rightarrow M$: $v = h(C, M, c, q, r, s)$	(c: microamount)
M: checks if $v = h(C, M, c, q, r, \overline{Mac_{k_B}(C,t)})$,	
keeps an account balance for the user and increases C's account by c, and (optionally) stores (t, q, r, v) if he is suspicious about this payment.	

From the response in message 3, M is able to determine whether C is in possession of the spending key s. If so M knows that C was authorised to make payments associated with the value t. Note that there is no mechanism to prevent C using s any number of times. In this sense SVP is a credit based scheme in which C is trusted to pay the bills accrued from use of s.

4.2 Payment clearing

The merchant regularly sends the broker a statement of the amount of money spent by consumers, and the broker monitors to check if the accounts are consistent. If not, the broker requests a valid proof (C, c, t, q, r, v) of payment from M. If it cannot be provided, the broker just refuses the payment and records that there must be a problem with C or M. *If such a proof is released, the broker pays and checks if (M, q, r) has already credited to M.* If it has, the broker suspects the merchant to be dishonest. If not, the broker stores (M, q, r) in the (C, t)-records.

4.3 Problems with the SVP scheme

We have identified a number of problems with the SVP scheme which affect both the security and efficiency of a practical implementation in the wireless environment.

- There is no signature from the user, and thereby the scheme does not provide non-repudiation (the merchant and/or the broker can generate all the security parameters). This is why the scheme depends heavily on the use of tamper resistant devices. The compromise of only one tamper resistant device in the merchant side enables an attacker to impersonate other consumers.

- The shared secret key k_B between the broker and all the merchants must be the same, because the user's spending key is a function of the key k_B. Every merchant (more precisely, its tamper-proof device) in transaction with the user must be able to compute the spending key. Stern and Vaudenay recognise this problem in their paper and suggested a solution in which the broker has several secret keys, a subset of which is shared with each merchant. Users must then obtain multiple spending keys and provide a corresponding subset of spending keys to the merchant. This solution adds storage and complexity to the scheme, while compromise of a set of merchants keys can still allow spending at different merchants.

- The user and the VASP must execute the three-way challenge-response protocol for every micropayment, which is inefficient compared with the hash chain approach exchanging only one message (preimage of a hash chain).

- There is no authentication of the merchant to the user. The cost of three-way moves for each microamount and still lack of mutual authentication may not be justified even for micropayment environment.

- *Weakness in the message freshness*: the mechanism of replay-detection against the merchant is vulnerable to the following attack scenario:
    - Broker resets all the account records periodically, e.g., every month.
    - Merchant reuses the old parameters (used in the previous months).
    - Broker checks if (M, q, r) has been used before, but the check cannot be applied to all the transaction out of the manageable period.

The last problem with regard to replay attack can be easily fixed by including date information (*yymmdd*) in the commitment computation procedure. This prevents a merchant from cheating the broker with old payment data received from the user previously. We have included such a mechanism in the new scheme described in Section 5.

5   A New Scheme Using Tamper-Resistant Devices

Exploiting the advantages of both the hash chain and the tamper-resistance schemes, we have designed a new scheme. We can avoid both the expensive asymmetric cryptography even for the payment initialization, and challenge-response for each payment of microamount.

Figure 5 shows the required setting of this scheme assuming the use of tamper-resistant devices. The role players in this scheme are taken by those in mobile environments: the user, the VASP, and the user's SP. There are three distinct kinds of shared secret keys:

- K_{us} between the user and the SP;
- K_{vs} between the VASP and the SP;
- K_{uv} between the user and the VASP.

In fact, the shared key K_{uv} is derived from the K_{vs} which is common to every VASP. Also, for simplicity of key management, the user-SP shared key K_{us} is computed using the user identity and a master key K_s of the SP.

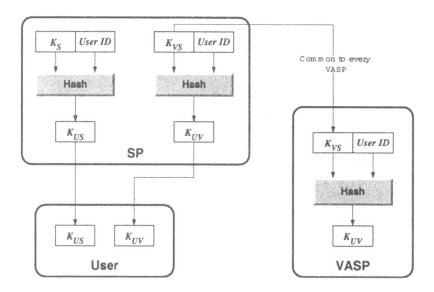

Figure 5: Enhanced Payment Scheme

The payment protocol assuming the use of hash chain is described in Figure 6. The user generates two separate commitments: one to the VASP, and another to the SP. The usage of pre-images of the hash chain is the same as in Figure 3. The computation of commitment for the VASP uses the shared key K_{uv}, the identities of the user and the VASP, random challenge r_v and time-stamp TS_v from the VASP, and the result of hashing c_0. This commitment value is, in turn, input, together with the shared key K_{us} between the user and the SP, to the commitment generation procedure for the SP. Note that by including the time-stamp which may be simply the date (yymmdd), this scheme is secure against the replay attack by the VASP which was possible in the scheme described in the previous section. The burden of computing the hash chain, if any, may be alleviated by reusing the previously generated hash chain in such a way that the remaining preimage with the smallest index is used for the commitment generation. Summarizing, the setting of this scheme basically comes from the SVP mechanism proposed by Stern and Vaudenay, and the actual payment protocol from the hash chain setting.

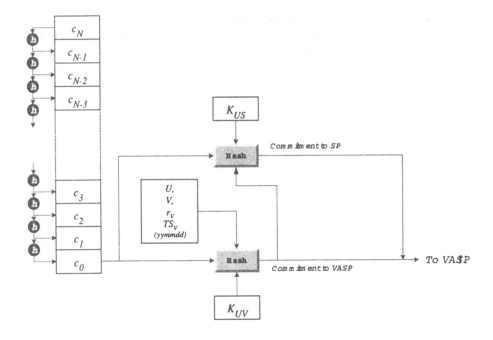

Figure 6: Payment Protocol

An example payment protocol set based on this enhanced scheme is shown in the following. We first summarise the goals of payment initialization protocol as follows.

- Mutual authentication between the user and the VASP
- Authentic and secure key establishment
- Mutual session key control
- Weak non-repudiation of the user to both the VASP and the SP based on shared keys
- Payment parameter initialisation

5.1 Payment initialization protocol

U: User, V: VASP, S: SP

1. $U \rightarrow V$: U, r_U
2. $U \leftarrow V$: $r_V, h(r_U, r_V, K_{UV}), ch\_data, TS_V$
3. $U \rightarrow V$: $c_0, commitment\_V, commitment\_S$

U: $commitment\_V = h(U, V, K_{UV}, r_U, r_V, TS_V, ch\_data, c_0)$
$commitment\_S = h(commitment\_V, K_{US})$

Protocol description

- *First message*: the user sends the VASP his identity U and a random challenge r_U.

- *Second message*: the VASP computes the common shared secret key K_{UV} using the user identity U and the share secret key K_{VS}, chooses a random challenge r_V and computes $h(r_U, r_V, K_{UV})$. It delivers to the user the random challenge r_V, $h(r_U, r_V, K_{UV})$, charging data ch_data and the time-stamp TS_V.

- *Third message*: upon receiving the second message from the VASP, the user computes $h(r_U, r_V, K_{UV})$, the value of which is compared with the received hash value from the VASP. The match of two values guarantees that the VASP has an authentic secret key K_{VS}. After that, the user computes the two commitments to the VASP and the SP using the secret keys K_{UV} and K_{VS}, respectively. The VASP checks the first commitment value by computing the same calculation as the user and comparing the result with the received value. If both values match, then it has confidence that the user has the correct shared secret key K_{UV}.

5.2   Payment protocol

U: User, V: VASP

1. $U \rightarrow V$: c_j $(j = 1,...,N)$

Protocol description

When the user and VASP need to exchange the actual payment data for a unit of charged service, the user sends the relevant preimage of the hash chain. Note that the three-way challenge-response messages are not used but a simple tick (a preimage of the hash chain) is delivered from the user to the VASP when required. Therefore this scheme achieves a significant improvement in terms of signalling load from the tamper-resistant device scheme proposed by Stern and Vaudenay.

5.3   Payment clearing

V: VASP, S: SP,

1. $V \rightarrow S$: $c_0, c_{j\_max}, j\_max, U, V, r_U, r_V, ch\_data, TS_V,$ commitment_V, commitment_S

Protocol description

After the transaction, the VASP stores the payment data for billing, which includes the user identity U, the user's commitments to the VASP and the SP, and the required data for the verification of the signature, i.e., r_U, r_V, ch_data, TS_V, c_0, the last received pre-image c_{j_max}, and the corresponding index value j_max, which equals the total number of ticks paid by the user in the transaction. The SP checks the *commitment_S* field by computing the same calculation as the user and comparing the result with the received value. The confidence gained by this check is to ensure that the commitment could not have been formed by the VASP, even if the tamper resistance of the VASP's device has been compromised.

5.4   Comparison with SVP

Compared to the original SVP protocol there are two main advantages that we can claim.

- Firstly, the interactive three move protocol for every micropayment has been avoided. This can be an important issue for mobile communications where call charges are still large in comparison with Internet based communications. It also reduces delay and removes the possibility of incomplete payment protocols due to communications failures.
- Secondly, mutual authentication between the user and the VASP is provided while keeping the protocol efficient in terms of computational and bandwidth requirements.

If we compare the disadvantages of the SVP scheme with our enhanced version we see that all disadvantages have been overcome except that there is still no signature to provide non-repudiation of user payments.

6   Conclusion

In the foreseeable future, mobile communication terminals will be a major method for electronic commerce, at least in transactions of small amounts. The well-studied and efficiency-proven hash chain scheme relies on the existence of digital signatures and an associated public key infrastructure. In this paper the alternative of using a tamper-resistant device has been explored. It has been shown that the SVP scheme has limitations for use in this environment. We have proposed an improved scheme which provides much reduced risk at no significant additional cost. We suggest that our enhanced scheme is suitable for implementation in current wireless communications systems.

References

[Ahuj96]   Vijay Ahuja, *Secure Commerce on the Internet*, Academic Press, 1996.

[AMS95]   Ross Anderson, Harry Manifavas, and Chris Sutherland, "A practical electronic cash system", *Personal Communication*, December 1995.

[ASPe96]   ASPeCT, *Initial Report on Security Requirements*, AC095/ATEA/W21/DS/P/02/B, February 1996.

[ASPe97]   ASPeCT, *Secure billing: evaluation report*, AC095/SAG/W25/DS/P/16/ 1, May 1997.

[Buha97]   K. Buhanal, et al., "IMT-2000: Service Providers' Perspective", *IEEE Personal Communications*, August 1997.

[GrFe99]   Marilyn Greenstein and Todd M Feinman, *Electronic Commerce: Security, Risk Management and Control*, McGraw-Hill, 1999, p. 2.

[HoPr98]   G. Horn and B. Preneel, "Authentication and payment in future mobile systems", *Computer Security - ESORICS'98*, Lecture Notes in Computer Science, 1485, 1998, pp. 277-293.

[HSW96]   Ralf Hauser, Michael Steiner, and Michael Waidner, *Micro-Payments based on iKP*, IBM Zurich Research Lab. Available as `http://www.zurich.ibm.ch/Technology/Security/publications/1996/HSW96.ps.gz`

[ITU97]   ITU, Recommendation ITU-R M.1223, *Evaluation of Security Mechanisms for IMT-2000*, 1997.

[Lamp81]   L. Lamport, "Password authentication with insecure communication", *Communications of the ACM*, 24(11), 770-771, Nov 1981.

[MaMi98]   K.M. Martin and C.J. Mitchell, "Evaluation of authentication protocols for mobile environment value-added services", *Journal of Computer Security*, to appear. Available online from `http://isg.rhbnc.ac.uk/cjm/Chris_Mitchell.htm`.

[Mehr97]   A. Mehrota, *GSM System Engineering*, Artech House, 1997.

[Mitc95]   C.J. Mitchell, "Security in Future Mobile Networks", *Second International Workshop on Mobile Multi-Media Communications (MoMuC-2)*, Bristol, April 1995. Also available online at `http://isg.rhbnc.ac.uk/cjm/SIFMW.ZIP`.

[Moha96]   Seshadri Mohan, "Privacy and Authentication Protocols for PCS", *IEEE Personal Communications*, October 1996, pp.34-38.

[MPMH98]   K.M. Martin, B. Preneel, C.J. Mitchell, H.J. Hitz, G. Horn, A. Poliakova, and P. Howard, "Secure billing for mobile information services in UMTS", in: S. Trigila, A. Mullery, M. Campolargo, H. Vanderstraeten and M. Mampaey (eds.), *Intelligence in Services and Networks: Technology for Ubiquitous Telecom Services* (Proceedings of the Fifth International Conference, IS&N 98, Antwerp, Belgium, May 1998), Springer-Verlag (LNCS 1430), pp.535-548.

[OjPr98]   T. Ojanpera and R. Prasad, "IMT-2000 Applications", in *Wideband CDMA for Third Generation Mobile Communication*, T Ojanpera and R. Prasad (ed.), Artech House, 1998, pp. 65-76.

[Pede95]   T. P. Pedersen, "Electronic payments of small amounts", DAIMI PB-495, Computer Science Department, Aarhus University, August 1995.

[Redl98]   S.M. Redl et al., *GSM and Personal Communications Handbook*, Artech House Publishers, 1998.

[RiSh96]   R. L. Rivest and A Shamir, "PayWord and MicroMint: Two simple micropayment schemes", *Cryptobytes*, Vol. 2, No. 1, May 1996, pp7-11. Available from `http://theory.lcs.mit.edu/~rivest`

[StVa97]   Jacques Stern and Serge Vaudenay, "SVP: a Flexible Micropayment Scheme", *Financial Crypto '97*, pp.161-171, Springer-Verlag, 1997.

[TAC99]   The Technical Advisory Committee to Develop a Federal Information Processing Standard for the Federal Key Management Infrastructure, *Requirements for Key Recovery Products*, available at `http://csrc.nist.gov/keyrecovery`.

[Wayn97]   Peter Wayner, *Digital Cash: Commerce on the Net*, Academic Press, 2nd Edition, 1997.

[Wien98]   M. Wiener, "Performance Comparison of Public-Key Cryptosystems", *Cryptobytes*, Vol. 1, No. 2, RSA Laboratories, 1998.

Cryptographic Applications of Sparse Polynomials over Finite Rings

William D. Banks[1], Daniel Lieman[2], Igor E. Shparlinski[3] and Van Thuong To[4]

[1] Department of Mathematics, University of Missouri
Columbia, MO 65211, USA
bbanks@math.missouri.edu
[2] Department of Mathematics, University of Georgia
Athens, GA 30602, USA
dlieman@math.uga.edu
[3] Department of Computing, Macquarie University
Sydney, NSW 2109, Australia
igor@comp.mq.edu.au
[4] Department of Computing, Macquarie University
Sydney, NSW 2109, Australia
tto@comp.mq.edu.au

Abstract. This paper gives new examples that exploit the idea of using sparse polynomials with restricted coefficients over a finite ring for designing fast, reliable cryptosystems and identification schemes.

1 Overview

The idea of using polynomials with restricted coefficients in cryptography, though fairly new, has already found several cryptographic applications such as the NTRU cryptosystem [10], the ENROOT cryptosystem [6], the PASS identification scheme [9, 11], and the SPIFI identification scheme [2]; see also [8].

In contrast to the constructions of NTRU and PASS, which consider classes of low-degree polynomials with many "small" nonzero coefficients, ENROOT and SPIFI are based on the use of polynomials of high degree that are extremely sparse. Although these latter constructions were originally considered only over finite fields, in this paper we improve and extend the ideas of [2, 6] and show that both ENROOT and SPIFI can be generalized to the setting of an arbitrary finite ring. In this generality, the user can be assured of an extra degree of security by selecting rings in which the problem of solving polynomial equations is notoriously difficult, as in the case of the residue ring for an RSA-modulus $M = pl$, where p and l are two privately held primes. In this paper, we have also introduced several new security features for the ENROOT and SPIFI protocols. Quite recently several powerful attacks on the original versions of ENROOT and SPIFI and some of their modifications have been presented in [1]. In particular the present version is a result of our iterative, thanks to many fruitful discussions with the authors of [1]) attempts to make ENROOT and SPIFI resistant to attacks of the types described in [1]. Although we believe that there is no immediate "danger", it still seems that these attacks still present a serious threat to ENROOT and SPIFI. Nevertheless, these objects, sparse polynomials, look too nice and natural (easy to evaluate but hard to invert) not to try to use them for a public key cryptography. We hope that our paper may help to bring more attention to this area.

2 Notation and Definitions

Throughout this paper, $\log z$ denotes the binary logarithm of z.

Let $\mathcal{R}$ be an arbitrary finite ring, and let N denote a fixed multiple of the exponent $\exp(\mathcal{R}^\times)$ of the multiplicative group of units $\mathcal{R}^\times$; thus we have $a^{N+1} = a$ for all $a \in \mathcal{R}^\times \cup \{0\}$.

To illustrate our ideas below, we will sometimes consider two important special cases, which we refer to as the "$\mathbb{F}_q$-case" and the "$\mathbb{Z}_M$-case." In the $\mathbb{F}_q$-case, $\mathcal{R}$ is the finite field $\mathbb{F}_q$ with q elements, and we can take $N = \exp(\mathcal{R}^\times) = q - 1$. In the $\mathbb{Z}_m$-case, $\mathcal{R}$ is the ring $\mathbb{Z}/M\mathbb{Z}$ of residue classes with respect to an RSA modulus $M = pl$, where p and l are primes. In this case, we can either take $N = \varphi(M) = (p-1)(l-1)$, where φ is the Euler function, or $N = \lambda(M) = \mathrm{lcm}(p-1, l-1)$ (λ is the *Carmichael function*). We remark that in all of these cases, we have $\log |\mathcal{R}| \approx \log |\mathcal{R}^\times| \approx \log N$,

We also assume that any element of $\mathcal{R}$ can be encoded by using about $\log |\mathcal{R}|$ bits.

Let d be a fixed positive integer. Given a set $S \subseteq \mathcal{R}$, we say that a polynomial $f(x_1, \ldots, x_d) \in \mathcal{R}[x_1, \ldots, x_d]$ is an *S-polynomial* if every coefficient of f belongs to S.

An expression of the form $ax_1^{e_1} \ldots x_d^{e_d}$ we call a *monomial* with the *coefficient* a and the *exponent* $(e_1, \ldots, e_d)$.

Finally, we say that a polynomial $f(x_1, \ldots, x_d) \in \mathcal{R}[x_1, \ldots, x_d]$ is *τ-sparse* if f has at most τ nonzero coefficients.

3 The SPIFI Identification Scheme

In this section, we describe a generalization of SPIFI (for *Secure Polynomial IdentiFIcation*; see [2]) for an arbitrary finite ring $\mathcal{R}$. For the sake of simplicity and practicality, we work only with polynomials of a single variable (that is, $d = 1$).

3.1 A Hard Problem

The hard problem underlying our one-way functions can be stated as follows:

Given 2m arbitrary elements $\alpha_1, \ldots \alpha_m, \beta_1, \ldots, \beta_m \in \mathcal{R}$ and a set $S \subseteq \mathcal{R}$ of small cardinality, it is not feasible to find a τ-sparse S-polynomial $f(x) \in \mathcal{R}[x]$ of degree $\deg(f) \le N$ with $f(\alpha_j) = \beta_j$ for each $j = 1, \ldots, m$, provided that N is of "medium" size relative to the choices of $m \ge 1$, the cardinality $|S|$, and $\tau \ge 3$.

More precisely, we expect that if one fixes the number of points m, the cardinality $|S|$, and the sparsity $\tau \ge 3$, then the problem requires exponential time as $N \to \infty$ (that is, exponential with respect to the bit length of N).

For example, let p be a prime, and consider the case where $\mathcal{R}$ is the finite field $\mathbb{F}_p$ with p elements. Let $a_{ij} \equiv \alpha_j^i \pmod{p}$ and $b_j \equiv \beta_j \pmod{p}$ be chosen so that $0 \le a_{ij}, b_j \le p - 1$ for $i = 0, \ldots, p - 1$ and $j = 1, \ldots, m$. Then in this simplified situation, the hard problem above is still equivalent to the problem of finding a feasible solution to the *integer programming problem*

$$\sum_{i=0}^{p-1} x_i \varepsilon_i a_{ij} + y_j p = b_j, \qquad j = 1, \ldots, m, \qquad \sum_{i=0}^{p-1} \varepsilon_i \le \tau,$$

where $y_j \in \mathbb{Z}$, $x_i \in S$, and $\varepsilon_i \in \{0, 1\}$ for all i and j.

3.2 Basic Idea

We fix the ring $\mathcal{R}$ and some integer parameters $k \ge 1$ and $r, s, t \ge 3$. This information is made *public*. The value of N may be kept *private*. Only *Alice* needs this value, so in this scenario the choice of the ring $\mathcal{R}$ (and the value of N) can be made by *Alice*.

In addition we require that $\mathcal{R}$ contains elements of multiplicative order in the interval $[0.5N^{1/4}, 2N^{1/4}]$. This certainly imposes some additional number theoretic requirements on N which in practice are easy to satisfy.

To create the identification message *Alice* uses the following algorithm, which we still denote by SPIFI.

Initial Set-up

Step 1
> Select at random k distinct elements $a_0, \ldots a_{k-1} \in \mathcal{R}^\times$ where a_0 of multiplicative order in the interval $[0.5N^{1/4}, 2N^{1/4}]$.

Step 2
> Select a random $\lceil t/2 \rceil$-sparse $\{0, 1\}$-polynomial $f_1(x) \in \mathcal{R}[x]$ with $\deg(f_1) \le N$ and $f_1(a_0) \in \mathcal{R}^\times$. Next, select a random $\lfloor t/2 \rfloor$-sparse $\{0, 1\}$-polynomial $f_2(x) \in \mathcal{R}[x]$ with $\deg(f_2) \le N$, $f_2(a_0) \ne 0$ and $f_2(a_0) \ne -f_1(a_0)$.

Step 3
> Compute $A = -f_2(a_0)f_1(a_0)^{-1}$ and put $f(x) = Af_1(x) + f_2(x)$. Then f is a t-sparse $\{0, 1, A\}$-polynomial with $\deg(f) \le N$, and $f(a_0) = 0$. The polynomial f is the *private key*.

Step 4

Compute $C_j = f(a_j)$ for $j = 1, \ldots, k-1$.

Step 5

Publish the set of values $\{A, a_0, \ldots a_{k-1}, C_1, \ldots, C_{k-1}\}$ as the **public key**.

To verify *Alice*'s identity, *Alice* and *Bob* use the following procedure.

Verification Protocol

Step 1

Alice selects a random r-sparse $\{0, 1\}$-polynomial $g(x) \in \mathcal{R}[x]$ with $\deg(g) \leq N$ and $g(0) = 0$ and random $\lfloor s/2 \rfloor$-sparse $\{0, 1, B\}$-polynomial $h_1(x) \in \mathcal{R}[x]$ of degree $\deg(h_1) \leq N$ with $B \neq 0, 1$ or A, computes

$$D_j = g(a_j), \qquad j = 1, \ldots, k-1,$$

and sends the sum $D = D_1 + \ldots + D_{k-1}$ to *Bob*.

Step 2

Bob selects a random $s - \lfloor s/2 \rfloor$-sparse $\{0, 1, B\}$-polynomial $h_2(x) \in \mathcal{R}[x]$ of degree $\deg(h_2) \leq N$ which has no common monomials with h_1, computes $h = h_1 + h_2$ and sends h to *Alice*.

Step 3

Alice verifies that h is an s-sparse $\{0, 1, B\}$-polynomial which contains all the terms of the polynomial h_1, computes

$$F(x) \equiv f(x)g(x)h(x) \pmod{x^{N+1} - x}$$

and sends the polynomial F and $\{D_1, \ldots, D_{k-1}\}$ to *Bob*.

Step 4

Bob computes

$$E_j = h(a_j), \qquad j = 1, \ldots, k-1,$$

and verifies that

$$D_1 + \ldots + D_{k-1} = D$$

and that $F(x)$ is an rst-sparse $\{0, 1, A, B, AB\}$-polynomial with $\deg(F) \leq N$, $F(a_0) = 0$, and

$$F(a_j) = C_j D_j E_j, \qquad j = 1, \ldots, k-1.$$

Of course, there is a chance that the constructed polynomial $F(x)$ is not a $\{0, 1, A, B, AB\}$-polynomial; however, if rst is substantially smaller than N, then this chance is negligible (and in this case, *Alice* and *Bob* can repeat the procedure).

3.3 Efficiency

The sparsity of the polynomials involved guarantees computational efficiency for this scheme. Using (naive) repeated squaring, one can compute the power a^e for any $a \in \mathcal{R}$ and $0 \leq e \leq N$ in about $2 \log N$ arithmetic operations in $\mathcal{R}$ in the worst case, or about $1.5 \log N$ arithmetic operations "on average"; see Section 1.3 of [3], Section 4.3 of [4], or Section 2.1 of [5]. Consequently, any τ-sparse polynomial $f(x) \in \mathcal{R}[x]$ of degree at most N can be evaluated at any point in about $O(\tau \log N)$ arithmetic operations in $\mathcal{R}$.

We recall that any element of $\mathcal{R}$ can be encoded by using about $\log |\mathcal{R}|$ bits.

Finally, we remark that if $0 \in \mathcal{S} \subseteq \mathcal{R}$, then any τ-sparse $\mathcal{S}$-polynomial $f(x) \in \mathcal{R}[x]$ of degree at most N can be encoded with about $\tau \log(N|\mathcal{S}| - N)$ bits. To do this, we have to identify at most τ positions at which f has a nonzero coefficient. The encoding of each position requires about $\log N$ bits, and for each such position, about $\log(|\mathcal{S}| - 1)$ bits are then required to determine the corresponding element of $\mathcal{S}$.

For example, the encoding of the values of B, $D_1, \ldots, D_{k-1}$ and the sum $D = D_1 + \ldots + D_{k-1}$ requires about $(k + 1) \log |\mathcal{R}|$ bits. The identification message must encode rst positions of the polynomial F (corresponding to its nonzero coefficients), which takes about $rst \log N$ bits. Each position requires two additional bits to distinguish between the possible nonzero coefficients 1, A, B and AB. Because of the same reason the encoding of the polynomial h_1 requires $\lfloor s/2 \rfloor \log(4N)$ bits. Hence the total number of bits which are sent from Alice to Bob is about $\lfloor s/2 \rfloor \log(4N) + (k + 1) \log |\mathcal{R}| + rst \log(4N)$.

Putting everything together, after simple calculations we derive that (using the naive repeated squaring exponentiation)

o the *initial set-up* takes $O(kt \log N)$ arithmetic operations in $\mathcal{R}$;

o the *private key size* is about $t \log(2N)$ bits;

o the *public key size* is about $k(\log |\mathcal{R}| + \log |\mathcal{R}^\times|)$ bits;

o the *identification message generation*, that is, computation of the polynomial F, elements D_j, $j = 1, \ldots, k - 1$, and their sum D, takes $O(rst)$ arithmetic operations with integer numbers in the range $[0, 2N]$ and $O((k - 1)r \log N)$ arithmetic operations in $\mathcal{R}$;

o the *identification message size* is about $rst \log(4N) + k \log |\mathcal{R}|$ bits;

o the *identification message verification*, that is, computation of $D_1 + \ldots + D_{k-1}$, $F(a_j)$ and the products $C_j D_j E_j$, $j = 1, \ldots, k - 1$, takes about $O(krst \log N)$ arithmetic operations in $\mathcal{R}$.

We remark that the practical and asymptotic performance of the SPIFI scheme can be improved if one uses more sophisticated algorithms to evaluate powers and sparse polynomials; see [3–5, 13, 15]. In particular, one can use pre-computation of certain powers of the a_j, $j = 1, \ldots, k - 1$, and several other clever tricks which we do not consider in this paper.

3.4 Possible Attacks

It is clear that recovering or faking the private key (that is, finding a t-sparse $\{0,1,A\}$-polynomial polynomial $\widetilde{f}(x) \in \mathcal{R}[x]$ with $\widetilde{f}(a_0) = 0$ and $\widetilde{f}(a_j) = C_j$ for $j = 1,\ldots,k-1$) or faking the identification message (that is, finding a rst-sparse $\{0,1,A,B,AB\}$-polynomial $\widetilde{F}(x) \in \mathcal{R}[x]$ with $\widetilde{F}(a_0) = 0$ and $\widetilde{F}(a_j) = C_j D_j E_j$ for $j = 1,\ldots,k-1$) are versions of the hard problem mentioned in Section 3.1 (with slightly different parameters).

We also remark that that without the reduction

$$f(x)g(x)h(x) \pmod{x^{N+1} - x},$$

one of the one possible attacks might be via polynomial factorization. In a practical implementation of this scheme, one should make sure that both f and g have terms of degree greater than $N/2$ so there are some reductions. Even without the reduction modulo $x^{N+1} - x$, the factorization attack does not seem to be feasible because of the large degrees of the polynomials involved; all known factorization algorithms (as well as their important components such as irreducibility testing and the greatest common divisor algorithms) do not take advantage of sparsity or any special structure of the coefficients; see [4, 14]. Moreover, the first factor that any of these algorithms will find would be the trivial one, that is, $(x - a_0)$. But the quotient $F(x)/(x - a_0)$ is most likely neither sparse nor an $\mathcal{S}$-polynomial for any small set $\mathcal{S}$. Finally, we remark that if one works in the setting of a ring $\mathcal{R}$ that is *not* a field (such as the $\mathbb{Z}_M$-case), then the problem of factorization becomes much more complicated, so this type of attack is even less likely to succeed.

It is possible that by using some "clever" choice of polynomials h, after several rounds of identification, *Bob* might be able to gain some information about f. But the polynomials g are specifically designed to prevent him from doing this. In Section 3.5 below, we present another idea which should render this attack completely infeasible, at least in the $\mathbb{F}_q$-case.

One might also consider lattice attacks. In particular, one can try to select a rt-sparse $\{0,1,A\}$-polynomial $e(X) \in \mathcal{R}[x]$ with $e(a_0) = 0$, compute

$$D_j = e(a_j)C_j^{-1}, \qquad j = 1,\ldots,k-1,$$

and then send these values together with

$$F(x) \equiv e(x)h(x) \pmod{x^{N+1} - x}$$

to the verifier. In principal this attack could succeed but finding such a polynomial e is kind of knapsack problem and since the dimension of the corresponding lattice would be equal to the (very large) degree N of the polynomials involved, any such attack seems completely infeasible at this time. With current technology, one can reduce lattices of degrees only in the hundreds, while in a practical implementation of this scheme our lattices will have dimension N of much large order of magnitude. Another attempt to construct such a polynomial e could be via solving the discrete logarithm in $\mathcal{R}$ to base a_0, see [1]. However because a_0 is selected to be of small order this attack is very unlikely to succeed either.

There is also another way to avoid the above attack via constructing a rt-sparse $\{0, 1, A\}$-polynomial $e(X) \in \mathcal{R}[x]$ with $e(a_0) = 0$. This way does not require any restrictions of the order of a_0 and thus can be used for arbitrary N. Namely we request that for each of the polynomials $f(x)$, $g(x)$ and $h(x)$ the sum of the degrees of the monomials is divisible by N. In this case the same condition also holds for $F(x)$. Indeed, if an s-sparse polynomial and a t-sparse polynomial have monomials of degrees $n_1, \ldots, n_s$ and $m_1, \ldots, m_t$, respectively, with

$$\sum_{i=1}^{s} n_i \equiv 0 \pmod{N} \qquad \text{and} \qquad \sum_{j=1}^{t} m_j \equiv 0 \pmod{N}$$

then their product has st monomials $n_i + m_j$, $i = 1, \ldots, s$, $j = 1, \ldots, t$ (unless a collision occurs which is very unlikely). Then

$$\sum_{i=1}^{s} \sum_{j=1}^{t} (n_i + m_j) = \sum_{i=1}^{s} \left(t n_i + \sum_{j=1}^{t} m_j \right)$$

$$= t \sum_{i=1}^{s} n_i + s \sum_{j=1}^{t} m_j \equiv 0 \pmod{N}$$

as claimed. Therefore the aforementioned discrete logarithm attack from [1] is very unlikely (with probability about $1/N$) to produce a polynomial $e(x)$ which besides the aforementioned conditions also has the sum of the degrees of the monomials which is divisible by N.

We remark that, although using composite moduli may add some additional security features, the security of the SPIFI scheme is not compromised even if the factorization of M is known. In fact, we believe that even in the case where the modulus is a (sufficiently large) prime (that is, in the $\mathbb{F}_q$-case), the scheme is still very secure.

It has turned out that *Alice* must make some commitment about the values of $D_1, \ldots, D_{k-1}$ before she receives the polynomial h from *Bob*, otherwise there is a very simple attack on this scheme. On the other hand, sending the whole set to *Bob* before he selects his polynomial h may open some ways of attacking for "cheating" verifier. Sending the sum $D = D_1 + \ldots + D_{k-1}$ is just one of many possible ways for *Alice* undertake some commitments about the values of $D_1, \ldots, D_{k-1}$ (just reducing the probability of the aforementioned "on-line" attack to $1/N$). Probably a more practical way would be just sending about a half of the bits of $D_1, \ldots, D_{k-1}$ at Step 1 (instead of computing and sending D) and then sending the rest of the bits at Step 3 (just reducing the probability of the aforementioned "on-line" attack to about $N^{-(k-1)/2}$).

Moreover, the SPIFI scheme is easily modified so that the value $N = \varphi(M)$ or $\lambda(M)$ (see Section 2) remains secret. Indeed, *Alice* can choose g in Step 1 of the verification protocol so that the reduction modulo $x^{N+1} - x$ that occurs in Step 3 produces a polynomial whose degree is not "too close" to N. In fact, "on average" it should be about $N(1-1/2rst)$ for the SPIFI scheme and about $N(1-1/2R)$ for the ENROOT scheme (see Section 4 below) since the corresponding polynomials are rst-sparse and R-sparse, respectively. Thus, in the case that $N = \varphi(M)$, the degree of F gives a worse approximation to N than the value of M itself, at least if $M = pl$ is a product of two primes of the same order.

3.5 Modification of the Basic Scheme

In this subsection, we consider only the case of a finite field $\mathcal{R} = \mathbb{F}_q$. It is very likely, however, that these ideas can be generalized to the setting of an arbitrary ring.

In order to prevent *Bob* from gaining any useful information about f by selecting certain special polynomials h, *Alice* can initially construct *two* polynomials f_1 and f_2, either one of which can serve as her private key. That is, for some $A, C_1, \ldots, C_{k-1} \in \mathbb{F}_q$ with $A \neq 0, 1$ and distinct $a_0, \ldots, a_{k-1} \in \mathbb{F}_q^\times$, *Alice* can find two t-sparse $\{0, 1, A\}$-polynomials $f_1(x), f_2(x) \in \mathbb{F}_q[x]$ of degree at most N that satisfy

$$f_1(a_j) = f_2(a_j) = C_j, \qquad j = 0, \ldots, k - 1, \tag{1}$$

for some $C_0, C_1, \ldots, C_{k-1} \in \mathbb{F}_q$.

To do this, *Alice* selects a certain parameter n and considers certain random $\{\pm 1, \pm A\}$-polynomials $\psi(x)$ of degree $4n$, looking for roots in $\mathbb{F}_q^\times$. For values of n of reasonable size this can be done quite efficiently, at least in probabilistic polynomial time; see [4, 14].

It follows from Theorem 3 of [12] that for sufficiently large q, the probability that a random monic polynomial of degree $4n$ over $\mathbb{F}_q$ will have $k + 1$ distinct roots in $\mathbb{F}_q$ is given by

$$P_{k+1}(4n, q) = \sum_{m=k+1}^{\infty} \binom{q}{m} q^{-m} \sum_{l=0}^{4n-m} (-1)^l \binom{q-m}{l} q^{-l}.$$

In particular,

$$\lim_{n \to \infty} \lim_{q \to \infty} P_{k+1}(4n, q) = \frac{1}{(k+1)! \, e^{k+1}}.$$

Letting A vary randomly over $\mathbb{F}_q/\{0, 1\}$, *Alice* considers $\{\pm 1, \pm A\}$-polynomials $\psi(x) \in \mathbb{F}_q[x]$ of degree $4n$ which have n coefficients of each type $1, -1, A$ or $-A$. Since

$$(q - 2) \frac{(4n)!}{(n!)^4 (k + 1)! \, e^{k+1}}$$

is large, after $O((k + 1)! e^{k+1})$ trials *Alice* will find with high probability such a polynomial with $k+1$ distinct roots $a_0, \ldots, a_k \in \mathbb{F}_q$. By reordering if necessary, we can assume that $a_0, \ldots, a_{k-1}$ are distinct elements in the multiplicative group

$\mathbb{F}_q^\times$. *Alice* now writes $\psi(x) = \varphi_1(x) - \varphi_2(x)$ where φ_1, φ_2 are $2n$-sparse $\{0, 1, A\}$-polynomials of degree at most $4n + 1$, and each φ_i has n coefficients of each type 1 or A. Moreover, $\varphi_1 \neq \varphi_2$, but clearly

$$\varphi_1(a_j) = \varphi_2(a_j), \qquad j = 0, \ldots, k - 1.$$

Now *Alice* selects a random $(\lceil t/2 \rceil - n)$-sparse $\{0, 1\}$-polynomial $\psi_1(x) \in \mathbb{F}_q[x]$ with $\deg(\psi_1) \leq q - 1$ and $\psi_1(a_0) \neq 0$. *Alice* then selects a random $(\lfloor t/2 \rfloor - n)$-sparse $\{0, 1\}$-polynomial $\psi_2(x) \in \mathbb{F}_q[x]$ with $\deg(\psi_2) \leq q - 1$. Assuming that ψ_1 and ψ_2 have been selected so that the non-constant monomials that occur in them have degree greater than $4n + 1$, *Alice* can now define

$$f_i(x) = A\psi_1(x) + \psi_2(x) + \varphi_i(x), \qquad i = 1, 2.$$

Then f_1 and f_2 are t-sparse $\{0, 1, A\}$-polynomials in the correct form for the SPIFI protocol, and they satisfy (1) some $C_0, C_1, \ldots, C_{k-1} \in \mathbb{F}_q$ We remark that in this case the value $C_0 = f_1(a_0) = f_2(a_0)$ must be published as well (although the scheme can easily be modified in such a way that as before $f_1(a_0) = f_2(a_0) = 0$ thus this value need not be sent).

Now *Alice* can alternate between f_1 and f_2 in a random order to confound *Bob*'s attempts to gain useful information about the private key.

It is easy to see that instead of the sum $A\psi_1(x) + \psi_2(x) + \varphi_i(x)$ for $i = 1, 2$, one can also consider more complicated expressions involving $\{0, 1\}$-polynomials. For example, one can put

$$f_i(x) = A\psi_1(x) + \psi_2(x) + \varphi_i(x)\varphi(x), \qquad i = 1, 2,$$

for appropriately chosen $\{0, 1\}$-polynomials $\psi_1(x), \psi_2(x)$ and $\varphi(x)$.

3.6 Remarks

It is natural to try to construct and utilize more than two t-sparse $\{0, 1, A\}$-polynomials that take the same values at k distinct points. However our approach of Section 3.5 does not seem to extend to this case.

Although we have not done so here, it can be interesting to extend our construction to multivariate polynomials.

4 The ENROOT Cryptosystem

In this section, we describe a generalization of ENROOT (for *ENcryption with ROOTs*; see [6]) for an arbitrary finite ring $\mathcal{R}$. We will now consider polynomials in $\mathcal{R}[x_1, \ldots, x_d]$, where $d \geq 2$ is fixed. Accordingly, we will often employ vector notation, writing $f(x)$ for $f(x_1, \ldots, x_d)$, $\mathcal{R}[x]$ for $\mathcal{R}[x_1, \ldots, x_d]$, etc.

4.1 Another Hard Problem

Our one-way functions are based on the following hard problem:

> Given the τ-sparse polynomials $f_1(x),\ldots,f_d(x) \subset \mathcal{R}[x]$ of degree at most N, it is not feasible to find an element $a = (a_1,\ldots,a_d) \in \mathcal{R}^d$ with $f_j(a) = 0$ for $j = 1,\ldots,d$, provided that N is sufficiently large relative to the choices of $d \geq 2$ and $\tau \geq 3$.

Again, we expect that if one fixes the number $d \geq 2$ and the sparsity $\tau \geq 3$, then the problem requires exponential time as $N \to \infty$ (see Section 4.4 below).

4.2 Basic Idea

We fix the ring $\mathcal{R}$ and the integers $d > \ell \geq 3$, $s_j, t_j \geq 3$, $j = 1,\ldots,d$ such that $t_1 = \ldots = t_\ell$. This information is made *public*. The value of N may be kept *private*. In fact, only *Bob* needs this value so in this scenario the choice of the ring $\mathcal{R}$ (and thus the value of N) is made by *Bob*.

The algorithm ENROOT can be described as follows.

ENROOT Algorithm

Step 1
   Alice selects d random elements $a_1,\ldots,a_d \in \mathcal{R}^\times$ which form her *private key*.
Step 2
   Alice selects d random polynomials $h_j(x) \in \mathcal{R}[x]$, of degree at most $|\mathcal{R}|$, containing at most $t_j - 1$ monomials, $j = 1,\ldots,d$, such that the first ℓ polynomials $h_1(x),\ldots,h_\ell(x)$ have the same set $\mathcal{E}$ of exponents of their monomials.
Step 3
   Alice publishes the polynomials $f_j(x) = h_j(x) - h_j(a)$ for $j = 1,\ldots,d$ as her *public key*, where a is the vector $(a_1,\ldots,a_d) \in (\mathcal{R}^\times)^d$.
Step 4
   To send a message $m \in \mathcal{R}$, *Bob* selects d random polynomials $g_j(x) \in \mathcal{R}[x]$ of degree at most N, containing at most $s_j - 1$ monomials such that one monomial has an exponent from the set $\mathcal{E}$ and having nonzero constant coefficients. *Bob* then computes the reduction $F(x)$ of the polynomial

$$m + f_1(x)g_1(x) + \ldots + f_d(x)g_d(x)$$

modulo the ideal in $\mathcal{R}[x]$ generated by $\{x_1^{N+1} - x_1,\ldots,x_d^{N+1} - x_d\}$, and he sends $F(x)$ to *Alice*.
Step 5
   To decrypt the message, *Alice* simply computes $F(a) = m$.

4.3 Efficiency

The sparsity of the polynomials involved again provides computational efficiency for this scheme. Using repeated squaring, one can compute the monomial $a_1^{e_1} \ldots a_d^{e_d}$ for any $(a_1, \ldots, a_d) \in \mathcal{R}^d$ and $0 \le e_j \le |\mathcal{R}|$, $j = 1, \ldots, d$, in about $O(d \log(2|\mathcal{R}|))$ arithmetic operations in $\mathcal{R}$. Consequently, any τ-sparse polynomial $f(x) \in \mathcal{R}[x]$ of degree at most $|\mathcal{R}|$ can be evaluated at any point in $\mathcal{R}^d$ in about $O(\tau d \log(2|\mathcal{R}|))$ arithmetic operations in $\mathcal{R}$.

We remark that any τ-sparse polynomial $f(x) \in \mathcal{R}[x]$ of degree at most $|\mathcal{R}|$ can be encoded with about $\tau((d+1) \log |\mathcal{R}|)$ bits. To do this, we have to identify at most τ monomials for which f has a nonzero coefficient. The encoding of each monomial $x_1^{e_1} \ldots x_d^{e_d}$ requires about $d \log |\mathcal{R}|$ bits, and for each such monomial about $\log |\mathcal{R}|$ bits are then required to encode the coefficient.

Let us set

$$T = \sum_{j=1}^{d} t_j, \qquad S = \sum_{j=1}^{d} s_j, \qquad R = \sum_{j=1}^{d} t_j s_j, \qquad Q = (d+1) \log |\mathcal{R}|.$$

Then after simple calculations we derive that (using the naive repeated squaring exponentiation)

- *generation of the public key*: to produce the vector a requires $O(d \log |\mathcal{R}|)$ random bits; to construct the polynomials $h_j(x)$ requires the generation of another $(T - d)Q$ random bits; the computation of the $h_j(a)$, $j = 1, \ldots, d$, takes $O(Td \log(2|\mathcal{R}|))$ arithmetic operations in $\mathcal{R}$;
- the *private key size* is about $d \log |\mathcal{R}| + (T - d)Q$ bits;
- the *public key size* is about TQ bits;
- *cost of encryption*: to construct the polynomials $g_j(x)$ requires the generation of about $d \log |\mathcal{R}| + (S - d)Q$ random bits; the computation of the polynomial $F(x)$ requires about R arithmetic operations in $\mathcal{R}$ plus Rd additions in $\mathbb{Z}/N\mathbb{Z}$;
- the *size of the encrypted message* is about RQ bits;
- the *cost of decryption*: the evaluation of $F(a) = m$ takes about $O(Rd \log(2N))$ arithmetic operations in $\mathcal{R}$.

In the $\mathbb{F}_q$-case, the above scheme can be accelerated if *Alice* sets $e_1 = 1$, selects a random element $a \in \mathcal{R}^\times$ and $d - 1$ random exponents $e_2, \ldots, e_d \in \mathbb{Z}/(q-1)\mathbb{Z}$, and defines a as $(a^{e_1}, \ldots, a^{e_d}) \in (\mathcal{R}^\times)^d$.

Again we mention that the performance of the ENROOT algorithm can be improved if one uses more sophisticated algorithms to evaluate powers and sparse polynomials; see [3–5, 13, 15].

Another possible way to improve performance is to use at Step 4 only $k < d$ randomly selected polynomials from the set $\{f_1, \ldots, f_d\}$. For the same level of security, there will be a trade-off between the complexity of Step 2 (hence the size of the private key) and the complexity of Step 4.

4.4 Security Considerations

The obvious way to attack the ENROOT cryptosystem is to try to find a simultaneous solution to the system of polynomial equations

$$f_j(x) = 0, \qquad j = 1, \ldots, d, \tag{2}$$

which amounts to solving the hard problem in Section 4.1. All known algorithms to solve systems of polynomial equations of total degree n require (regardless of sparsity) an amount of time that is polynomial in n (that is, exponential time with respect to the bit length of n); see [7, 14]. Since the degrees of the polynomials in (2) will be very large in practical implementations (about the size of N), this attack is totally infeasible.

Another possibility is to simply "guess" a solution. One expects that a system of τ-sparse polynomial equations in d variables of high degree over $\mathcal{R}$ will have very few zeros if $d \geq 2$, even though the number of zeros of a polynomial over an arbitrary ring is not necessarily bounded by the degree. Working heuristically, if we view the vector of polynomials $f(x) = (f_1(x), \ldots, f_d(x))$ as defining a "random" map $f : \mathcal{R}^d \to \mathcal{R}^d$, then the expected number of roots common to all of the polynomials $f_j(x)$ (that is, the cardinality of the kernel of f) is given by

$$\frac{1 - |\mathcal{R}|^{d(1-|\mathcal{R}|^d)}}{1 - (1 - |\mathcal{R}|^{-d})^{|\mathcal{R}|^d}} \approx \frac{1}{1 - e^{-1}} \approx 1.5819,$$

hence this brute force attack will take roughly $0.245|\mathcal{R}|^d$ trials "on average" to succeed. For arbitrary rings, we expect the choice $d \geq 2$ will provide the 2^{90} level of security against this attack if N is at least 50 bits long.

Although it is tempting to choose $d = \ell = 1$, in this case there are more sophisticated attacks that provide some information about *Alice*'s private key. One of these is based upon consideration of the difference set of the powers of monomials occurring in the polynomial $F(x)$. Indeed, if

$$f(x) = \sum_{i=1}^{t} A_i x^{n_i} \qquad \text{and} \qquad g(x) = \sum_{j=1}^{s} B_i x^{m_j}$$

are the polynomials selected by *Alice* and *Bob*, respectively, with $n_1 = m_1 = 0$ (and such that the sets $\{n_1, \ldots, n_t\}$ and $\{m_1, \ldots, m_s\}$ have a reasonably large intersection) then $F(x)$ contains at most $\tau \leq st$ monomials $C_\nu x^{r_\nu}$, where

$$r_\nu \equiv n_i + m_j \pmod{N}, \qquad \nu = 1, \ldots, \tau,$$

for some $i = 1, \ldots, t$ and $j = 1, \ldots, s$.

Consequently, finding the repeated elements in the difference set

$$\Delta = \{r_\nu - r_\eta \pmod{N} \mid \nu, \eta = 1, \ldots, \tau\},$$

may reveal some information about the polynomial $f(x)$.

In addition, if $d = 1$, one can also compute the greatest common divisor of $f(x)$ with $x^{N+1} - x$. Since this polynomial will have lower degree than f in general, it would be easier to find a root from a theoretical standpoint. Although it is not clear how to do this in an amount of time that is polynomial in the sparsity rather than in the degree of $f(x)$, it remains a potential threat.

On the other hand, these attacks on ENROOT fail when $d > \ell \geq 2$. Indeed, the first attack may help to gain some information about the total set of monomials in all of the polynomials $f_1, \ldots, f_d$, but it does not provide any information about the individual polynomials since it is impossible to determine which monomial comes from which product $f_j g_j$, $j = 1, \ldots, d$. In order to try all possible partitions into d groups of $s_j t_j$ monomials, $j = 1, \ldots, d$, needs to examine

$$\mathcal{P} = \frac{R!}{(s_1 t_1)! \ldots (s_d t_d)!} \tag{3}$$

total combinations. In particular, the most interesting case is when $s_1, \ldots, s_d$ are of approximately the same size as well as $t_1, \ldots, t_d$, that is, when $s_j \sim s$ and $t_j \sim t$ for $j = 1, \ldots, d$. In this case

$$\log \mathcal{P} \sim st(d \log d),$$

hence the number $\mathcal{P}$ of combinations to consider grows exponentially with respect to all parameters, provided that $d > \ell \geq 2$.

The second attack fails as well since the notion of greatest common divisor for multivariate polynomials is not defined, and taking resolvents to reduce to one variable is too costly.

However the case $d = 2$ is still not secure. Indeed, in this case we have either $d = \ell = 2$ or $\ell = 1$. In either case there are very simple linear algebra attacks which do not apply when $d > \ell \geq 2$ which we believe to be completely secure against the aforementioned attacks. There are some other alternative ways to guarantee that there are sufficiently many common elements in the sets of exponents of monomials of $f_1, \ldots, f_d$. In particular, the first few monomials of each polynomial $h_1, \ldots, h_d$ can be selected the same exponents.

Finally, we remark that the ENROOT cryptosystem is probably secure against lattice attacks for the same reason that the SPIFI scheme is secure (see Section 3.4): most lattice attacks would necessarily be based on lattices of dimension equal to the cardinality of $|\mathcal{R}|$, and in practical implementations this number would be very large.

4.5 Remarks

The ENROOT cryptosystem is well-suited to private key sharing among multiple parties. For parameter choices and approximate runtimes in the $\mathbb{F}_q$-case, we refer the reader to [6]. The main inherent weakness of this cryptosystem is its high message expansion cost. It is likely that working over certain noncommutative rings or rings that are not principal ideal domains may improve the overall security, so that smaller rings could be employed. Working over these rings, it would be interesting to have a more thorough security analysis than we have attempted here.

Acknowledgments. The authors are grateful to Claus Schnorr for pointing out an error in the initial version of the SPIFI scheme. The authors also thank the authors of [1] for very interesting discussions of other attacks on SPIFI and ENROOT and generally the perspectives of sparse polynomial based cryptography.

Thanks also go to Kwangjo Kim and Arjen Lenstra for their assistance in contacting with the authors of [1].

This work was done during a series of visits by W. B. and D. L. to Macquarie University, whose hospitality and support are gratefully acknowledged.

Work also supported in part, W. B. by NSF grant DMS-0070628, for D. L. by NSF grant DMS-9700542, and a Big 12 Faculty Fellowship from the University of Missouri and for I. S. by ARC grant A69700294.

References

1. F. Bao, R. H. Deng, W. Geiselmann, C. Schnorr, R. Steinwandt and H. Wu, 'Crytoanalysis of two sparse polynomial based cryptosystems', *Proc. Int. Conf. on Public Key Cryptography, PKC'2001, Lect. Notes in Comp. Sci.*, Springer-Verlag, Berlin, 2001, to appear.

2. W. Banks, D. Lieman and I. E. Shparlinski, 'An identification scheme based on sparse polynomials', *Lect. Notes in Comp. Sci.*, Springer-Verlag, Berlin, **1751**, 68–74.

3. H. Cohen *A course in computational algebraic number theory*, Springer-Verlag, Berlin, 1997.

4. J. von zur Gathen and J. Gerhard, *Modern computer algebra*, Cambridge Univ. Press, Cambridge, 1999.

5. D. M. Gordon, 'A survey of fast exponentiation methods', *J. Algorithms*, **27** (1998), 129–146.

6. D. Grant, K. Krastev, D. Lieman and I. E. Shparlinski, 'A public key cryptosystem based on sparse polynomials', *Proc. International Conference on Coding Theory, Cryptography and Related Areas, Guanajuato, 1998*, Springer-Verlag, Berlin, 2000, 114–121.

7. M.-D. A. Huang and Y.-C. Wong, 'Solving systems of polynomial congruences modulo a large prime', *Proc. 37 IEEE Symp. on Found. of Comp. Sci.*, 1996, 115–124.

8. J. Hoffstein, B. S. Kaliski, D. Lieman, M. J. B. Robshaw and Y. L. Yin, 'A new identification scheme based on polynomial evaluation', *US Patent*, No. **No. 6076163**, 2000.

9. J. Hoffstein, D. Lieman and J. H. Silverman, 'Polynomial Rings and Efficient Public Key Authentication', *Proc. the International Workshop on Cryptographic Techniques and E-Commerce*, City University of Hong Kong Press, to appear.

10. J. Hoffstein, J. Pipher and J. H. Silverman, 'NTRU: A ring based public key cryptosystem', *Lect. Notes in Comp. Sci.*, Springer-Verlag, Berlin, **1433** (1998), 267–288.

11. J. Hoffstein and J. H. Silverman, 'Polynomial rings and efficient public key authentication II', *Proc. the International Workshop on Cryptography and Computational Number Theory, Singapore, 1999*, Birkhäuser, 2001, 269–286.

12. A. Knopfmacher and J. Knopfmacher, 'Counting polynomials with a given number of zeros in a finite field', *Linear and Multilinear Algebra*, **26** (1990), 287–292.

13. N. Pippenger, 'On the evaluation of powers and monomials', *SIAM J. Comp.*, **9** (1980), 230–250.

14. I. E. Shparlinski, *Finite fields: Theory and computation*, Kluwer Acad. Publ., Dordrecht, 1999.

15. A. C.-C. Yao, 'On the evaluation of powers', *SIAM J. Comp.*, **5** (1976), 100–103.

Efficient Anonymous Fingerprinting of Electronic Information with Improved Automatic Idetification of Redistributors [†]

Chanjoo Chung*, Seungbok Choi*, Youngchul Choi**, Dongho Won*

* School of Electrical and Computer Engineering Sungkyunkwan University, Korea
{cjchung, sbchoi, dhwon}@dosan.skku.ac.kr
** BCQRE Co., Ltd, Korea
ycchoi@bcqre.com

Abstract. The proposed scheme by Domingo in Electronic Letters presented the first anonymous fingerprinting scheme in which help of a registration center is not required in order to identify redistributors. However, registration protocol of this scheme is 4-pass and identification process also required many exponential operations. In this paper, we propose more efficient protocol than the scheme by Domingo, which require 2-pass in registration protocol and need only 3+1 times exponential operation in identification process. In the electronic commerce of digital contents, registration protocol is more efficient than the previous scheme introduced. Moreover, computational complexity is diminished since identification process requires only 3+1 times exponential operations as the previous one requires 3+N/2 exponential operations. We now show efficient anonymous fingerprinting of electronic information with improved automatic identification of redistributors.

1. Introduction

Fingerprinting schemes do not prevent illegal copies; however, they do not deter people from illegally redistributing digital contents by enabling the original merchant of the contents to identify the original buyer of a redistributed copy. General methods of copyright protection previously proposed are to use the encryption and access control techniques. But those techniques are capable of illegal copying after acquisition of permission to the digital contents. The newly proposed method of copyright protection is copyright marking techniques. Copyright marking is the technique that an ownership information is embedded into digital contents to prevent illegal copies. Methods of copyright marking techniques are generally to use digital watermarking and digital fingerprinting techniques. Digital watermarking is the method which same authentication codes are embedded into same contents; whereas digital fingerprinting is different form it, in that different authentication codes are

[†] This paper is supported by Korea Science and Engineering Foundation(KOSEF) grant 97-01-00-13-01-5.

embedded into same contents. Since different authentication codes should identify redistributor of illegal copy.

The subordinate classification of digital fingerprinting is traitor tracing that provides identification with used keys. Traitor tracing techniques are to detect redistributor of used keys at an image decryption in broadcasting system such as pay-TV.

Before the emergence of computers, only physical fingerprinting had been studied and developed. With the increasing importance of digital contents, there is strong desire to use digital fingerprinting to protect intellectual properties, because it requires light-weight cryptographic capability but satisfies the purpose. Examples of digital contents to be fingerprinted include documents, images, movies, sounds and so on.

Fingerprinting has a problem about collusions. Suppose digital contents are distributed with different fingerprints. If a collusion group that got those contents compare their copies, they can easily discover all the fingerprints. Therefore collusion group can remove these fingerprints, interpolate gaps, and resell the digital contents without worrying about being traced.

This problem of collusions was first discussed by Blakley et al. [2] and solution against larger collusions was presented by Boneh and Shaw [3]. Low and Maxemchuk [4] presented a collusion analysis in their model for general multiparty cryptographic protocols. Explicit collusion-tolerant constructions were given in [5, 6].

Traitor tracing is the equivalent of fingerprinting for cryptologic keys. It was introduced by Chor et al. [7] for broadcast encryption. For example, when digital movies are broadcasted in encrypted form, and only the decryption keys are sold, a different key is sold to each pay-TV subscriber. Furthermore, the encryption scheme is adapted so that all keys can be used to decrypt the same cipher-contents. Since decryption key is different for each subscriber, the pay-TV company can trace the redistributor who made illegal copies of his key. Naor et al. [8] introduced threshold traitor tracing a scheme. Boneh et al. [9] and Fiat et al. [10] introduced efficient public key and dynamic traitor tracing scheme.

Recently, several studies enhance the functionality of fingerprinting scheme in various ways. Asymmetric fingerprinting was introduced by Pfitzmann and Schunter [11]. Unlike conventional fingerprinting schemes, only the buyer of a fingerprinted object knows the contents with the fingerprints. When a merchant finds the illegal copy, he can nevertheless identify the buyer and prove to third parties that this buyer bought the copy from him. Pfitzmann [12] also proposed a traitor tracing scheme using asymmetric fingerprinting. Anonymous fingerprinting was introduced by Pfitzmann and Waidner [13] as an analogy of the blind signature for fingerprinting. It uses a trusted third party, called the registration center, to identify buyers suspected of behaving illegally. Thus the merchant is not able to identify him without help of the registration center. Coin-based anonymous fingerprinting was introduced by Pfitzmann and Sadeghi [14]. Automatic identifying redistributor when fingerprinted contents are illegally redistributed without help of the registration center was introduced by Domingo [1]. Figure 1 is presented by classification of copyright marking system [19].

The remainder of the paper is organized into the following sections: Section 2 describes the classification of fingerprinting techniques. Section 3 illustrates the requirement of digital fingerprinting system by contrary to digital watermarking system. Section 4 overviews Domingo scheme [1]. Section 5 shows an efficient anonymous fingerprinting scheme of electronic information with improved automatic identification of redistributors. Section 6 describes security of proposed system. Conclusion is given in Section 7.

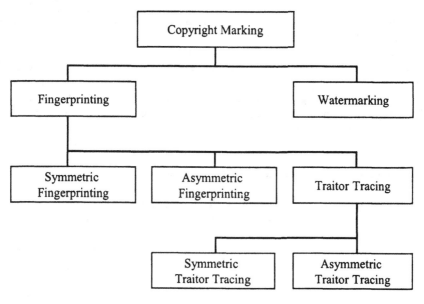

Figure 1. Classification of Copyright Marking System

2. Classification of Digital Fingerprinting

Fingerprinting schemes are technical means to discourage people from illegally redistributing the digital contents they have legally purchased. Fingerprinting schemes can be classified by five criterions like as follows: the objects to be fingerprinted, detection sensitivity, fingerprinting methods, generated fingerprints and disclosure of identification information to be fingerprinted. These five categories are not exclusive [18].

2.1 Object-based classification

The nature of the objects is a basic criterion, since it may provide a customized way to fingerprint the object. Object-based classification is divided into two categories. One is physical fingerprinting(e.g. human fingerprints, iris patterns and so on). The other is digital fingerprinting. If an object to be fingerprinted is digital format type so that computer can process fingerprints, we call it digital fingerprinting.

2.2 Detection sensitivity based classification

The sensitivity level of a fingerprinting scheme against illegal use is another criterion. Based on the detection sensitivity against violation, we classify fingerprinting into three categories. One is perfect fingerprinting. This is any alteration to the objects, which makes the fingerprint to be unrecognizable and also does the object unusable. Another is statistical fingerprinting. This is that if misused objects to examine the fingerprint generators was very much sufficiently given, we can gain any desired degree of confidence that they have correctly identified the compromised user. The other is threshold fingerprinting. Threshold fingerprinting is a hybrid type of the above two fingerprinting. It allows a certain level of illegal use, but it identifies the illegal copy when the threshold is reached. It is allowed to make copies of an object less than the threshold, and copies are not detected at all.

2.3 Fingerprinting method-based classification

Basic methods for fingerprinting such as recognition, deletion, addition and modification have also been used as another classification criterion, Recognition type fingerprinting is consists of recognizing and recording fingerprints that are already part of the object. Deletion type fingerprinting is that some legitimate portion of the original object is deleted. If some new portion is added to the object, it is of addition type fingerprinting. These additional parts can be sensible or meaningless. If a change to some portion of the object is made, it is called modification type fingerprinting.

2.4 Fingerprint-based classification

There are two categories in fingerprint-based classification. One is discrete fingerprinting. This is that a generated fingerprint has a finite value of discontinuous numbers. Many of digital fingerprints are included in this category. The other is continuous fingerprinting. This is that a generated fingerprint has a continuous value and essentially there is no limit to the number of possible values. Most physical fingerprint are of this type.

2.5 Identification information-based classification

There are two categories in identification information-based classification. One is symmetric fingerprinting, which digital content to be fingerprinted is known to buyer and merchant. Compared to the symmetric cryptography, encryption key is known to sender and receiver. The other is asymmetric fingerprinting. This is that digital content to be fingerprinted is only known to buyer. Compared to the asymmetric cryptography, secret key is only known to receiver.

3. Requirements of fingerprinting systems

Requirements of fingerprinting system are similar to the requirements of digital watermarking system, but there are a few different requirements. That is, while digital watermarking authenticates ownership of digital content, digital fingerprinting authenticates ownership of the digital contents and also provide identification function [18].

Requirement 1. Collusion tolerance
Attacker should not be able to find, generate, or delete the fingerprint by comparing the copies, even if they have access to a certain number of copies. In particular, the fingerprints must have a common intersection.

Requirement 2. Object quality tolerance
This is similar to perceptibility of the digital watermarking. The fingerprints have not to significantly decrease the usefulness of quality of digital contents.

Requirement 3. Object manipulation tolerance
If an attacker tampers the digital contents, the fingerprint should still be negotiable, unless there is so much noise that makes the digital contents useless. In particular, the fingerprint should tolerate lossy data compression. This is in order to trace illegal distributor after digital contents manipulated.

4. Overview of previous scheme

Domingo [1] described a construction for anonymous fingerprinting in which, on finding a fingerprinted copy, the merchant needs no help to identify the dishonest buyer. The role of the registration center is limited to buyer registration. In addition, the redistribution fraud can be proven to third parties. In the proposed scheme before Domingo suggested, on finding a fingerprinted copy, the merchant needs the help of a registration center to identify the redistributor.

4.1 System set up

Let p be a large prime such that $q = (p-1)/2$ is also prime. Let G be a group of order $p-1$, and let g be a generator of G such that computing discrete logarithms to the base g is difficult. Assume that both the buyer B and the registration center R have ElGalmal-like public-key pairs [15]. The buyer's secret key is x_g and his public key is $y_g \equiv g^{x_g} \bmod p$. The registration center R uses its secret key to issue certificates that can be verified using R's public key. All public keys are assumed to be known and certified.

4.2 Buyer's registration protocol

(i) The registration center R chooses a random secret nonce $x_r \in Z_p$ and sends $y_r \equiv g^{x_r} \bmod p$ to buyer B.

(ii) Buyer B chooses secret random x_1 and x_2 in Z_p such that $x_1 + x = x_B \in Z_p$ and sends $S_1 \equiv y_r^{x_1} \bmod p$ and $S_2 \equiv y_r^{x_2} \bmod p$ to registration center R. The buyer B convinces registration center R of zero-knowledge of possession of x_1 and x_2. The proof given in [16] for showing possession of discrete logarithms may be used here. The buyer B computes an ElGamal public key $y_2 \equiv g^{x_2} \bmod p$ and sends it to registration center R. In fingerprinting protocol, S_1 will act as a pseudonym.

(iii) The registration center R checks that $S_1 S_2 \equiv y_B^{x_r} \bmod p$ and $y_2^{x_r} \equiv S_2 \bmod p$. The registration center R returns to B certificates $Cert(y_r \| S_1)$, $Cert(y_r \| S_2)$ and the value x_r . The certificates are linked by y_r and state the correctness of S_1 and S_2. $Cert(x)$ is generated using registration center's secret key x_r.

By going through the above registration procedure several times, a buyer can obtain several different pseudonym pairs (S_1, S_2).

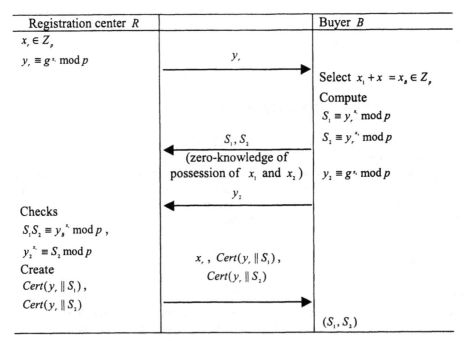

Figure 2. Buyer registration protocol

4.3 Fingerprinting protocol

(i) The buyer B sends $y_1, y_2, [S_1, Cert(y, \| S_1)]$ and *text* to merchant M, where *text* is a string identifying the purchase. The buyer B computes an ElGamal signature *sig* on *text* with the secret key x_2. *sig* is not sent to the merchant M.

(ii) The merchant M verifies the certificate on S_1

(iii) The buyer B and merchant M enter a secure two-party computation [17]. The merchant M's inputs are y_1, y_2, *text* and *item*, where *item* is the original information to be fingerprinted. The buyer B's inputs are x_2, sig, S_2 and $Cert(y, \| S_2)$. The computations performed are:

(a) $ver_1 = Verify(text, sig, y_2)$. The signature *sig* on *text* is verified using the public key y_2. The output ver_1 is a boolean variable only seen by the merchant M which is true if and only if the signature verification succeeds.

(b) $ver_2 = Verify(S_2, Cert(y, \| S_2), x_2, y_1, y_2)$. First, the certificate on S_2 is verified. Secondly, it checks that $g^{x_2} \equiv y_2 \bmod p$ and $y_2^{x_2} \equiv S_2$. Thirdly, it verifies that the value of y_1 in the certificate on S_1 previously verified by the merchant M. The output ver_2 is boolean variable only seen by the merchant M which is true if and only if the three aforementioned checks succeed.

(c) $item^* = Fing(item, emb)$. A classical fingerprinting algorithm is used to embed *emb* into the original information *item*, where

$$emb = text \| sig \| y_2 \| x_2 \| y_1 \| S_2 \| Cert(y, \| S_2)$$

The fingerprinted information $item^*$ is obtained as output and is only seen by the buyer B. In the above two-party computation, the merchant M obtains outputs first, and unless ver_1 and ver_2 are both true, the merchant M does not obtain the output $item^*$.

Buyer B		Merchant M
	$y_1, y_2, [S_1, Cert(y, \| S_1)]$, *text*	
Select $k \in Z_p$	⟶	Verify
$r \equiv g^k \bmod p$		$Cert(y, \| S_1)$
$text \equiv x_2 \cdot r + k \cdot s \bmod p - 1$		Record
$s = \dfrac{text - x_2 \cdot r}{k} \bmod p - 1$		$[S_1, Cert(y, \| S_1)]$
$sig = (text, r, s)$		

Figure 3. Buyer authentication process of fingerprinting protocol

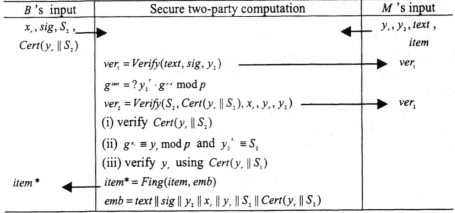

Figure 4. Secure two-party computation of fingerprinting protocol

4.4 Identification process

On finding a redistributed copy, merchant M only extract emb . The extracted information contains the values specified by $emb = text \| sig \| y_2 \| x_, \| y_, \| S_2 \| Cert(y_, \| S_2)$ and is combined by M with the purchase record $[S_1, Cert(y_, \| S_1)]$ to construct a redistribution proof.

(i) The signature sig on the $text$ is verified using anonymous public key y_2 .

(ii) The value $y_,$ links the certificates on S_1 and S_2. Moreover $y_,$ cannot be altered since it is part of the certificates.

(iii) The value $x_,$ proves that the owner of the key y_2 is the same as the owner of S_2. This is so because, according to the registration protocol, the registration center R only reveals $\log_g y_, = x_,$ to B after B has provided y_2 such that $y_2^{x_,} \equiv S_2 \bmod p$. Now if the Diffie-Hellman key exchange is secure, the buyer B cannot produce a correct y_2 without knowing $\log_g y_2 = x_2 = \log_{y_,} S_2$.

(iv) Finally, to identify a buyer, the merchant M raises the public keys of buyers to $x_,$ until a public key y_B is found such that $S_1 S_2 \equiv y_B^{x_,} \bmod p$. The dishonest buyer B has been identified. Note that, since $y_,$ is certified $x_,$ cannot be forged by the merchant M to unjustly accuse a buyer.

5. Proposed schemes

In Domingo's scheme, registration protocol is 4-pass. The buyer brings random nonce $x_,$ from registration center in order to buyer's identity is blinding. It is not efficient in electronic commerce since communication pass number is increased. In our proposed schemes, registration protocol is 2-pass. And his identification process needs many exponential operations, since the merchant M raises the public keys in public key directory of buyers to x_r. It needs much computational quantity. In our proposed scheme, exponential operation is only 3+1 times. And our scheme's identification is improved more than his scheme since many exponential computations replace 3+1 time exponential operations. We now show efficient and improved schemes.

Our scheme has three sub-protocol and one process. Those are registration protocol, fingerprinting protocol, identification process, and dispute protocol. System set up is the same as Domingo scheme. Our scheme is constructed following.

5.1 Buyer's registration protocol

(i) The buyer B chooses secret random x_1 and x_2 in Z_p such that $x_1 \cdot x_2 = x_B \in Z_p$ and sends $t \equiv g^{x_1} \bmod p$ and encrypting x_2 using registration center's public key pk_R such as $E_{pk_R}(x_2)$ to registration center R. t is anonymous public key of buyer's B. The buyer B convinces registration center R of zero-knowledge of possession of x_1. The proof given in [16] for showing possession of discrete logarithms may be used here. The buyer B's public key is an ElGamal public key y_B.

(ii) Registration center R first decrypt $E_{pk_R}(x_2)$ using its secret key sk_R and checks that $t^{x_2} \equiv y_B \bmod p$. If verification is successful, then the registration center R chooses a random secret nonce $x_r \in Z_p$ and computes $T \equiv g^{x_r \cdot x_2} \bmod p$. T will act as a pseudonym of the buyer B. The registration center R returns to B certificates $Cert$ $(t \| x_r)$, $Cert(T)$ and the value x_r and state the correctness of t and T.

By going through the above registration procedure several times, a buyer can obtain several different pseudonyms pairs (t, T).

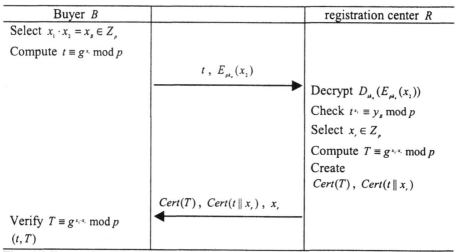

Buyer B		registration center R
Select $x_1 \cdot x_2 = x_B \in Z_p$		
Compute $t \equiv g^{x_1} \bmod p$		
	t, $E_{pk_R}(x_2)$ $\longrightarrow$	Decrypt $D_{sk_R}(E_{pk_R}(x_2))$
		Check $t^{x_2} \equiv y_B \bmod p$
		Select $x_r \in Z_p$
		Compute $T \equiv g^{x_r \cdot x_2} \bmod p$
		Create
		$Cert(T)$, $Cert(t \| x_r)$
	$Cert(T)$, $Cert(t \| x_r)$, x_r $\longleftarrow$	
Verify $T \equiv g^{x_r \cdot x_2} \bmod p$		
(t, T)		

Figure 5. Proposed registration protocol

5.2 Fingerprinting protocol

(i) The buyer B sends $t, [T, Cert(T)]$ and *text* to M, where *text* is a string identifying the purchase. The buyer B computes an ElGamal signature *sig* on *text* with the random secret key x_1. The signature *sig* is not sent to the merchant M.

(ii) The merchant M verifies the certificate on T.

(iii) The buyer B and merchant M perform a secure two-party computation [17]. The merchant M's input are $T, t, text$ and *item*, where *item* is the original information to be fingerprinted. The buyer B's input are x_1, sig, x_2 and $Cert(t \| x_2)$. The computations performed are:

(a) $ver_1 = Verify(text, sig, t)$. The signature *sig* on *text* is verified using the anonymous public key t. The output ver_1 is a boolean variable only seen by the merchant M which is true if and only if the signature verification succeeds.

(b) $ver_2 = Verify(t, Cert(t \| x_2), x_2, x_1, T)$. First, the certificate on t is verified. Secondly, it checks that $T \equiv g^{x_1 x_2} \bmod p$. The output ver_2 is a boolean variable only seen by the merchant M which is true if and only if the two aforementioned checks succeed.

(c) $item^* = Fing(item, emb)$. A classical fingerprinting algorithm is used to embed *emb* into the original information *item*, where

$$emb = text \| sig \| t \| x_1 \| x_2 \| T \| Cert(t \| x_2)$$

The fingerprinted information $item^*$ is obtained as output and is only seen by the buyer B. In the above two-party computation, through the merchant M obtains outputs ver_1 and ver_2 first if they are both true, it does not obtain the output $item^*$.

5.3 Identification process

On finding a redistributed copy, the merchant M extract *emb*. The extracted information contains the values specified by $emb = text \| sig \| t \| x_1 \| x_2 \| T \| Cert(t \| x_2)$ and is combined by the merchant M with the purchase record $[T, Cert(T)]$ to construct a redistribution proof.

Buyer B		Merchant M
	$t, [T_1, Cert(T)], text$	
Select $k \in Z_p$		Verify $Cert(T)$
$r \equiv g^k \bmod p$		Record $[T, Cert(T)]$
$text \equiv x_1 \cdot r + k \cdot s \bmod p - 1$		
$s = \dfrac{text - x_1 \cdot r}{k} \bmod p - 1$		
$sig = (text, r, s)$		

Figure 6. Buyer authentication process of proposed fingerprinting protocol

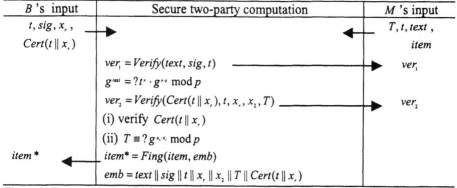

Figure 7. Proposed secure two-party computation of fingerprinting protocol

(i) The signature sig on the $text$ is verified using anonymous public key t.

(ii) The value x_r links the certificates on T and t. Moreover x_r cannot be altered, since it is part of the certificates.

(iii) The value x_2 proves that the owner of the key t is the same as the owner of T. This is so because, according to the registration protocol, the registration center R only reveals x_r to B after B has provided T such that $T \equiv g^{x_r \cdot x} \bmod p$. Now if the Diffie-Hellman key exchange is secure, B cannot produce a correct T.

(iv) Finally, to identify a buyer, the merchant M raises the public keys of buyers to x_2 such that $t^{x_2} \bmod p = id$. The merchant M searches for $id = public - key$ in the public key directory. The dishonest buyer B has been identified. Note that, since T and t are certified, x_r cannot be forged by M to unjustly accuse a buyer.

5.4 Disputation protocol

This protocol is optional. If the merchant M shows redistribution proof to third party, this protocol is performed. The merchant M sends proof (purchase record $[T, Cert(T)]$ and emb) to a judge. Judge verifies the proof. First, certificate $[T, Cert(T)]$ is verified by the registration center R's public key. Finally, checks $t^{x_2} \bmod p = y_B$. If above four verification is valid then owner of the public key y_B is accused. If registration record is necessary to prove guilty of buyer then judge ask registration record to registration center.

5.5 Comparison with previous scheme

We now compare our scheme with previous it. According to the computation quantity point of view, we contrast above three protocols(registration, fingerprinting, identification).

Table 1. Comparison with previous scheme

Protocol	Compare object	Previous scheme	Our scheme
Registration	Exponential operation	6	7
	Multiplication operation	1	2
	Pass number	4	2
Fingerprinting	Exponential operation	5	4
	Multiplication operation	5	6
	Pass number	6	Same
Identification	Exponential operation	3+N/2 (Average)	3+1 (Any case)
	Multiplication operation	2	3
	Comparison operation	N/2	Same

N : Number of public key in directory

In registration protocol, multiplication operation of our scheme and exponential operation are on the increase 1 time, but pass number are on the decrease 4 to 2. Because pass numbers are 2-pass, our scheme is efficient in electronic commerce in digital contents.

In fingerprinting protocol, multiplication operation of our scheme is on the increase 1 time, but pass number is same, and exponential operation is on the decrease 5 to 4.

In identification protocol, multiplication operation and comparison operation are same, but exponential operation is the decrease of 3+N/2 times to 3+1 times. And our scheme is more improved in automatic identification of redistributor.

6. Security of proposed scheme

We show security of our scheme. Security is not decrease than previous scheme, since our scheme is the same basis problem as previous one.

Proposition 1 (*Buyer's Registration security*) : Registration protocol provides buyer authentication without compromising the private key x_{B} of the buyer.

Proof : Registration center sees t, x_2 and zero-knowledge proofs. The latter leaks no information. If we don't consider the zero-knowledge proofs, then registration center must need no knowledge of x_1 to find the value t' such that $t'^{x_1} \bmod p \equiv y_{B}$. If we consider the zero-knowledge proofs, then impersonator not knowing x_{B} can compute t, x_2 such that $t^{x_1} \bmod p \equiv y_{B}$. Hence the impersonator can compute the discrete logarithm x_{B}. If impersonation is feasible, so is computing discrete logarithm problem. In general, discrete logarithm problem is hard, so registration center does not make t' such that $t'^{x_i} \bmod p \equiv y_s$.

Proposition 2 (*Buyer anonymity*) : An honest buyer who follows fingerprinting protocol will not be identified if computing discrete logarithm is hard and secure two-party computation is feasible.

Proof : In fingerprinting protocol, merchant M knows $t, [T, Cert(T)]$ and his outputs of a secure two-party computation that ver_1 and ver_2 . Finding y_s would require knowledge of x_s . However, if secure two-party computation is feasible, the only way for merchant to find x_2 is to compute $\log_g T$ using $Cert(T)$. As illustrated in security of registration protocol, discrete logarithm $x_2 \cdot x_r$ should be computed such as $T \equiv g^{x_2 x_r} \bmod p$. But, polynomial algorithm proving discrete logarithm problem does not exist, so attacker do not compute $x_2 \cdot x_r$. And, buyer anonymity is guaranteed.

7. Conclusion

We presented an efficient and improved identification asymmetric anonymous fingerprinting solution to redistribution of electronic information problem. Our construction is based on discrete logarithm problem. Security of our scheme is the same alike previous scheme, but efficiency is increased and computation quantity is decreased. According to the growth of electronic commerce, more efficient scheme is necessary. And we mention an application of our scheme to defend against the piracy of many software and multimedia contents.

Reference

1. Josep Domingo-Ferrer, "Anonymous Fingerprinting of Electronic Information with Automatic Identification of Redistributors", Electronics Lettes 34/13, 1998, pp. 1303-1304.
2. Blakley, G. R., C. Meadows, and G. B. Purdy, "Fingerprinting Long Forgiving Messages", in Advances in Cryptology, Proceedings of CRYPTO '85, vol. 218 of Lecture Notes in Computer Science, Springer-Verlag, 1986, pp.180-189.
3. Boneh, D., and J. Shaw, "Collusion-secure Fingerprinting for Digital Data", in Advances in Cryptology, Proceedings of CRYPTO '95, vol. 963 of Lecture Notes in Computer Science, Springer-Verlag, 1995, pp.452-465.
4. Low, S. H., and N. F. Maxemchuk, "Modeling Cryptographic Protocols and their Collusion Analysis", in Information Hiding : First International Workshop, Proceedings, vol. 1174 of Lecture Notes in Computer Science, Springer-Verlag, 1996, pp.169-184.
5. Pfitzmann, B., and Waidner, M., "Asymmetric Fingerprinting for Larger Collusions", 4th ACM Conference on Computer and Communications Security, acm press, 1997, pp.151-160.
6. Biehl, I. and Meyer, B., "Protocol for Collusion-Secure Asymmetric Fingerprinting", STACS 97, vol. 1200 of Lecture Notes in Computer Science, Springer-Verlag, 1997, pp.399-412.

7. Chor, B., A. Fiat, and M. Naor, "Tracing Traitors", in Advances in Cryptology, Proceedings of CRYPTO '94, vol. 839 of Lecture Notes in Computer Science, Springer-Verlag, 1994, pp.257-270.

8. Naor, M., and Pinkas, B., "Threshold Traitor Tracing", in Advances in Cryptology, Proceedings of CRYPTO '98, vol. 1462 of Lecture Notes in Computer Science, Springer-Verlag, 1998, pp.502-517.

9. Boneh, D., and Franklin, M., "An Efficient Public Key Traitor Tracing Scheme", in Advances in Cryptology, Proceedings of CRYPTO '99, vol. 1666 of Lecture Notes in Computer Science, Springer-Verlag, 1999, pp.338-353.

10. Fiat, A., and Tassa, T., "Dynamic Traitor Tracing", in Advances in Cryptology, Proceedings of CRYPTO '99, vol. 1666 of Lecture Notes in Computer Science, Springer-Verlag, 1999, pp.354-371.

11. Pfitzmann, B., and M. Schunter, "Asymmetric Fingerprinting", in Advances in Cryptology, Proceedings of EUROCRYPT '96, vol. 1070 of Lecture Notes in Computer Science, Springer-Verlag, 1996, pp.84-95.

12. Pfitzmann, B., "Trials of Traced Traitors", in Information Hiding: First International Workshop, Proceedings, vol. 1174 of Lecture Notes in Computer Science, Springer-Verlag, 1996, pp.49-64.

13. Pfitzmann, B., and M. Waidner, "Anonymous Fingerprinting", in Advances in Cryptology, Proceedings of EUROCRYPT '97, vol. 1233 of Lecture Notes in Computer Science, Springer-Verlag, 1997, pp.88-102.

14. Pfitzmann, B., and A. Sadeghi, "Coin-Based Anonymous Fingerprinting", in Advances in Cryptology, Proceedings of EUROCRYPT '99, vol. 1592 of Lecture Notes in Computer Science, Springer-Verlag, 1999, pp.150-164.

15. ElGamal, T., "A public-key cryptosystem and signature scheme based on discrete logarithms", IEEE Trans. Inf. Theory, 1985, IT-31, pp.469-472.

16. Chaum, D., Evertse, J.-H., and Van De Graaf, J., "An improved protocol for demonstrating possession of discrete logarithms and some generalizations", in Advanced in Cryptology, Proceedings of EUROCRYPT '87, vol 304 of Lecture Notes in Computer Science, Springer-Verlag, 1987, pp.127-141.

17. Chaum, D., Damgaard, I. B., and Van De Graaf, J., "Multiparty computation ensuring privacy of each party's input and correctness of the result", in Advanced in Cryptology, Proceedings of CRYPTO '87, vol 293 of Lecture Notes in Computer Science, Springer-Verlag, 1987, pp.87-119.

18. Katzenbeisser, S., Petitcolas, F. A. P, "Information Hiding Techniques for Steganography and Digital Watermarking", Artech House, 2000.

19. Peticolas, F. A. P., R. J. Anderson, and M. G. Kuhn, "Information Hiding-A Survey", Proceedings of the IEEE, vol. 87, no. 7, 1999, pp.1062-1078.

Hash to the Rescue:
Space Minimization for PKI Directories

Adam Young[1]   and   Moti Yung[2]

[1] Columbia University, New York, NY, USA. Email: ayoung@cs.columbia.edu
[2] CertCo, New York, NY, USA. Email: moti@cs.columbia.edu

Abstract. In this paper we investigate the notion of space efficient public-key infrastructure (PKI) directories. The area of PKI is relatively young and we do not know yet the long term implications of design decisions regarding PKI and its interface with applications. Our goal is to study mechanisms for networks and systems settings where the size of directories is a significant resource (due to space restrictions).

Naturally, the tools we employ are cryptographic hashing techniques combined with the tradeoffs of public storage and computation. Our mechanisms are quite simple, easy to implement and thus practical, yet they are quite powerful in making the operation substantially less costly (mainly) storage-wise and in trading storage for computation. In the past, tree based mechanisms were considered extensively to improve the complexity of PKI directories. We show that hashing techniques provide various advantages as well.

1   Introduction

We are currently at the point, due to the enormous surge of Internet use, where large-scale Public Key Infrastructures (PKI) are being deployed. A primary usage of PKI's, which we deal with herein, is for representing and authenticating digital identities via the use of digital signatures. In this case the public keys available are used for signature verification. This setting enables authenticated channels (on top of which privacy can be established as well, via authenticated key exchange). It also enables undisputed "evident collection" in audits and logs, since unforgeable digital signatures authenticate the message and its originator in a non-repudiated fashion. PKIs are employed in the area of "network security" to assure smooth, robust and secure operation of Intranets and inter-organization communication [KPS95]. PKIs will also be used in the growing area of distributed web-based electronic-commerce applications as the basic enabling technology [FB97].

Digital signature schemes can be used directly between users who know in advance each other's public signature verification functions. However, for larger scale interactions one needs a PKI setting. A very basic PKI configuration consists of a Certificate Authority (CA) which publishes signature verification keys in a public file [Koh78] (also called "trusted directory" [M98] and "certificate directory" in [MOV97]). Users may query the file to find bindings of public keys to their owners (*pull model*, see [MOV97] chapter 13.6.3). This basic setting is equivalent to a "white pages directory" (it does not deal with revocations of certificates and, in fact, the SET infrastructure for credit cards has adopted this model since credit cards are revoked at a different level of management, namely in the physical world [FL98]). A full fledge PKI, in turn, also includes a Repository (distributed directories) where Certificate Revocation Lists (CRL's) include lists of revoked public keys, and users may query the CA and the repository for status checks of keys (or may be notified of these status checks). CRL can also work (using Online/Real-time Certificate Status Checking Protocols ([OCSP-ietf]) as being defined by the IETF) with the *push model* ([MOV97]), where each user possesses his/her certificate and presents it upon usage.

PKI is a new area and we lack the long term understanding of its requirements regarding aspects like: space minimization issues, temporal (history-related) issues, validity determination issues, or application and device integration issues. We believe that we need to start thinking about such aspects as PKI becomes a reality.

Here we deal with cases where storage required by signatures' directories, such as Certification Authorities (CAs) and Revocation servers (CRLs), cashes of signature values, time stamping servers, etc. needs to be minimized. Space minimization is motivated by certain environments; in particular, it is significant for application layer scenarios involving small devices (mobiles and PDAs, set tops, network computers, etc.) and network and application layer security mechanisms involving logging large audit information (in firewalls and gateways), and caching large server data bases, as well as for small business directories.

In certain network contexts, digital signatures and certificates (which themselves include digital signatures) may be stored in numerous places. A signature of a message may be used for an intermediate node to have irrefutable evidence of the fact that a certain packet passed through that node. This will create authenticated logs which are basic components of network security. In fact, firewalls may store such logs. Another scenario where signatures are stored is when signatures and certificate directories are cached by end-nodes for e-commerce applications. In scenarios like the above, we realize that it is possible to have an extensive duplication of storage when the operations become complex and involve many parties (say, in a multicast scenario or a multi-party protocol). We were therefore motivated to reduce the storage spent in the network on "signatures and certificate" while not losing their functional power (non-repudiated evidence collection). We believe that our work can provide further options to designers of PKI and of architectures which exploit digital signatures (e.g., IETF, X509 etc.) when considering operation under certain (storage) constraints.

A public key certificate can be approximately 256 bytes in length (by an underestimating conservative calculation that assumes a DSA signature of 320 bits and a public key of 1024 bits as well as 88 bytes for alphanumeric information: dates and identities, not including the user ID). A system of 100 million users (e.g., the subscribers of the US postal service) would therefore require the CA to have at least 25.6 gigabytes of storage for all of the certificates. In this paper we describe a method that allows the CA to store merely 10-15 bytes of information for each user (excluding the user's name and information related to the file access method: hash or inverted list, etc.), instead of 256 bytes. This gives a 16 to 25-fold decrease in the storage requirements of the CA directory, reducing the size of the database to 1-1.5 gigabyte. (We note that the certificate size is conservative and assumed to be optimized, a typical X.509 certificate can contain much more information).

What we present is the following:

1. We first identify cases where one does not need to store the signature itself to fulfill its function; rather, a "signature digest" suffices. This simple demonstration is merely our starting point, stimulating the possibility of reduced storage and tradeoffs.

2. Sometimes, storage saving in directories is an outcome of using small size signature values. We therefore deal with further compressing signature size of existing methods. We show how to simply modify the DSA signature scheme to enable smaller sized signatures (we start with DSA which is already a small size signature; note that we do not consider multi-variate polynomial based schemes). This modified scheme preserves the algebraic properties of DSA, and at the same time yields signatures of size $|q| + K$ bits long, where K is the base-2 logarithm of the maximum value in the domain of a one-way hash function H used in computing the signature. The size of the signature is made comparable to Schnorr signatures while being based on the freely available DSA. This reduction is the starting point of our techniques and is based on hashing.

3. We then show how, using hashing, any signature scheme can be represented in a public file in a much shorter record. This demonstrates the tradeoff of verification time and signature size vs. the public storage size.

4. Then we describe how in certain PKI contexts, the techniques can save on storage and communication between users and CAs/ directories. Some of the saving implies saving of computation time as well.

5. Using accumulated hashing, we show how to minimize the task of verifying membership of certificates in lists (e.g. CRL).

Our work represent a few simple steps, perhaps some of which have been noticed earlier, yet it is effective and practical; it can be readily employed in various setting which are different than the X.509 standard one (we believe that non-standard solution will be needed and applicable in many crucial cases).

Related Work

The notion of a digital signature was put forth in [DH76] and the first realization of them was the RSA scheme [RSA78] which is based on the problem of factoring. Since then many digital signature schemes have been proposed with various properties. The first public signature scheme making use of the discrete log problem was ElGamal [ElG85]. Schnorr, then, proposed a signature scheme based on the problem of computing discrete logs in a subgroup of order q, where $q|p-1$, with q and p prime. With $|q| = 140$ bits, a Schnorr signature is about 212 bits long, which is much smaller than an RSA signature. The Digital Signature Algorithm due to Kravitz [DSS91] is a variant of ElGamal and Schnorr. Like Schnorr, DSA makes use of the problem of computing discrete logs in a subgroup. A DSA signature is 320 bits in length.

Many issues regarding PKI and trust have been dealt with in the last few years. Notions of management of public key directories making them current and optimizing the sizes of proofs to users of a certificate status are dealt with in [FL98,M98,K98,R98,M96,NN98,ALO98]. We are not aware of attempts to optimize storage of directories in models where we need to save on it. The issue of compressing and optimizing the usage of signatures in networking scenario was dealt with, in the case of bulk (flows of messages and multicasts) e.g., in [WL98], our setting is different (we deal with individual signature as a block) and it specializes to the PKI setting, though we show how to employ their method as well in combination with our techniques.

2 Minimizing Signature Size

We skip the definition of a signature scheme in this version, a definition is given e.g., [MOV97] Chapter 11. A signature scheme involves a signing function which is private and a corresponding public verification function associated with the user owning the signing function. The verification is efficient while forging a signature on unsigned messages is hard and considered impossible for all practical purposes. Given a message m the signing algorithm generates $\sigma(m)$ which is the signature. If m is recoverable from the signature, the scheme is called a "message recovery scheme". The typical signature methods are RSA [RSA78] (based on the hardness of factoring) and DSA [DSS91] (based on the hardness of extracting discrete logarithm).

2.1 A very simple starting point: Signature Digest

In network applications, often the signature is kept for the sake of having a "record." This is typical in a scenario where an intermediary (third party which is neither the sender nor the receiver(s)) is performing an "audit function". In many cases, the signature and the message are kept in storage. This may be unnecessary. In many cases, when a dispute occurs over the authenticity of messages, either the receiver or the sender(s) present the message and the signature. In this case the intermediary need only store a digest of the signature and the message, or just of the signature in case it is a "message recovery signature." Given a signature $\sigma(m)$ and a message m, the stored value can be $H(\sigma(m), m)$ where H is a collision-intractable hash function (a hash where collision finding is hard). In practice, the SHA-1 message digest function [SHA1] suffices (or SHA-2). When presented with the signature and the message the third party can recompute the digest and look and compare with his/her storage. (The storage is assumed to be organized so that access based on the hash value is efficient; such storage schemes are common).

2.2 On Smaller Signatures: Size Efficient DSA Variant

The above signature storage technique is a standard one, yet one may want to "digest" the signature at its creation so that the digest be sufficient for validation of the signature by the verification algorithm. Such digest methods were originated by Schnorr, and used in a variant in the DSA signature suggested by NIST for general use. One advantage of the former is that it may be shorter while the later is free to be used by the public. Here we combine both advantages into a third variant which we present in this subsection. This is yet a bit less simpler starting point to our needs.

First, let us review the Digital Signature Algorithm. Let p be a large prime, such that $q|p-1$, where q is a 160 bit prime. Let g be an element with order q mod p. The private signing key of a user is x, where $x \in_R Z_q$. The public verification key of the user is $y = g^x \bmod p$. To sign a message m, the user computes the following:

1. $k \in_R Z_q$
2. $r = (g^k \bmod p) \bmod q$
3. $s = k^{-1}(H(m) + xr) \bmod q$
4. output the signature (r, s)

The function H used here is SHA. To verify the signatures authenticity, the verifier checks that $r = (g^{H(m)s^{-1}} y^{rs^{-1}} \bmod p) \bmod q$. Note that $|q| = 160$ bits, hence the signature is 320 bits long. The question we ask is, "why is $g^k \bmod p$ reduced modulo q?".

To answer this, let us consider DSA and it's origins. After all, DSA was designed by the NSA, and many questioned whether or not it has a back-door. It may seem that by reducing $g^k \bmod p$ by q we conceal $g^k \bmod p$. But, this is not the case. It was shown in e.g. [NR94,YY97] that DSA can and in fact does give away $g^k \pmod p$, and can be (ab)used for key exchange and encryption. This attack exploits the fact that $g^k \bmod p$ is in fact readily computable from (r, s). To see this, note that $g^{H(m)s^{-1}} y^{rs^{-1}} \bmod p$ is $g^k \bmod p$ for valid signatures. Thus, reducing mod q does not hide $g^k \bmod p$ at all. What then is the purpose of reducing modulo q to derive r? To help answer this, let us review the Schnorr digital signature scheme [Sc89] which is another small space signature which is covered by a patent [Sc89-p].

Let s be the private signing key chosen randomly mod q. Let $v = g^{-s} \bmod p$ be the corresponding public key. Here p, q, and g are the same as in DSA. To sign m the signer does the following:

1. $k \in_R Z_q$
2. $t = g^k \bmod p$
3. $e = H(m, t)$

4. $u = k + se \bmod q$
5. output the signature (e, u)

To verify the signature, the verifier computes $z = g^u v^e \bmod p$ and checks that $e = H(m, z)$. Note that in Schnorr, $g^k \bmod p$ is hashed. Indeed this is also possible in DSA. So, the following is our modified DSA signing algorithm:

1. $k \in_R Z_q$
2. $r = H(g^k \bmod p)$
3. $s = k^{-1}(H(m) + xr) \bmod q$
4. output the signature (r, s)

To verify the signature, the verifier checks that $r = H(g^{H(m)s^{-1}} y^{rs^{-1}} \bmod p)$. So, the only change made to DSA is that we hash $g^k \bmod p$ rather than reduce it modulo q. Hence, by making DSA more like Schnorr, we end up with a faster, more space efficient signature scheme. It is faster since H can be SHA, MD5, etc., and hashing is faster than modular reduction. It is more efficient since the range of H can be narrower than $|Z_q|$ for any 160 bit prime q nowadays, and in the future if we use larger p's and q's (say 2048 bit p's and 320 bit q's), the range of H may become even a larger improvement compared with the size of q

.

Speculating Remark: Why then was DSA not defined this way? It does not seem that it was designed this way for the purposes of implementing a backdoor in DSA. It was perhaps designed this way to look less like Schnorr to avoid patent conflicts. However, this is pure speculation.

2.3 Size Efficient Generic Public Verification Key

The above can be viewed as a method which hashes the random key used within an individual signature operation. We now show that such hashing can be applied to the permanent verification key as well.

Assume now the setting where a user publishes his public key (in his home page or in a white page directory or in a PGP system [Zim92]). Let a signature scheme in use have a private signing key s_U for user U, and its corresponding public verification key v_U. In a typical public key system v_U gets published. We modify the system to publish $h_U = H(v_U)$ which is made as U's public key. Here H is a one-way (collision intractable) hash function with, for instance, a range of $\{0, 1\}^{80}$; in practice the first 10 (or 15) bytes of SHA can be used (public keys are fixed size strings which are very structured (we may add an error correction code field to the signature value to add redundancy), and collisions will be still hard to find even when using a 10-15 byte hash function size).

To sign, the user signs using s_U and also sends an appendix v_U. The verifier, in turn, first verifies that the hash of the appendix matches the publicly available h_U, and then uses the appendix to verify the signature itself.

The above simple hashing idea demonstrates a tradeoff between public key size and the size and time of signature verification (but the time does not grow by more than twice). The fact that the hash function is collision intractable assures the unique binding of v_U and the publicly available h_U[1]. Of course, a tradeoff between space and security (collision probability) is immediate.

We note that the above method is a hybrid which combines the "pull model" (where the verifier has access to the signature scheme) with the "push model" (where the signature scheme is presented to the verifier). The appendix may include the certification chain, which may be checked up to the required point by the receiver.

3 Space Efficient PKI Directories

Next we exploit the techniques above and directory organization methods to reduce space and processing time in the context of PKI. The methods described here are an alternative to organize PKI when storage is the limiting constraint.

3.1 Efficient White-Pages Based Signature PKI

In this subsection we will explain how to implement a size efficient public file which is maintained by a CA (CA directory), in a basic digital signature public key system (a white pages system). Note that a white page design is not what we suggest here, but it is a construction known in the literature (where certificates are backed up or maintained centrally).

In a typical public key system, the CA signs public keys and issues the resulting digital certificates to users; in each signature verification a CA signature needs to be checked as well. In our model, we have a white pages (trusted file) and we can save space of certificates. In addition we will save on signature verification time.

In this subsection, let C_U denote the (lean) digital certificate of user U which is a usual certificate but without the CA signature on the user's public key. Recall that the usual certificate information typically includes: user (entity) ID, validity period, serial number, verification-key description, issuer ID, information about the entity, information about the key, information about the usage. The CA typically includes this information and its own signature when making a certificate. In our scenario, the CA makes $h_U = H(C_U)$ available in the public file.

To sign a message, U signs the message using his private key as in a normal public key system. However, C_U (the lean certificate) is included with the message and the signature. The 'new signature' then, is the signature from the relevant digital signature scheme plus the lean digital certificate of the user.

---

[1] assuming 2^{28} non-maliciously chosen certificate for all US citizens, there are 2^{56} pairs while 2^{80} (or 2^{120}) random hash values are possible (assuming SHA is a good ideal random hash, as done in many analyses) which makes the probability of a "collision pair" very small.

A verifier verifies a signature by first hashing the lean certificate and comparing to h_U from the white pages file. This is accomplished by taking U's public key from C_U and performing signature verification.

This system provides an enormous storage savings for the CA. In a typical public key system, the CA stores C_U for each user, which is approximately 256 bytes of information (under a very conservative estimate). By having the CA store $H(C_U)$, the CA only stores 10 bytes per user, giving the CA a 25-fold storage savings. Also, we take advantage of the trusted white pages file to save time in verifying the certificate (essentially cutting the verification time by half).

Note that the white pages file itself can be signed (and verified by a user) once by means of a signature on it by a CA when the file is published (loaded by a user). In these case, the CA's key is used minimally, using the white pages model which is relatively static.

In fact, various blocks can be signed separately, to allow monotonic increase of the directory, and speeding up the verification of the integrity of the directory (which can be done block by block whenever the block is in use). Aggregated "groups of pages" can be signed to balance the checking of validity and size. In fact a method like the one by Wong and Lam [WL98] can be exploited here for signing bulk data efficiently. (We will explain this method which combines message authentication and signature to achieve more efficient signature, in more details in the next version).

We also note that the hash values can be used for file organization and access methods as well. Further, another organization method allows the file server to keep the certificate value off-line (to be retrieved if needed) while the on-line organization is as above. The off-line "full-fledge certificate" can be used rarely depending on the application, in which case it can be retrieved from secondary storage (server) rather than using the available data (or cached data).

3.2 Size Efficient CRL Repositories in PKI's

Now we show the functionality of our technique in the context of a Public Key Infrastructure where the CA signs certificates and distribute to users (what is known as the push model originally adopted by X.509).

In this setting it is a "repositories' job" to maintain a list of revoked public keys for each user of a Public Key Infrastructure (CRLs). In the event that a signing private key is lost or stolen, the corresponding public key and/or digital certificate is stored by the repository for that user. From then on, when user A verifies a signature of user B, user A has the option of checking the repository to see if the version of user B's public key that A has, has been revoked. If the public key had been revoked, after the query, user A will be informed of this, and the authenticity of the message could then be brought into question. This capability is important because a malicious user could use the lost and or stolen private key and try to sign a message, impersonating user B. The attacker may be aware that user A is storing the corresponding outdated public key, and may hope that user A doesn't query the repository.

A repository typically stores either the public key or the digital certificate of lost or stolen public keys. However, hashing can be used to greatly reduce the repository's storage requirements. By having the repository store $v_U = H(C_U)$ instead of the certificate C_U, the repository allocates merely, say, 10 bytes for each lost or stolen public key. A user checks to see if a certificate C_U has been revoked by checking that $v_U = H(C_U)$ is in a CRL. This saves space and communication (queries can be based on the hash value in the "pull model", and list distribution in the "push model" can also be based on hash values which can help in accessing certificates). These methods are also applicable to PKI's that are used for encryption and decryption (rather than signing and verification).

3.3   Proving Revocation/ Active Certificate

Here the tradeoff will be between the server work and the user's time and resources of checking.

If a user needs a "proof of revocation" (does not trust the directory), one can hash the certificate hash values using a accumulated hash (as given in [BdM94,Ny96]), and the CA signs this value. If the list is: $\{L_1, L_2, ..., L_m\}$, $H(L) = H(L_1(H(L_2(...(H(L_m))...))$ The CA signs $H(L)$, to prove revocation the CA present to the user three values the hash of the values till L_i, L_i and hash of the values larger than L_i. From this three values the user can reconstruct (for each i) the value $H(L)$ and can check the signature (if this was not done before). The list is harder to maintain by the CA (keeping hashes of sublists will help). As the list changes $H(L)$ changes, periodically. Or a list of lists is maintained.

This method implies short proofs to users of the fact that a key is a member of a list (CRL). the user's task now is independent of the size of the CRL (which may be attractive for small energy devices).

Further trust may be implied by having distributed directories (in which case storing hashed values is important since there is no need to replicate longer values when a shorter digest can be employed and is sufficient).

In [GGM00], it was suggested to have am active certificate list, similar to a revocation list where the CA signs the root, and a CA can send the path to the root in the CRL. This can also be replaced by an accumulated hash as above.

4   Conclusions

Scenarios for minimizing the signature storage and representation have been investigated. These were motivated by the needs of certain systems and communication settings. In some settings directory size issues are crucial and standards may not provide a feasible solution.

We discussed various space minimization issues and tradeoffs between space and other resources in various potential PKI components.

A modification of DSA was presented that allowed for variable sized DSA-like signatures. Not only does this scheme provide for smaller signatures with nearly the same level of security (for use with the directories), but the signing algorithm is also faster. We showed how hashing can be used to decrease the storage requirements of public verification keys by making the signature larger. This is used to implement a trusted public key file in a PKI which also decreases the verification time. The same method was then applied to reducing the storage/communication requirements of a directory or CRL repository in various PKI settings.

We believe that determining the right components of PKI is an open issues which will evolve as we gain experience with PKI and its applications. The interaction between cryptographic issues and traditional data base issues in the case of PKI directories presents potentially interesting open issues.

References

[ALO98] W. Aiello, S. Lodha and R. Ostrovsky. Fast Digital Identity Revocation. In *Advances in Cryptology—Crypto'98*, pages 137–152.

[BdM94] J. Benaloh and M. de Mare, One-Way Accumulators: A Decentralized Alternative to Digital Signatures, In Advances in Cryptology—EUROCRYPT 93.

[DH76] W. Diffie, M. Hellman. New Directions in Cryptography. In volume IT-22, n. 6 of IEEE Transactions on Information Theory, pages 644–654, Nov. 1976.

[DSS91] Proposed Federal Information Processing Standard 186 for Digital Signature Standard (DSS). In volume 56, n. 169 of *Federal Register*, pages 42980–42982, 1991.

[SHA1] Proposed Federal Information Processing Standard 180-1 for Secure Hash Standard, 1995.

[ElG85] T. ElGamal. A Public-Key Cryptosystem and a Signature Scheme Based on Discrete Logarithms. In CRYPTO '84, pages 10–18.

[FB97] W. Ford and M. Baum, Secure Electronic Commerce: Building the Infrastructure for Digital Signature and Encryption. Prentice Hall, 1997.

[FL98] B. Fox and B. LaMacchia. Certificate Revocation: Mechanisms and Meaning. In *Financial Cryptography 98*, pages 158–164, 1998.

[GGM00] I. Gasseko, P. Gemmel and P. MacKenzie, Efficient and Fresh Certification. In *Public Key Cryptography: PKC 2000*, LNCS 1751.

[OCSP-ietf] On-line Certificate Status Checking Protocol, IETF.

[KPS95] C. Kaufman, R. Perlman and M Speciner, Network Security: Private Communication in a Public World, Prentice Hall, 1995.

[K98] P. Kocher. On Certificate Revocation and Validation. In *Financial Cryptography 98*, pages 172–178, 1998.

[Koh78] L. Kohnfelder. A Method for Certification. MIT Lab. for Computer Science, Cambridge Mass., May 1978.

[MOV97] Alfred J. Menezes, Paul C. van Oorschot and Scott A. Vanstone. Handbook of Applied Cryptography, 1997. CRC Press LLC.

[M96] S. Micali. Efficient Certificate Revocation. In *MIT Tech. Report.* 1996.

[M98] M. Myers. Revocation: Options and Challenges. In *Financial Cryptography 98*, pages 165–171, 1998.

[NN98] M. Naor and K. Nissim. Certificate Revocation and Certificate Update. In *7-th USENIX Security Symp.*, 1998.

[Ny96] K. Nyberg. Fast Accumulated Hashing. In *Fast Software Encryption* 96, pages 83–87.

[NR94] K. Nyberg, R. Rueppel. Message Recovery for Signature Schemes Based on the Discrete Logarithm Problem. In *Advances in Cryptology—Eurocrypt '94*, pages 182–193, 1994. Springer-Verlag.

[R98] R. Rivest. Can We Eliminate Certificate Revocation Lists? In *Financial Cryptography 98*, pages 178–183, 1998.

[RSA78] R. Rivest, A. Shamir, L. Adleman. A method for obtaining Digital Signatures and Public-Key Cryptosystems. In *Communications of the ACM*, volume 21, n. 2, pages 120–126, 1978.

[Sc89] C. P. Schnorr. Efficient Signature Generation for Smart Cards. In *Advances in Cryptology—CRYPTO '89*, pages 239-252, 1990. Springer-Verlag.

[Sc89-p] C. P. Schnorr. Method for Identifying subscribers and for generating and verifying electronic signatures in a data exchange system In U.S. Patent 4,995,082, 19 Feb. 1991. Springer-Verlag.

[YY97] A. Young, M. Yung. The Prevalence of Kleptographic Attacks on Discrete-Log Based Cryptosystems. In CRYPTO '97, pages 264–276. Springer-Verlag.

[WL98] C. K. Wong and S. Lam, Digital Signatures for Flows and Multicasts, IEEE ICNP'98, 1998.

[Zim92] Phil Zimmerman. PGP User's Guide, 4 Dec. 1992.

A Design of the Security Evaluation System for Decision Support in the Enterprise Network Security Management

Jae Seung Lee[1], Sang Choon Kim[1], and Seung Won Sohn[1]

[1] Information Security Technology Division, ETRI
161 Kajong-Dong, Yusong-Gu, Taejon, 305-350, Korea
{jasonlee, kimsc, swsohn}@etri.re.kr

Abstract. Security Evaluation System is a system that evaluates the security of the entire enterprise network domain consists of various components and that supports a security manager or a security management system in making decisions about security management of the enterprise network based on the evaluation. It helps the security manager or the security management system to make a decision about how to change the configuration of the network to prevent the attack due to the security vulnerabilities of the network. Security Evaluation System checks the "current status" of the network, predicts the possible intrusion and supports decision-making about security management to prevent the intrusion in advance. In this paper we analyze the requirements of the Security Evaluation System that automates the security evaluation of the enterprise network consists of various components and that supports decision-making about security management to prevent the intrusion, and we propose a design for it which satisfies the requirements.

1 Introduction

1.1 Overview of the Security Evaluation System

Recent explosive increase in the use of the network makes the security of the network that consists of various components much more important than that of individual hosts. In addition to the Firewall and the Intrusion Detection System [11], we need a tool that can check the current security status of the enterprise network. To maintain the security of the enterprise network, the tool must have the capability of evaluating the security of the entire enterprise network and the capability of decision support for the network security management to prevent the intrusion caused by the security vulnerabilities of the network.

Security evaluation requires various kinds of test and analysis, and it is hard for a network security manager to perform all the tests and analyses of the large enterprise network for himself. Therefore, a tool is needed which automates the tests and analyses. The tool must have the capability of evaluating the security of the entire enterprise network domain as well as the capability of the security analysis of the individual host or subnetwork. It must support a security manager or a security management system in making decisions about security management of the network based on the security evaluation result to prevent intrusion caused by the vulnerability of the network.

Security Evaluation System is a system that evaluates the security of the entire enterprise network domain consists of various components using automated analysis tools and that supports decision-making about security management based on the evaluation to prevent the intrusion in advance. It helps security managers or security management systems to make a decision about how to change the configuration of the network to prevent the attack due to the security vulnerabilities of the network. The Intrusion Detection System [11] detects an attack when the intrusion really happens, but Security Evaluation System checks the "current security status" of the network, predicts the possible intrusion and helps security managers to prevent the intrusion in advance.

1.2 Necessity of the Research on the Security Evaluation System

So far there is no formal and standardized technology for Security Evaluation System, and vulnerability assessment tools developed by individual software companies are used for security evaluation of the network instead.

Hacking tools such as sscan, SATAN [8], SAINT, ISS are widely used for vulnerability assessment but the individual tools are not integrated and it is hard to get integrated security evaluation report from the vulnerabilities found by each tool. The tools have to be updated whenever a new vulnerability is found and a new version of the tool is available reflecting the vulnerability. They just provide a report on several well-known vulnerabilities from a network scanning.

Commercial vulnerability assessment tools such as AXENT NetRecon [3], ISS Internet Scanner, System Scanner, Database Scanner [4] and RSA Kane Security Analyst [5] are used for security evaluation but these tools just provide several host or network scanning functionality.

Security Evaluation System has to be evolved to provide automatic update of the search capability for the new vulnerability check, vulnerability analysis using hacking simulation, and analysis of the possible intrusion due to a specific vulnerability. It also has to provide integration with the Security Policy Server, vulnerability recovery functionality integrated with the Security Management System, decision support for the security management and scalability for the large enterprise network. But so far the research on the Security Evaluation System is not enough, and there is no tool that supports all the functions described above. Therefore, further research and development are needed for the progress of the Security Evaluation System Technology.

In this paper we analyze the existing tools for security evaluation. Then we analyze the requirements of the Security Evaluation System that automates the security evaluation of the enterprise network and that supports decision-making about security management to prevent the intrusion, and we propose a design for it that satisfies the requirements.

2 Vulnerability Assessment Tools

Security evaluation requires various kinds of test and analysis, and it is hard for a network security manager to perform all the tests and analyses of the large enterprise network for himself. Therefore, a vulnerability assessment tool is needed which automates the tests and analyses. Network-based scanners and host-based scanners [1] are the typical vulnerability assessment tools.

2.1 Network-Based Scanner

A network-based scanner performs quick, detailed analyses of an enterprise's critical network and system infrastructure from the perspective of an external or internal intruder trying to use the network to break into systems. Features of the network-based scanners are as follows [1].

- A network-based scanner analyzes network-based devices on the network and quickly provides repair reports to allow quick corrective action.
- It can be set up and used quickly because it requires no host software to be installed on the system being scanned.
- It assesses network-based vulnerabilities by replicating techniques that intruders use to exploit remote systems over the network.
- Many network-based vulnerabilities are more efficiently investigated over the network.
- It can test vulnerabilities of critical network devices that don't support host-scanning software, including routers, switches and printers.

2.2 Host-Based Scanner

Host-based scanning's strengths lie in direct access to low-level details of a host's operating system, specific services, and configuration details. While a network-based scanner emulates the perspective that a network-based intruder would have, a host-based scanner can view a system from the security perspective of a user who has a local account on the system [1, 2].

- A host-based scanner can identify the risky user activities.
- It detects signs that an intruder has already infiltrated a system. It also detects any unauthorized changes in critical system files.

- It detects signs that an intruder is still active on a system, including locating "sniffer [8]" programs.
- It detects well-known hacker back-door programs such as "Back Orifice" and local host services vulnerable to "local buffer overflow".
- It is ideal for performing resource-intensive file system checks, which are impractical with network-based scanners.

2.3 Vulnerability Assessment Tools in Use

AXENT NetRecon [3], ISS Internet Scanner, System Scanner, Database Scanner [4] and RSA Kane Security Analyst [5] are famous vulnerability assessment tools in use. They provide network-based and host-based scanning functionality.

They provide Graphic User Interface and some product can analyze the communication devices such as router. Some product suggests corrective actions when the vulnerability is found. Table 1 summarizes the major vulnerability assessment tools [6].

Table 1. Vulnerability Assessment Tools

	NetRecon	Internet Scanner	System Scanner	Database Scanner	Kane Security Analyst
Product	AXENT	ISS	ISS	ISS	RSA
Target	analyzes the network and host vulnerabilities	analyzes the network vulnerabilities	analyzes the system vulnerabilities	analyzes the database vulnerabilities	analyzes the vulnerabilities of Winows NT
Features	suggests the corrective action for the vulnerability	integrated assessment tool for intranet, firewall and web server	analyzes the low-level vulnerabilities using agents	analyzes the vulnerabilities related to DB	analyzes the system configuration of Windows NT
Platform	Windows NT	Windows NT	UNIX, Windows 95/98/NT	Windows 95/98/NT	Windows NT

3 Design of the Security Evaluation System

In this section, we analyze the requirements of the scalable Security Evaluation System that automates the security evaluation of the enterprise network consists of various components, and that supports decision-making about security management of the network to prevent the intrusion caused by the vulnerability of the network. Then we propose a design for it which satisfies the requirements.

3.1 Requirements of the Security Evaluation System

The requirements of the Security Evaluation System are as follows.

- It must support the security evaluation of the enterprise network consists of various components and it must be scalable. Its performance should not be decreased even if the size of the network grows.
- It must be able to analyze the vulnerability of a host. For example, it must be able to check system configuration error, Trojan horses and system file integrity. It also must be able to trace signs of the hacker's intrusion.
- It must perform the vulnerability checks and security analyses of each subnet. For example, it must be able to analyze the vulnerability of network services.
- It must provide the function of generating the security evaluation reports. It also must support a security manager or a security management system in making decisions about security management of the network based on the security evaluation result to prevent intrusion caused by the vulnerability of the network.
- It must analyze whether the various components of the enterprise network obey the security policy of the network domain or not using the Security Policy Server.
- It must support the remote and central administration. It must provide the central view of the security status of the entire network.
- It must provide the vulnerability recovery functionality using Security Management System.
- It must be able to get new vulnerability information from directory services frequently and apply it to the security evaluation.

3.2 Architecture of the Security Evaluation System

Fig. 1 illustrates the architecture of the Security Evaluation System that satisfies the requirements described above. The major components of the Security Evaluation System are Agents, Subnet Analyzers, a Domain Analyzer, a Security Evaluation Rule Manager and a Manager Tool.

3.2.1 Agent

Agents are executed in each host and they analyze the security of the host in detail. They access the low-level detail of a host's operating system, specific services, and configuration details. Each Agent analyzes the system configuration error, signs of the hacker's intrusion, system file integrity and the existence of the Trojan horses or viruses. Each Agent then generates a security evaluation report of the host and sends it to the Subnet Analyzer that is in the same subnet.

3.2.2 Subnet Analyzer

Subnet Analyzers evaluate the security of each subnet from the viewpoint of network user. Each Subnet Analyzer analyzes vulnerability of the network services and daemons. It analyzes the host security evaluation reports from each Agent in the subnet and finds out vulnerability which can be only checked with the information of several hosts that are members of the distributed environments. A subnet security evaluation report is sent to the Domain Analyer.

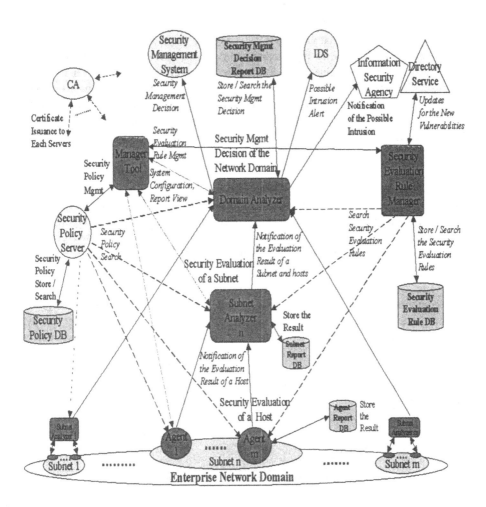

Fig. 1. Architecture of the Security Evaluation System

3.2.3 Domain Analyzer

Domain Analyzer evaluates the security of the entire enterprise network domain based on the security evaluation results from each Agents and Subnet Analyzers. It analyzes the reports from each analyzer and finds out correlation between the vulnerability. It predicts the possible intrusion due to the vulnerability analyzed so far. Then it makes decisions about security management to prevent the possible intrusion and report them to the security manager. It also sends the notification mail to the system manager of the host that has the security weaknesses. It notifies the decisions about security management to the Security Management System or to the Intrusion Detection System [11] and helps them to prevent the possible intrusion.

3.2.4 Security Evaluation Rule Manager

Security Evaluation Rule Manager helps the security manager to generate and manage the evaluation rules that are needed for security evaluation. Security evaluation rules consist of vulnerability information related to a specific host or network resource, possible intrusion due to the vulnerability, and countermeasures for it. Security evaluation of Agents, Subnet Analyzers, and Domain Analyzer is based on the security evaluation rules that are managed by the Rule Manager. Rule Manager gets new vulnerability information from the external trusted directory service frequently and updates the security evaluation rule based on it. Therefore Security Evaluation System can cope with the new vulnerability rapidly.

3.2.5 Manager Tool

Manager Tool helps the administration of Security Evaluation System. A security manager can configure each component of Security Evaluation System remotely. He can centrally manage the security evaluation rules and security policy. Manager Tool helps the security manager to view the evaluation reports in various formats.

3.2.6 Etc

Each component of Security Evaluation System such as Agents, Subnet Analyzers, Domain Analyzer, Manager Tool and Security Evaluation Rule Manager encrypts the data that are confidential such as security evaluation reports. For the authentication between the components, they use the certificates issued by the trusted CA (Certificate Authority).

Security Policy Server is the server that manages the security policy of the network domain. Security Evaluation System always references the security policy when it performs the security evaluation and checks whether the network resource obeys the security policy or not. It makes the components of the enterprise network follow the consistent security policy.

3.2.7 Features of the System Design

Security Evaluation System proposed in this paper evaluates the current security status of the enterprise network and supports decision-making about security management to remove the vulnerability of the network and to prevent the possible intrusion. Security evaluation performed by each analyzer module (Agent, Subnet Analyzer, Domain Analyzer) is based on the security evaluation rules that are managed by Rule Manager. Rule Manager gets the new vulnerability information from the external trusted directory service frequently and updates the security evaluation rule based on it. Security Evaluation System can cope with the new vulnerability rapidly by this feature.

In this design, Security analysis processes are distributed over several analyzers. That is, an Agent analyzes the security of a host, a Subnet Analyzer analyzes the security of a subnet and Domain Analyzer analyzes the results from all analyzers. Therefore the processing load of each analyzer is relatively small. If the number of hosts in the network domain increase, overall evaluation performance will be decreased. But the overall performance can be preserved from decrease by adding new Agents and Subnet Analyzers to the network domain, since the processing load can be distributed over new analyzers. That is, Security Evaluation System is scalable. Security Evaluation System requires relatively small network traffic because not all the host or network information is sent to the Domain Analyzer. Agents or Subnet Analyzers gather all the information about hosts or subnet, analyzes the hosts or subnet information, and they send the security evaluation reports to the Domain Analyzer. The reports contain only information related to the security evaluation result of the host or the subnet. It also greatly reduces the amount of data that the Domain Analyzer has to analyze. Therefore, Security Evaluation System is suitable for large enterprise network.

Each analyzer module in Security Evaluation System references the security policy of the enterprise network domain and checks whether the components of the network obey the security policy whenever the analyzer performs the security evaluation of the component. It makes the components of the enterprise network follow the consistent security policy.

3.3 Detail Architecture and Processing Flow of the Security Evaluation System

An outline of the processing flow of Security Evaluation System is as follows. Update of the Security Evaluation Rule DB is performed frequently independent of each analyzer in the Security Evaluation System.

Security Evaluation of each host by Agents	→
Security Evaluation of each subnet by Subnet Analyzers	→
Decision-making about Security Management of the network	
by Domain Analyzer	

An outline of the processing flow of each analyzer module (Agent, Subnet Analyzer, Domain Analyzer) is as follows.

> *Analysis of the host or network resources based on Security*
>
> *Evaluation Rules and Security Policy*   →
>
> *Generation of the Security Evaluation Result Report*   →
>
> *Notification of the Evaluation Result to the higher analyzer module*

Both the security evaluation rule and the security policy have to be referenced during security evaluation. To do this, the analyzer selects the rules and policy that are suitable to the target hosts or network environment from the Security Evaluation Rule DB and the Security Policy DB, and stores them in Customized Rule DB. Customized rules in the DB are referenced for the security evaluation. If the customized rules already exist and if they are made for the same host or network environment from the same security evaluation rules and security policy, the analyzer omits the Rule DB and Policy DB access and uses the customized rules that already exist. This reduces the DB access and search time. If there is any conflict between security evaluation rules and security policy, security policy has the preference.

Detailed architecture and processing flow of the Security Evaluation System are as follows.

3.3.1 Generation of the Security Evaluation Rules by the Security Evaluation Rule Manager

Fig. 2 illustrates the architecture of the Security Evaluation Rule Manager.

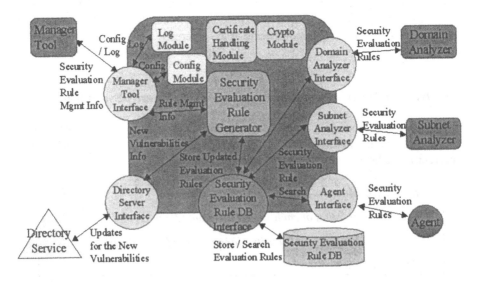

Fig. 2. Architecture of the Security Evaluation Rule Manager

Security Evaluation Rule Manager gets the new vulnerability information from the external trusted directory service frequently. Security Evaluation Rule Generator generates the security evaluation rules from the vulnerability information by security manager's command from the Manager Tool, and stores them in the Security Evaluation Rule DB. The security manager generates, updates and deletes the security evaluation rules.

Security Evaluation Rule Manager helps the security manager to generate and manage the evaluation rules. Security evaluation of Agents, Subnet Analyzers, and Domain Analyzer is based on the security evaluation rules. If an analyzer request the security evaluation rule, Rule Manager searches for the rule in the Rule DB and return it using the analyzer interface.

3.3.2 Security Evaluation of a Host Using an Agent

Fig. 3 illustrates the architecture of the Agent.

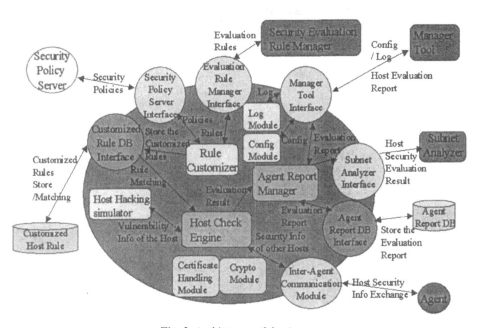

Fig. 3. Architecture of the Agent

Rule Customizer of each Agent selects the rules and policy that are suitable to the target host from the Security Evaluation Rule Manager and the Security Policy Server, and stores them in the Customized Rule DB. If the Customized Host Rules already exist and if they are made for the same host environment from the same security evaluation rules and Security Policy, the Agent omits the Rule DB and Policy DB access and uses the customized rules that already exist.

Host Check Engine evaluates the security of the target host based on the Customized Host Rules. It may exchange the host security information with other Agents to find out the vulnerability that can be checked using the information of other hosts.

It checks the vulnerability of the target host using Host Hacking Simulator. In this case, Host Hacking Simulator has to leave a message to the host to make it distinguish the test from real hacking. Host Check Engine accesses the low-level detail of the target host's operating system, specific services, and configuration details. It analyzes the system configuration error, bug patch status, signs of the hacker's intrusion, system file integrity and the existence of the Trojan horses, viruses or sniffers. It also checks whether the host obeys the security policy or not.

Agent Report Manager generates the host security evaluation result report that describes the information related to the vulnerability based on the security evaluation result from Host Check Engine. It stores the report in the Agent Report DB and sends the report to the Subnet Analyzer that is in the same subnet.

3.3.3 Security Evaluation of a Subnet Using a Subnet Analyzer
Fig. 4 illustrates the architecture of the Subnet Analyzer.

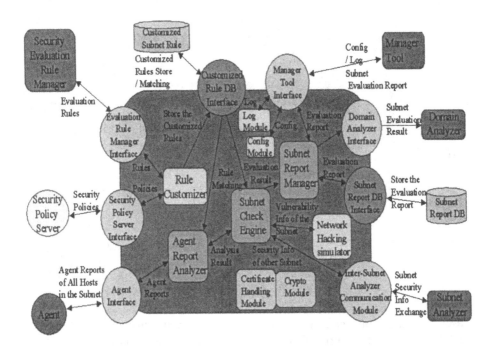

Fig. 4. Architecture of the Subnet Analyzer

If the Customized Subnet Rule does not exist in the Customized Rule DB, Rule Customizer of the Subnet Analyzer selects the rules and policy that are suitable to the target subnet from the Security Evaluation Rule Manager and the Security Policy Server. It stores them in the Customized Subnet Rule DB.

Subnet Check Engine evaluates the security of the target subnet based on the Customized Subnet Rules. It may exchange the subnet security information with other Subnet Analyzers to find out the vulnerability that can be checked using the information of other subnet.

It checks the vulnerability of the remote hosts using Network Hacking Simulator. In this case, Network Hacking Simulator has to leave a message to the remote host to make them distinguish the test from real hacking.

Subnet Check Engine evaluates the security of a subnet from the viewpoint of a network user. Each Subnet Analyzer analyzes vulnerability of the network services and daemons. It analyzes the host security evaluation reports from each Agent in the subnet and finds out the vulnerability which can be only checked with the information of several hosts that are part of the distributed environments using Agent Report Analyzer. It also checks whether the subnet obeys the security policy or not.

Subnet Report Manager generates the subnet security evaluation result report that describes the information related to the vulnerability based on the security evaluation result from Subnet Check Engine. It stores the report in the Subnet Report DB and sends the report to the Domain Analyzer.

3.3.4 Decision-Making about Security Management of the Enterprise Network Domain Using Domain Analyzer

Fig. 5 illustrates the architecture of the Domain Analyzer.

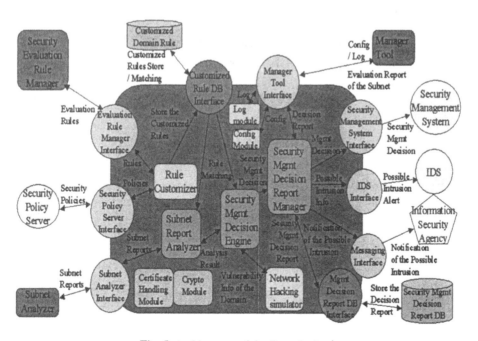

Fig. 5. Architecture of the Domain Analyzer

Rule Customizer of the Domain Analyzer selects the rules and policy that are necessary for Domain Analyzer from Security Evaluation Rule Manager and Security Policy Server, and stores them in the Customized Domain Rule DB.

Security Management Decision Engine evaluates the security of the entire enterprise network domain based on the Customized Domain Rules and security evaluation results from each Agents and Subnet Analyzers. It analyzes the subnet security evaluation reports from each subnet in the domain and finds out the vulnerability that can be checked with the information of several subnet using Subnet Report Analyzer. It also finds out the correlation between the vulnerability. It combines the duplicated results and removes wrong results from the evaluation results. It predicts the possible intrusion due to the vulnerability analyzed so far. Then it makes decision about security management to prevent the possible intrusion.

Security Management Decision Report Manager generates the Security Management Decision Report based on the security management decisions from Security Management Decision Engine. It stores the report in the Security Management Decision Report DB. A security manager can view the report using Manager Tool and use it for decision-making about security management of the network.

Domain Analyzer notifies the decisions about security management to the Security Management System or to the Intrusion Detection System and helps them to prevent the possible intrusion. It requests proper system configuration changes to the Security Management System and it notifies the possible intrusion to the Intrusion Detection System.

If signs of the serious hacking have been found, it sends a notification mail to the Information Security Agency. It also sends a notification mail to the system manager of the host that has the security weaknesses.

3.3.5 Report View and System Management Using Manager Tool

Fig. 6 illustrates the architecture of the Manager Tool.

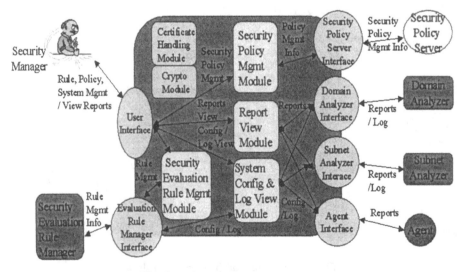

Fig. 6. Architecture of the Manager Tool

A security manager can configure each component of Security Evaluation System remotely using Manager Tool. He can centrally manage Security Evaluation Rules and Security Policy. Manager Tool helps the security manager to view the evaluation reports in various formats.

User Interface calls the proper module to process the manager's requests and displays the result to the manager. A security manager can add, delete, or update the security policy of the Security Policy Sever using Security Policy Management Module. He can view the reports in various formats using Report View Module. He also can add, delete, or update the security evaluation rules of the Security Evaluation Rule Manager. He can configure the system or view log using System Configuration & Log View module.

4. Conclusions

In this paper we analyze the requirements of the Security Evaluation System that automates the security evaluation of the enterprise network consists of various components and that supports decision-making about security management to prevent the intrusion, and we propose a design for it which satisfies the requirements.

In our design, Security Evaluation System gets new vulnerability information from directory services frequently and updates the security evaluation rule based on it. Therefore Security Evaluation System can cope with the new vulnerability rapidly. The system analyzes the security of each host and subnet that are components of the enterprise network and it evaluates the security of the entire enterprise network domain based on the analysis result of each component. Security Evaluation System then supports decision-making about security management based on the evaluation result to prevent the intrusion in advance.

In the system design, Security analysis processes are distributed over several analyzers such as Agents and Subnet Analyzers. Therefore the processing load of each analyzer is relatively small. Overall performance of Security Evaluation System can be preserved from decrease by adding new Agents and Subnet Analyzers to the network domain even though the number of hosts in the network domain increases. That is, Security Evaluation System is scalable. Security Evaluation System requires relatively small network traffic because not all the host or network information is sent to the Domain Analyzer. Agents or Subnet Analyzers send security evaluation reports to the Domain Analyzer and they contain only information related to the security evaluation result of the host or the subnet. It also greatly reduces the amount of data that the Domain Analyzer has to analyze. Therefore, Security Evaluation System is suitable for large enterprise network.

Security Evaluation System evaluates the security of the network based on the security policy of the Security Policy Server. Therefore, it is easy to check whether the various components of the enterprise network obey the security policy of the network domain or not.

Security Evaluation System references the security policy for the security evaluation but so far the research on the security policy system is not enough and further research is required. Artificial Intelligence technology such as Expert System has to be considered for the accurate and intelligent decision support for network security management. AI technology is also needed for the prediction of the intrusion caused by specific vulnerability.

References

1. ISS, "Network and Host-based Vulnerability Assessment,"
   http://documents.iss.net/whitepapers/nva.pdf
2. ISS, "Securing Operating Platforms: A solution for tightening system security," January 1997.
3. AXENT Home Page, http://www.axent.com
4. ISS Home Page, http://www.iss.net
5. Kane Security Analyst Product Home Page, http://www.mantech.co.kr/ksa.html
6. J.S. Lee, S.C. Kim, J.T. Lee, K.B. Kim and S.W. Sohn, "Design of the Security Evaluation System for the prevention of hacking incidents under large-scale network environment," *Proceedings of the 12th Workshop on Information Security and Cryptography*, pp. 160-176, Chun-An, 2000.9.
7. J.S. Lee, S.C. Kim, K.B. Kim and S.W. Sohn, "Design of the Security Evaluation System for the automatic security analysis of the large-scale network," *Proceedings of the 5th Conference on Communication Software*, pp. 172-176, Sok-Cho, 2000.7.
8. Larry J. Hughes, Jr., *Actually Useful Internet Security Techniques*, New Riders Publishing, 1995.
9. S. J. Shin, J. W. Yoon and B. M. Lee, "A Prototype Design of Expert System for Automated Risk Analysis tool," *Proceedings of the 10th Workshop on Information Security and Cryptography*, pp. 383-395, 1998.
10. S.W. Kim, H. J. Jang and B. Park, "Dynamic Monitoring based on Security Agent," *Proceedings of the 10th Workshop on Information Security and Cryptography*, pp. 518-530, 1998.
11. Sundaram. Aurobindo, "An Introduction to Intrusion Detection," ACM CROSSROADS Issue 2.4, 1996.4.
12. Simson Garfinkel & Gene Spafford, *Practical UNIX & Internet Security*, O'REILLY, Second Edition, April 1996.

Author Index

Lecture Notes in Computer Science

For information about Vols. 1–1924
please contact your bookseller or Springer-Verlag